T0301988

LEARNING CURVES

Theory, Models, and Applications

Industrial Innovation Series

Series Editor

Adedeji B. Badiru

Department of Systems and Engineering Management
Air Force Institute of Technology (AFIT) – Dayton, Ohio

LEARNING CURVES

Theory, Models, and Applications

Edited by
Mohamad Y. Jaber

CRC Press
Taylor & Francis Group
Boca Raton   London   New York

CRC Press is an imprint of the
Taylor & Francis Group, an **informa** business

CRC Press
Taylor & Francis Group
6000 Broken Sound Parkway NW, Suite 300
Boca Raton, FL 33487-2742

First issued in paperback 2017

© 2011 by Taylor & Francis Group, LLC
CRC Press is an imprint of Taylor & Francis Group, an Informa business

No claim to original U.S. Government works

ISBN 13: 978-1-4398-0738-5 (hbk)
ISBN 13: 978-1-138-07201-5 (pbk)

Visit the Taylor & Francis Web site at
http://www.taylorandfrancis.com

and the CRC Press Web site at
http://www.crcpress.com

Dedication

To the soul of my father,
and to my wife and sons.

Contents

PART I   Theory and Models

PART II   Applications

Preface

Early investigations of the learning phenomenon focused on the behavior of individual subjects who were learning-by-doing. These investigations revealed that the time required to perform a task declined, but at a decreasing rate as experience with the task increased. Such behavior was experimentally recorded and its data then fitted to an equation that adequately describes the relationships between the learning variables – namely that the performance (output) improves as experience (input) increases. Such an equation is known as a "learning curve" equation.

The learning curve has more general applicability and can describe the performance of an individual in a group, a group in an organization, and of an organization itself. Learning in an organization takes a more complex form than learning-by-doing. Learning in an organization occurs at different levels, involving functions such as strategic planning, personnel management, product planning and design, processes improvement, and technological progress. Many experts today believe that for an organization (manufacturing or a service) to sustain its competitive advantage, it has to have a steeper learning curve than its competitors. If the organization fails to have this, then it will forget and decay. As predicted by Stevens (*Management Accounting 77*, 64–65, 1999), learning curves continue to be used widely today due to the demand for sophisticated high-technology systems and the increasing interest in refurbishment to extend asset life. Therefore, understanding and quantifying the learning process can provide vital means to observe, track, and continuously improve processes in organizations within various sectors.

Since the seminal review paper of Yelle (*Decision Sciences 10*(2), 302–328, 1979), two books have been published on learning curves. These books treated the learning curve, as a forecasting tool with applications to accounting (A. Riahi-Belkaoui, 1986, *The learning curve: A management accounting tool*, Quorum Books: Westport, CT.), and as an industrial engineering tool (E. Dar-El, 2000, *Human learning: From learning curves to learning organizations*, Kluwer: Dordrecht, the Netherlands), respectively. For the past decade or so, some research has focused on opening the black box of learning curves in order to understand how learning occurs within organizations in many sectors. Some of these research studies have been the result of the careful examination of organizational systems in the manufacturing and service sectors. Recent studies show that applications of learning curves extend beyond engineering to include healthcare, information technology, technology assessment, postal services, military, and more.

This book is a collection of chapters written by international contributors who have for years been researching learning curves and their applications. The book will help draw a learning map that shows the reader where learning is involved within organizations and how it can be sustained, perfected, and accelerated. The book is a suitable reference for graduate students and researchers in the areas of operations research/management science, industrial engineering, management (e.g., healthcare, energy), and social sciences. It is divided into two parts. The first part, consisting of

Chapters 1–13, describes the theory and models of learning curves. The second part, consisting of Chapters 14–23, describes applications of learning curves.

During the preparation of this book, one of the contributors (Professor) Leo Schrattenholzer, sadly and unexpectedly passed away. Leo had initially asked Clas-Otto Wene and Bob van der Zwaan to be the co-authors of his book chapter and, thankfully and honorably, Bob and Clas-Otto carried on with the task of contributing Chapter 23 in memory of Leo.

I would like to thank all those who have encouraged me to edit this book; in particular Professor A. Badiru and Professor M. Bonney. Finally, I would also like to thank my wife for her continued support, which made completing this book possible.

Mohamad Y. Jaber
Ryerson University
Toronto, ON, Canada

Editor

Mohamad Y. Jaber is a professor of Industrial Engineering at Ryerson University. He obtained his PhD in manufacturing and operations management from The University of Nottingham. His research expertise includes modeling human learning and forgetting curves, workforce flexibility and productivity, inventory management, supply chain management, reverse logistics, and the thermodynamic analysis of production and inventory systems. Dr. Jaber has published extensively in internationally renowned journals, such as: *Applied Mathematical Modeling, Computers & Industrial Engineering, Computers & Operations Research, European Journal of Operational Research, Journal of Operational Research Society, International Journal of Production Economics*, and *International Journal of Production Research*. His research has been well cited by national and international scholars. Dr. Jaber's industrial experience is in construction management. He is the area editor—logistics and inventory systems—for *Computers & Industrial Engineering*, and a senior editor for *Ain Shams Engineering Journal*. He is also on the editorial boards for the *Journal of Operations and Logistics, Journal of Engineering and Applied Sciences* and the *Research Journal of Applied Sciences*. Dr. Jaber is the editor of the book, *Inventory Management: Non-Classical Views,* published by CRC Press. He is a member of the European Institute for Advanced Studies in Management, European Operations Management Association, Decision Sciences Institute, International Institute of Innovation, Industrial Engineering and Entrepreneurship, International Society for Inventory Research, Production & Operations Management Society, and Professional Engineers Ontario.

Contributors

Michel J. Anzanello
Department of Industrial Engineering
Federal University of Rio Grande do Sul
Rio Grande do Sul, Brazil

Adedeji B. Badiru
Department of Systems and
    Engineering Management
Air Force Institute of Technology
Wright-Patterson Air Force Base, Ohio

Charles D. Bailey
School of Accountancy
University of Memphis
Memphis, Tennessee

Roger Bohn
School of International Relations and
    Pacific Studies
University of California
San Diego, California

Maurice Bonney
Nottingham University Business School
University of Nottingham
Nottingham, United Kingdom

Tonya Boone
Mason School of Business
College of William and Mary
Williamsburg, Virginia

Roger Chow
Dymaxium Inc.
Toronto, Ontario, Canada

Guido Fioretti
Department of Management Science
University of Bologna
Bologna, Italy

Flavio S. Fogliatto
Department of Industrial Engineering
Federal University of Rio Grande do Sul
Rio Grande do Sul, Brazil

Ram Ganeshan
Mason School of Business
College of William and Mary
Williamsburg, Virginia

Christoph H. Glock
Faculty of Economics
University of Wuerzburg
Wuerzburg, Germany

Aziz Guergachi
Ted Rogers School of Management
Ryerson University
Toronto, Ontario, Canada

Ramsey F. Hamade
Department of Mechanical Engineering
American University of Beirut
Beirut, Lebanon

Robert L. Hicks
Department of Economics and Thomas
    Jefferson School in Public Policy
College of William and Mary
Williamsburg, Virginia

Mohamad Y. Jaber
Department of Mechanical and
    Industrial Engineering
Ryerson University
Toronto, Ontario, Canada

Corinne M. Karuppan
Department of Management
Missouri State University
Springfield, Missouri

Mehmood Khan
Department of Mechanical and
    Industrial Engineering
Ryerson University
Toronto, Ontario, Canada

Hemant V. Kher
Alfred Lerner College of Business and
    Economics
University of Delaware
Newark, Delaware

Michael A. Lapré
Owen Graduate School of Management
Vanderbilt University
Nashville, Tennessee

Jean-Philippe Laurenceau
Department of Psychology
University of Delaware
Newark, Delaware

Edward V. McIntyre
Department of Accounting
Florida State University
Tallahassee, Florida

Walid F. Nasrallah
Faculty of Engineering and Architecture
American University of Beirut
Beirut, Lebanon

W. Patrick Neumann
Department of Mechanical and
    Industrial Engineering
Ryerson University
Toronto, Ontario, Canada

Ojelanki Ngwenyama
Ted Rogers School of Management
Ryerson University
Toronto, Ontario, Canada

Margaret Plaza
Ted Rogers School of Management
Ryerson University
Toronto, Ontario, Canada

Sverker Sikström
Department of Psychology
Lund University
Lund, Sweden

Timothy L. Smunt
Sheldon B. Lubar School of Business
University of Wisconsin - Milwaukee
Milwaukee, Wisconsin

Sunantha Teyarachakul
ESSEC Business School
Paris, France

Daphne Diem Truong
Royal Bank of Canada
Toronto, Ontario, Canada

J. Deane Waldman
Health Sciences Center
University of New Mexico
Albuquerque, New Mexico

Clas-Otto Wene
Wenergy AB
Lund, Sweden

Steven A. Yourstone
Anderson School of Management
University of New Mexico
Albuquerque, New Mexico

Bob van der Zwaan
Policy Studies Department
Energy research Center of the
    Netherlands
Amsterdam, The Netherlands

Part I

Theory and Models

1 Learning Curves: The State of the Art and Research Directions

Flavio S. Fogliatto and Michel J. Anzanello

CONTENTS

INTRODUCTION

Several authors have investigated in various industrial segments the way in which workers improve their performance as repetitions of a manual-based task take place; e.g., Anderson (1982), Adler and Clark (1991), Pananiswaml and Bishop (1991), Nembhard and Uzumeri (2000a), Nembhard and Osothsilp (2002), Vits and Gelders (2002), Hamade et al. (2007). A number of factors may impact the workers' learning process, including: (1) task complexity, as investigated by Pananiswaml and Bishop (1991) and Nembhard and Osothsilp (2002); (2) structure of training programs (Terwiesch and Bohn 2001; Vits and Gelders 2002; Serel et al. 2003; Azizi et al. 2010); (3) workers' motivation in performing the tasks (Kanfer 1990; Eyring et al. 1993; Natter et al. 2001; Agrell et al. 2002); and (4) prior experience with the task (Nembhard and Uzumeri 2000a, 2000b; Nembhard and Osothsilp 2002). Other studies have focused on measuring knowledge and dexterity retention after task interruption; e.g., Dar-El and Rubinovitz (1991), Wickens et al. (1998), Nembhard and Uzumeri (2000b), and Jaber and Guiffrida (2008). Analyses presented in the works listed above were carried out by means of mathematical models suitable to describe the workers' learning process.

Learning curves (LCs) are deemed to be efficient tools in monitoring workers' performance in repetitive tasks, leading to reduced process loss due to workers'

3

inability in the first production cycles, as reported by Argote (1999), Dar-El (2000), Salameh and Jaber (2000), and Jaber et al. (2008). LCs have been used to allocate tasks to workers according to their learning profiles (Teplitz 1991; Uzumeri and Nembhard 1998; Nembhard and Uzumeri 2000a; Anzanello and Fogliatto 2007), to analyze and control productive operations (Chen et al. 2008; Jaber and El Saadany 2009; Janiak and Rudek 2008; Wahab and Jaber 2010), to measure production costs as workers gain experience in a task (Wright 1936; Teplitz 1991; Sturm 1999), and to estimate costs of consulting and technology implementation (Plaza and Rohlf 2008; Plaza et al. 2010).

In view of its wide applicability in production systems and given the increasing number of publications on the subject, we discuss, in this chapter, the state of the art in relation to LCs, covering the most relevant models and applications. Mathematical aspects of univariate and multivariate LCs are discussed, describing their applications, modifications to suit specific purposes, and limitations.

The chapter is divided into three sections including the present introduction. "A Review of Learning and Forgetting Models" section presents the main families of LC models and their mathematical aspects. "Research Agenda" section closes the chapter by presenting three promising research directions on LCs.

A REVIEW OF LEARNING AND FORGETTING MODELS

An LC is a mathematical description of workers' performance in repetitive tasks (Wright 1936; Teplitz 1991; Badiru 1992; Argote 1999; Fioretti 2007). Workers are likely to demand less time to perform tasks as repetitions take place due to increasing familiarity with the operation and tools, and because shortcuts to task execution are found (Wright 1936; Teplitz 1991; Dar-El 2000).

LCs were empirically developed by Wright (1936) after observing a decrease in the assembly costs of airplanes as repetitions were performed. Based on such empirical evidence, Wright proposed a rule, widely applied in the aeronautical industry of the time, according to which cumulative assembly costs were reduced on average by 20% as the number of units manufactured was duplicated (Teplitz 1991; Cook 1991; Badiru 1992; Argote 1999; Askin and Goldberg 2001).

Measures of workers' performance that have been used as dependent variables in LC models include: time to produce a single unit, number of units produced per time interval, costs to produce a single unit, and percentage of non-conforming units (Teplitz 1991; Franceschini and Galetto 2002). LC parameters may be estimated through a non-linear optimization routine aimed at minimizing the sum of squares error. The model's goodness of fit may be measured through the coefficient of determination (R^2), the sum of squares error, or the model adherence to a validation sample.

The wide range of applications conceived for LCs yielded univariate and multivariate models of varying complexity, which enabled a mathematical representation of the learning process in different settings. The log-linear, exponential, and hyperbolic models are the best known univariate models. These are described in the sections to follow.

LOG-LINEAR MODEL AND MODIFICATIONS

Wright's model, also referred to as the "log-linear model," is generally viewed as the first formal LC model. Its mathematical representation is given in Equation 1.1.

$$y = C_1 x^b, \tag{1.1}$$

where y is the average time or cost per unit demanded to produce x units, and C_1 is the time or cost to produce the first unit. Parameter b, ranging from -1 to 0, describes the workers' learning rate and corresponds to the slope of the curve. Values of b close to -1 denote high learning rate and fast adaptation to task execution (Teplitz 1991; Badiru 1992; Argote 1999; Dar-El 2000).

The following modification on Wright's model enables the estimation of the total time or cost to produce x units:

$$y_{1 \to x} = C_1 x^{b+1}. \tag{1.2}$$

The time or cost required to produce a specific unit i may be determined by further modifying the model in Equation 1.1 as follows:

$$y_i = C_1 \left[i^{b+1} - (i-1)^{b+1} \right]. \tag{1.3}$$

Numerical results from Equations 1.1 through 1.3 are summarized in tables for different learning rates (Wright 1936; Teplitz 1991), enabling prompt estimation of the time required to complete a task.

The log-linear model has several reported applications in the literature. For example, estimation of the time to task completion (Teplitz 1991), estimation of a product's life cycle (Kortge et al. 1994), evaluation of the effect of interruptions on the production rate (Jaber and Bonney 1996; Argote 1999), and assessment of the production rate as product specifications are changed through the process (Towill 1985). These applications are detailed in the paragraphs to follow.

Some industrial segments are well known for applying log-linear LCs and modifications to model the workers' learning process. Examples include the semiconductor industry (Cook 1991; Gruber 1992, 1994, 1996, 1998), electronic and aerospace components manufacturers (Garvin 2000), the chemical industry (Lieberman 1984), automotive parts manufacturers (Baloff 1971; Dar-El 2000), and truck assemblers (Argote 1999). Use of the log-linear LC for cost monitoring is reported by Spence (1981), Teng and Thompson (1996), Teplitz (1991), and Rea and Kerzner (1997).

The log-linear curve is the most used LC model for predicting the production rate in repetitive operations (Blancett 2002, Globerson and Gold 1997). Globerson and Levin (1987) and Vits and Gelders (2002) state that although presenting a non-complex mathematical structure, the log-linear model describes most manual-based operations with acceptable precision. Blancett (2002) applied the model in several sectors of a building company, evaluating workers' performance from product development to final manufacturing. With similar purposes, Terwiesch and Bohn (2001) analyzed the learning effect throughout the production process of a new product model. Finally, productivity in different cellular layouts was compared in Kannan and Polacsay (1999), using modifications of the log-linear LC.

Production planning activities may also benefit from applications of the log-linear LC, as reported by Kopcso and Nemitz (1983), Muth and Spremann (1983), Salameh et al. (1993), Jaber, Rachamadugu and Tan (1997), Jaber and Bonney (1999, 2001, 2003), and Pratsini (2000). These authors investigated the impact of workers' learning on inventory policies, optimal lot size determination, and other production planning activities.

The integration of log-linear LCs to tools designed to assist production control has also been reported in the literature. Yelle (1980, 1983), Kortge (1993), and Kortge et al. (1994) combined LCs and product life cycle models aiming at improving production planning (for a review of product life cycle models see Cox [1967] and Rink and Swan [1979], among others). Pramongkit et al. (2000) proposed the combination of LC and the Cobb-Douglas function to assess how specific factors such as invested capital and expert workforce affected workers' learning in Thai companies. Similarly, Pramongkit et al. (2002) used a log-linear LC associated with the total factor productivity tool to assess workers' learning in large Thai companies. Finally, Karaoz and Albeni (2005) integrated LCs and indices describing technological aspects to evaluate workers' performance in long production runs.

The combination of log-linear-based LCs and quality control techniques was suggested by Koulamas (1992) to evaluate the impacts of product redesign on process quality and cost. Tapiero (1987) established an association between learning process and quality control in production plants. Teng and Thompson (1996) assessed the way workers' learning rate influences the quality and costs of new products in automotive companies. Further, Franceschini and Galetto (2002) used LCs to estimate the reduction of non-conformities in a juice production plant as workers increased their skills. Jaber and Guiffrida (2004) proposed modifications on Wright's LC for processes generating defects that required reworking; the resulting model was named the "quality learning curve" (QLC). Jaber and Guiffrida (2008) investigated the QLC under the assumption that production is interrupted for quality maintenance aimed at bringing it back to an in-control state. Finally, Yang et al. (2009) proposed a quality control approach integrating on-line statistical process control (SPC) and LCs.

Applications of log-linear LCs have also been reported in the service sector. Chambers and Johnston (2000) applied LC modeling in two service providers: a large air company and a small bank. Saraswat and Gorgone (1990) evaluated the performance of software installers in companies and private residences. Sturm (1999) verified a 15% cost reduction in the process of filling out clinical forms as the number of forms doubled, roughly adhering to Wright's rule.

Log-linear LCs have been thoroughly investigated regarding their limitations and modifications for specific purposes (Baloff 1966; Zangwill and Kantor 1998, 2000; Waterworth 2000). Modifications generally aim at eliminating inconsistencies in the mathematical structure of the log-linear model.

Hurley (1996) and Eden et al. (1998) state that Wright's model yields execution times asymptotically converging to zero as a function of the number of repetitions, which is not verified in practice. To overcome this, the authors propose the inclusion of a constant term in Wright's model. Another drawback of Wright's model is pointed out by Globerson et al. (1989). They claim that the model does not take into account workers' prior experience, which clearly impacts on production planning and workforce allocation.

Another limitation of Wright's LC is related to inconsistencies in definition and inferences regarding LC outputs. Towill (1985, 1990) and Waterworth (2000) claim that many applications consider the mean execution time until unit x and the specific execution time of unit i as analogous. To correct this, Smunt (1999) proposed an alternative definition of repetition based on the continuous learning theory.

A factor that may undermine the fit of LC models is the variability in performance data collected from a given process (Yelle 1979). Globerson (1984) and Vigil and Sarper (1994) state that imprecise estimation of the learning parameter b jeopardizes the LC's predictive ability. They suggest using confidence intervals on the response estimates for predicting a process production rate. Globerson and Gold (1997) developed equations for estimating the LC's variance, coefficient of variation, and probability density function. Finally, Smunt (1999) proposed modifications on Wright's model in order to embrace situations where parameter b changes as the process takes place, while Smunt and Watts (2003) proposed the use of data aggregation techniques to reduce variance of LC-predicted values.

The use of cumulative units as independent variables has also been investigated in the LC literature. Fine (1986) argues that the number of produced units may hide learning deficiencies, since they do not take into account the units' quality. To overcome this, the author modified the LC to consider only data from conforming units. Li and Rajagopalan (1997) extended Fine's (1986) idea to include both conforming and non-conforming data in the LC model. Finally, Jaber and Guiffrida (2004) proposed modifications in Wright's model that were aimed at monitoring processes with a high percentage of non-conforming and reworked units.

Modifications in Wright's model were initially proposed to adapt the equation to specific applications, and then to become recognized as alternative models. An example is the Stanford-B model, presented in Equation 1.4, which incorporates workers' prior experience.

$$y = C_1(x+B)^b. \tag{1.4}$$

Parameter B, corresponding to the number of units of prior experience, shifts the LC downward with respect to the time/unit axis (Teplitz 1991; Badiru 1992; Nembhard and Uzumeri 2000a). The model was tested in the assembly stages of the Boeing 707, as well as in improvement activities performed later in the product (Yelle 1979; Badiru 1992; Nembhard and Uzumeri 2000a).

DeJong's model, presented in Equation 1.5, incorporates the influence of machinery in the learning process.

$$y = C_1[M + (1-M)x^b], \tag{1.5}$$

where M ($0 \leq M \leq 1$) is the incompressibility factor that informs the fraction of the task executed by machines (Yelle 1979; Badiru 1992); $M = 0$ denotes a situation where there is no machinery involved in the task, while $M = 1$ denotes a task fully executed by machinery where no learning takes place (Badiru 1992).

The S-curve model aims at describing learning when machinery intervention occurs, and when the first cycles of operation demand in-depth analysis. The model is a combination of DeJong's and Stanford-B's models, as presented in Equation 1.6. Parameters in the model maintain their original definitions (Badiru 1992; Nembhard and Uzumeri 2000a).

$$y = C_1[M + (1 - M)(x + B)^b].$$ (1.6)

The plateau model in Equation 1.7 displays an additive constant, C, which describes the steady state of workers' performance. The steady state is reached after learning is concluded or when machinery limitations block workers' improvement (Yelle 1979; Teplitz 1991; Li and Rajagopalan 1998).

$$y = C + C_1 x^b.$$ (1.7)

A comprehensive comparison of several of the LCs discussed above is reported in Hackett (1983).

In addition to the LC models presented above, other less cited log-linear-based LCs are proposed in the literature. Levy's (1965) adapted function is one such model:

$$My = \left[\frac{1}{\beta} - \left(\frac{1}{\beta} - \frac{x^b}{C_1} \right) k^{-kx} \right]^{-1},$$ (1.8)

where β is a task-defined production coefficient for the first unit, and k is the workers' performance in steady state. Remaining parameters are as previously defined.

Focused on production runs of long duration, Knecht (1974) proposed an alternative adapted function model that allows evaluating the production rate as parameter b changes during the production run (see Equation 1.9). Parameters are as previously defined.

$$y = \frac{C_1 x^{b+1}}{(1 + b)}.$$ (1.9)

A summation of LCs characterized by n different learning parameters b, proposed by Yelle (1976), is given in Equation 1.10. The resulting model could be applied to production processes comprised of n different tasks. However, Howell (1980) claims that the model in Equation 1.10 leads to imprecise production rate estimates.

$$y = C_1 x_1^{b1} + C_2 x_2^{b2} + \cdots + C_n x_n^{bn}.$$ (1.10)

Alternative LC models were developed following the log-linear model's principles, although relying on more elaborate mathematical structures to describe complex production settings. We refrain from exposing the mathematical details of those models in order to avoid overloading the exposition; only their purpose is presented. Klenow (1998) proposed an LC model to support decisions on updating production technology. Demeester and Qi (2005) developed an LC customized to situations in which two generations of the same product (i.e., two models) are simultaneously being produced. Their LC helps identify the best moment to allocate learning resources (e.g., training programs and incentive policies) to produce the new model.

Mazzola et al. (1998) developed an LC-based algorithm to synchronize multiproduct manufacturing in environments characterized by workers' learning and forgetting processes. Gavious and Rabinowitz (2003) proposed an approach using an LC to evaluate the training efficiency of internal resources in comparison with that of outsourced resources. Similarly, Fioretti (2007) suggested a disaggregated LC model to analyze complex production environments in terms of time reduction for task completion. Finally, Park et al. (2003) proposed a multiresponse LC aimed at evaluating knowledge transference at distinct production stages in a liquid display (LCD) factory.

The integration of an LC and scheduling techniques was introduced by Biskup (1999), analyzing the effect of learning on the position of jobs in a single machine. Mosheiov and Sidney (2003) extended that approach by combining job-dependent LCs (which are LCs with a different parameter for each job) to programming formulations aimed at minimizing flow time and makespan in a single machine, as well as flow time in unrelated parallel machines.

Hyperbolic Models

An LC model relating the number of conforming units to the total number of units produced is reported in Mazur and Hastie (1978). In that model, x describes the number of conforming units, and r is the number of non-conforming units; thus, y corresponds to the fraction of conforming units multiplied by a constant k. The model is called "2-parameter hyperbolic curve" and is represented by:

$$y = k\left(\frac{x}{x+r}\right). \tag{1.11}$$

For learning modeling purposes, y describes the workers' performance in terms of number of items produced after x units of operation time ($y \geq 0$ and $x \geq 0$), k quantifies the maximum performance level ($k \geq 0$), and r denotes the learning rate, given in time units (Nembhard and Uzumeri 2000a).

A more complete model can be generated by including workers' prior experience in executing the task. For that matter, Mazur and Hastie (1978) suggested the inclusion of parameter p in Equation 1.11, leading to the 3-parameter hyperbolic LC in Equation 1.12. In that equation, parameter p refers to workers' prior experience evaluated in time units ($p \geq 0$).

$$y = k\left(\frac{x+p}{x+p+r}\right). \tag{1.12}$$

Uzumeri and Nembhard (1998) and Nembhard and Uzumeri (2000a) improved the definition of parameter r, associating it with the time required to achieve production level $k/2$, which is half the maximum performance level k. A worker presenting high values of r requires much practice in order to achieve k, thus displaying slow learning. The authors also state that r acts as the shape factor in the hyperbolic model, leading to three possible learning profiles: (1) $r > 0$—the curve presents an increasing profile until k, representing the typical behavior of workers performing new tasks; (2) $r \to 0$—the curve follows a horizontal pattern, denoting absence of workers' improvement; and (3) $r < 0$—the curve follows a decreasing performance pattern, usually associated with fatigue or forgetting.

The 3-parameter hyperbolic model presented remarkable performance when compared to ten other LC models in terms of efficiency, stability, parsimony, and the ability to model scenarios with forgetting (Nembhard and Uzumeri 2000a). Further, Anzanello and Fogliatto (2007) found the hyperbolic model to be more robust in comparison with the 3-parameter exponential and time constant models.

A major application of the 3-parameter hyperbolic LC is related to the allocation of tasks to workers aimed at improving production systems. Uzumeri and Nembhard (1998) and Shafer et al. (2001) applied the model in a population of workers assigned to new tasks. The authors concluded that fast learners (workers presenting low values of r) tend to achieve lower maximum performance k if compared with slow learners (workers presenting high values of r). Therefore, they propose allocating fast learners to short duration tasks, while slow learners should perform tasks of long duration, given their final performance. With similar objectives, Nembhard and Osothsilp (2002) evaluated the effects of task complexity on the allocation of tasks to workers, while Nembhard and Uzumeri (2000b) evaluated distinct workers' profiles in terms of ability gaining and retention under different tasks. Anzanello and Fogliatto (2007) used the 3-parameter hyperbolic model to allocate tasks to workers according to the duration of production runs in a shoe manufacturing process.

Other applications of the 3-parameter hyperbolic model include Anzanello and Fogliatto (2010), where workers' processing times are estimated using the model and integrated in several heuristics for task scheduling, and Anzanello (2010), where a variable selection method using hyperbolic LC parameters as variables for clustering in mass customized scenarios affected by workers' learning is proposed.

EXPONENTIAL MODELS

One of the first applications of hte exponential LC is reported in Knecht (1974), who integrated exponential and log-linear functions to improve predictions in long duration production runs. The proposed model is given in Equation 1.13.

$$y = C_1 x^b e^{cx}, \tag{1.13}$$

where c is a second constant; other parameters are as previously defined.

Although several exponential LC models were proposed in the literature, three are frequently discussed: the 3-parameter exponential, the 2-parameter exponential, and the constant time models. The first is given in Equation 1.14, with parameters defined as those in the 3-parameter hyperbolic model.

$$y = k(1 - e^{-(x+p)/r}). \tag{1.14}$$

The LC in Equation 1.14 was tested in Mazur and Hastie (1978); the model provided a poor fit to processes characterized by workers allocated to complex and demanding new tasks. By contrast, the model fitted well in scenarios characterized by high levels of workers' prior experience. The elimination of parameter p in Equation 1.14 gives rise to the 2-parameter exponential LC model. As expected, such a simplified model offers poorer fit to performance data if compared with the 3-parameter exponential LC (Mazur and Hastie 1978).

A comparison between exponential models and the 3-parameter hyperbolic was performed by Mazur and Hastie (1978). They found that parameters p and r assume similar values, while parameter k is generally underestimated by the exponential model. In addition, the 3-parameter hyperbolic model presented better fit to data collected from several processes, based on the coefficient of determination (R^2).

An alternative exponential model is the time constant model proposed by Towill (1990). It is structurally similar to the 3-parameter exponential LC, as shown below:

$$y = y_c + y_f(1 - e^{-t/\tau}), \tag{1.15}$$

where y_c corresponds to workers' initial performance (in number of items produced per time), and y_f is the maximum performance when workers' learning has been completed, given in the same units. Variable t is the cumulative operation time (analogous to x in the previous models), which enables easier estimation of the time required to achieve a certain performance level. Towill (1990) applied the time constant model in a process where performance data collection had started after a short adaptation of workers to the task.

Modifications on the time constant model were proposed by Naim and Towill (1990); these authors added trigonometric functions to the model with the aim of better describing scenarios where cyclical variations in performance are verified. Further, Howell (1990) evaluated the impact of inaccurate parameter inputs in the model's predictive ability and proposed approaches to achieve convergence in complex modeling situations. Finally, Dardan et al. (2006) applied the time constant model in a hardware company aimed at evaluating the relationship between workers' learning process and the duration of technological investments.

Multivariate Models

Scenarios where quantitative and qualitative factors affect the learning process require extensions of the traditional LC models (Badiru 1992). For that matter, multivariate LCs (i.e., LCs relying on two or more independent variables) have been developed. Such models often display the following generic structure:

$$C_x = K \prod_{i=1}^{n} c_i x_i^{b_i}, \qquad (1.16)$$

with K as the performance (cost) to produce the first unit, and c_i as the coefficient for the independent variable i; other parameters are as previously defined.

Gold (1981) and Camm et al. (1987) used relations similar to that in Equation 1.16 to monitor production costs as a function of the following independent variables: number of items produced, production rate, duration and cost of training programs, and task complexity. Alternative multivariate LCs based on integration procedures were proposed by McIntyre (1977) and Womer (1979); such models, however, present limited application due to their complexity. Recently Hamade et al. (2009) used univariate and multivariate LCs to analyze CAD (computer-aided design) procedural and cognitive data describing the performance of trainees.

A comprehensive comparison between univariate and multivariate LCs' predictive ability is presented in Badiru (1992). The author states that multivariate models provide relevant information on variables' interactions, but the addition of non-significant variables jeopardizes the model's quality. In addition, the estimation of regression coefficients in multivariate models can become numerically unstable and can be affected by multicollinearity among variables, thereby yielding low-quality models. Thus, univariate LCs should be preferred where the importance of additional independent variables is not clear. The adherence of the multivariate model to data can be evaluated using traditional criteria; e.g., R^2 and the sum of squares error.

Forgetting Models

Modifications in product specifications and interruptions in the production process impose additional challenges to workers in terms of resuming activities (Jaber 2006). Forgetting is a major consequence of interruptions and becomes evident by (1) reduction in production rate after an inactive period, and (2) manufacturing of lower quality products compared to those produced during continuous operation, especially in the first cycles after the process is resumed. Thus, predicting workers' production rate at resuming tasks enables better production planning and precise resource allocation, as reported in Globerson et al. (1989), Argote (1999), Dar-El and Rubinovitz (1991), Dar-El (2000), and Bailey and McIntyre (1997, 2003).

The forgetting process and its consequences on workers' performance have been widely studied in many fields. Globerson et al. (1989) found that a log-linear model properly described workers' forgetting process, indicating that learning and forgetting occur in similar ways. Jaber and Bonney (1996) and Bailey and McIntyre (2003) also modeled the forgetting process using log-linear-based models. Jaber and Kher (2004) extended the studies above to evaluate the variability on workers' time to achieve a complete forgetting status. Their models also enable estimation of the time required to produce the first unit, once the task is resumed. Further, Jaber and Bonney (1997) proposed three mathematical models for describing both learning and forgetting processes, while Bailey and McIntyre (1997) introduced a relearning curve based on the log-linear model, recommending it to model processes where interruptions in production are frequent. Another interesting study on the issue was conducted by Jaber et al. (2003), who described factors that influence workers' forgetting within industrial environments and assesses the efficiency of existing mathematical structures in modeling the forgetting process. Additional studies of forgetting and learning were carried by Jaber and Guiffrida (2007), Jaber and Sikström (2004), and Zamiska et al. (2007).

The 3-parameter hyperbolic LC has also been used to model forgetting processes as reported in Nembhard and Uzumeri (2000a), Shafer et al. (2001) and Nembhard and Osothsilp (2002). Further, Davidovitch et al. (2008) estimated the efficiency of the learning–forgetting–relearning process using simulation; they analyzed how different interruption lengths affect workers' performance once the activity is resumed.

The impact of forgetting on production planning activities such as optimal lot size definition and inventory level decisions has also received attention (Salameh et al. 1993; Bailey 1989; and Jaber et al. 2009). Jaber and Bonney (1996, 2003) found that the optimal lot size decreases as workers continuously perform the task without major interruptions. In addition, Alamri and Balkhi (2007) assessed the effects of learning and forgetting on lot size problems considering an infinite planning horizon. Other studies relating to learning, forgetting, and the lot size determination problem are reported by Jaber and Bonney (2007).

Similar to interruptions in production, product redesign may disrupt workers' performance (Yelle 1979). Such modifications are usually related to customers' requirements or product customization to new markets, and demand workers' adaptation as they are submitted to wholly new tasks (see Eden et al. 1998). Finally, Lam et al. (2001) applied LC to evaluate the effects of forgetting and product modifications in the construction segment, and estimated their impact on productivity indices.

RESEARCH AGENDA

LCs re-emerge as an important research topic in industrial engineering and operations management motivated by two main events: The first one is the increasing popularity of mass customization (MC) as a production strategy in manufacturing and service industries. Mass customized products are tailored to the individual needs and expectations of customers, implying a large variety of product models and small lot sizes. LC modeling enables the assignment of tasks to individuals or teams, allowing the economically feasible production of small lots. The second event is the greater availability of automated data collection devices, which enables easy data collection and storage. This opens an opportunity to propose and explore multivariate LC models, thus expanding the scarce literature on the subject. We now propose three research directions on LCs related to the two motivating events referred to above.

The first research direction is related to job scheduling. Job scheduling theory aims at allocating jobs to resources such that an objective (e.g., minimization of completion time) is optimized (Pinedo 2008). However, it seems that understanding the impacts of workers' learning on the scheduling problem has only recently become a research topic. This issue becomes critical in mass customized productive environments where a large range of items are produced following a small lot size policy.

Seminal studies regarding the effects of learning on scheduling were proposed by Biskup (1999) and Mosheiov (2001). Further, Mosheiov and Sidney (2003) integrated LCs and scheduling formulations aimed at minimizing flow time and makespan in single and unrelated parallel machines. More recently, Anzanello and Fogliatto (2010) proposed an approach to minimize total weighted earliness and tardiness in scenarios characterized by workers' learning; job completion times were estimated using LC profiles, and were then inputted into scheduling algorithms.

Estimates of job completion times are highly affected by the LC goodness of fit. Since imprecise estimates lead to unreliable scheduling results, we propose the use of LC goodness of fit indices to adjust poorly estimated completion times. Such indices could be used to reorganize jobs in a scheduling procedure such that completion times obtained from well-fitted LCs are prioritized. In addition, multivariate LC models that take into account the effect of other independent variables in addition to time should also be investigated. A potentially relevant independent variable could be the training procedure.

The second research topic we propose is related to job rotation. Job rotation makes it possible to observe workers performing different tasks, and to determine which task is best suited for each worker. Although usually dominant, the productivity criterion is not the only one motivating job rotation. Rotation may benefit workers ergonomically by reducing exposure to work-related musculoskeletal disorders due to repetitive postures, as investigated by Kuijer et al. (1999) and Frazer et al. (2003). In addition, production managers may benefit from the flexibility provided by multi-task workers, which is a natural outcome of job rotation.

A few authors have worked on the association of job rotation and learning. Jovanovic (1979) was one of the first to formally investigate job rotation associated with workers' learning. Although not explicitly using LC modeling, the author proposes a regression model in which a worker's contribution to the total company output is dependent on the amount of time the worker is functional and on independent parameters related to the worker–job match. Ortega (2001) extends the propositions in Jovanovic (1979) to investigate whether job rotation policies are an efficient way to learn about workers within a company. To accomplish this, data are collected from different predetermined worker–job matches. Two rotation policies are investigated: specialization and job rotation.

None of the works above make direct use of LC modeling to determine job rotation strategies. A promising way to address the problem would be through the introduction of sequence-dependent learning models, in which workers' performance on a new model depends on the model previously assigned and its characteristics. A job rotation scheme that minimizes productivity losses would be pursued.

The last research direction is related to LC clustering. In assembly lines, a critical task can be performed by several workers to balance production capacity. Workers are usually selected based on their average time in executing the task once the performance steady state is reached; workers with similar time averages are assigned to the task. This procedure, however, does not take into account the time required by workers to learn the task, which becomes significant when dealing with small size lots.

Methods for clustering functional data aim at grouping information following a pattern instead of a single measure (Heckman and Zamar 2000; Tarpey and Kinateder 2003). Such patterns are usually time dependent, as is the LC itself. We envision LC clustering as a promising research issue, since it would enable the optimal grouping of workers based on their performance immediately after task start-up, and not only after performance steady state is achieved. Productive scenarios characterized by reduced lot sizes could benefit from this, since workers with similar learning profiles would be part of the same group and line balance would thus be ensured.

REFERENCES

Adler, P.S., and Clark, K.B., 1991. Behind the learning curve: A sketch of the learning process. *Management Science* 37(3): 267–281.

Agrell, P.J., Bogetoft, P., and Tind, J., 2002. Incentive plans for productive efficiency, innovation and learning. *International Journal of Production Economics* 78(1): 1–11.

Alamri, A., and Balkhi, Z., 2007. The effects of learning and forgetting on the optimal production lot size for deteriorating items with time varying demand and deterioration rates. *International Journal of Production Economics* 107(1): 125–138.

Anderson, J.R., 1982. Acquisition of cognitives. *Psychological Review* 89(4): 369–406.

Anzanello, M.J., 2010. Selecting relevant clustering variables in mass customization scenarios characterized by workers learning. In *Mass customization: Engineering and managing global operations*, eds. F.S. Fogliatto and G. Da Silveira. Springer, 303–318.

Anzanello, M.J., and Fogliatto, F.S., 2007. Learning curve modeling of work assignment in mass customized assembly lines. *International Journal of Production Research* 45(13): 2919–2938.

Anzanello, M.J., and Fogliatto, F.S., 2010. Scheduling learning dependent jobs in customized assembly lines. *International Journal of Production Research* 48(22): 6683–6699.

Argote, L., 1999. *Organizational learning: Creating, retaining and transferring knowledge.* New York: Springer.

Askin, R., and Goldberg, J., 2001. *Design and analysis of lean production systems.* New York: John Wiley.

Azizi, N., Zolfaghari, S., and Liang, M., 2010. Modeling job rotation in manufacturing systems: The study of employee's boredom and skill variations. *International Journal of Production Economics* 123(1): 69–85.

Badiru, A.B., 1992. Computational survey of univariate and multivariate learning curve models. *IEEE Transactions on Engineering Management* 39(2): 176–188.

Bailey, C., 1989. Forgetting and the learning curve: A laboratory study. *Management Science* 35(3): 340–352.

Bailey, C.D., and McIntyre, E.V., 1997. The relation between fit and prediction for alternative forms of learning curves and relearning curves. *IIE Transactions* 29(6): 487–495.

Bailey, C.D., and McIntyre, E.V., 2003. Using parameter prediction models to forecast post-interruption learning. *IIE Transactions* 35(12): 1077–1090.

Baloff, N., 1966. The learning curve – Some controversial issues. *The Journal of Industrial Economics* 14(3): 275–282.

Baloff, N., 1971. Extension of the learning curve – Some empirical results. *Operational Research Quarterly* 22(4): 329–340.

Biskup, D., 1999. Single-machine scheduling with learning considerations. *European Journal of Operational Research* 115(1): 173–178.

Blancett, R.S., 2002. Learning from productivity learning curves. *Research-Technology Management* 43(3): 54–58.

Camm, J.D., Evans, J.R., and Womer, N.K., 1987. The unit learning curve approximation of total costs. *Computers and Industrial Engineering* 12(3): 205–213.

Chambers, S., and Johnston, R., 2000. Experience curves in services: Macro and micro level approaches. *International Journal of Operations and Production Management* 20(7): 842–859.

Chen, C., Lo, C., and Liao, Y., 2008. Optimal lot size with learning consideration on an imperfect production system with allowable shortages. *International Journal of Production Economics* 113(1): 459–469.

Cook, J.A., 1991. Competitive model of the Japanese firm. *Journal of Policy Modeling* 13(1): 93–114.

Cox, W.E., 1967. Product life cycles as marketing models. *Journal of Business* 40(4): 375–384.

Dardan, S., Busch, D., and Sward, D., 2006. An application of the learning curve and the non-constant-growth dividend model: IT investment valuation at Intel Corporation. *Decision Support Systems* 41(4): 688–697.

Dar-El, E.M., 2000. *Human learning: From learning curves to learning organizations.* Boston: Kluwer Academic.

Dar-El, E.M., and Rubinovitz, J., 1991. Using learning theory in assembly lines for new products. *International Journal of Production Economics* 25(1–3): 103–109.

Davidovitch, L., Parush, A., and Shtub, A., 2008. Simulation-based learning: The learning–forgetting–relearning process and impact of learning history. *Computers & Education* 50(3): 866–880.

Demeester, L., and Qi, M., 2005. Managing learning resources for consecutive product generations. *International Journal of Production Economics* 95(2): 65–283.

Eden, C., Willians, T., and Ackermann, F., 1998. Dismantling the learning curve: The role of disruptions on the planning of development projects. *International Journal of Project Management* 16(3): 131–138.

Eyring, J.D., Johnson, D.S., and Francis, D.J., 1993. A cross-level units-of-analysis approach to individual differences in skill acquisition. *Journal of Applied Psychology* 78(5): 805–814.

Fine, C.H., 1986. Quality improvement and learning in productive systems. *Management Science* 32(10): 1301–1315.

Fioretti, G., 2007. The Organizational learning curve. *European Journal of Operational Research* 177(3): 1375–1384.

Franceschini, F., and Galetto, M., 2002. Asymptotic defectiveness of manufacturing plants: An estimate based on process. *International Journal of Production Research* 40(3): 537–545.

Frazer, M., Norman, R., Wells, R., and Neumann, P., 2003. The effects of job rotation on the risk of reporting low back pain. *Ergonomics* 46(9): 904–919.

Garvin, D., 2002. *Learning in action – A guide to put the learning organization to work.* Boston: Harvard Business School Press.

Gavious, A., and Rabinowitz, G., 2003. Optimal knowledge outsourcing model. *Omega* 31(6): 451–457.

Globerson, S., 1984. The deviation of actual performance around learning curve models. *International Journal of Production Research* 22(1): 51–62.

Globerson, S., and Gold, D., 1997. Statistical attributes of the power learning curve model. *International Journal of Production Research* 35(3): 699–711.

Globerson, S., and Levin, N., 1987. Incorporating forgetting into learning curves. *International Journal of Production Management* 7(4): 80–94.

Globerson, S., Levin, N., and Shtub, A., 1989. The impact of breaks on forgetting when performing a repetitive task. *IIE Transactions* 21(4): 376–381.

Gold, B., 1981. Changing perspectives on size, scale, and returns: An interpretive survey. *Journal of Economic Literature* 19(1): 5–33.

Gruber, H., 1992. The learning curve in the production of semiconductor memory chips. *Applied Economics* 24(8): 885–894.

Gruber, H., 1994. The yield factor and the learning curve in semiconductor production. *Applied Economics* 26(8): 837–843.

Gruber, H., 1996. Trade policy and learning by doing: The case of semiconductors. *Research Policy* 25(5): 723–739.

Gruber, H., 1998. Learning by doing and spillovers: Further evidence for the semiconductor industry. *Review of Industrial Organization* 13(6): 697–711.

Hackett, E., 1983. Application of a set of learning curve models to repetitive tasks. *Radio and Eletronic Engineer* 53(1): 25–32.

Hamade, R., Artail, H., and Jaber, M., 2007. Evaluating the learning process of mechanical CAD students. *Computers & Education* 49(3): 640–661.

Hamade, R., Jaber, M., and Sikstrom, S., 2009. Analyzing CAD competence with univariate and multivariate learning curve models. *Computers and Industrial Engineering* 56(4): 1510–1518.

Heckman, N.E., and Zamar, R.H., 2000. Comparing the shapes of regression functions. *Biometrika* 87(1): 135–144.

Howell, S., 1980. Learning curves for new products. *Industrial Marketing Management* 9(2): 97–99.

Howell, S., 1990. Parameter instability in learning curve models – Invited comments on papers by Towill and by Sharp and Price. *International Journal of Forecasting* 6(4): 541–547.

Hurley, J.W., 1996. When are we going to change the learning curve lecture? *Computers Operations Research* 23(5): 509–511.

Jaber, M.Y., 2006. Learning and forgetting models and their applications. In *Handbook of industrial and systems engineering,* ed. A.B. Badiru, 30.1–30.24. Boca Raton: CRC Press.

Jaber, M.Y., and Bonney, M., 1996. Production breaks and the learning curve: The forgetting phenomenon. *Applied Mathematics Modeling* 20(2): 162–169.

Jaber, M.Y., and Bonney, M., (1997). A comparative study of learning curves with forgetting. *Applied Mathematics Modeling* 21(8): 523–531.

Jaber, M.Y., and Bonney, M., 1999. The economic manufacture/order quantity (EMQ/EOQ) and the learning curve: Past, present, and future. *International Journal of Production Economics* 59(1–3): 93–102.

Jaber, M.Y., and Bonney, M., 2001. Economic lot sizing with learning and continuous time discounting: Is it significant? *International Journal of Production Economics* 71(1–3): 135–143.

Jaber, M.Y., and Bonney, M., 2003. Lot sizing with learning and forgetting in set-ups and in product quality. *International Journal of Production Economics* 83(1): 95–111.

Jaber, M.Y., and Bonney, M., 2007. Economic manufacture quantity (EMQ) model with lot-size dependent learning and forgetting rates. *International Journal of Production Economics* 108(1–2): 359–367.

Jaber, M.Y., Bonney, M., and Moualek, I., 2009. Lot sizing with learning, forgetting and entropy cost. *International Journal of Production Economics* 118(1): 19–25.

Jaber, M.Y., and El Saadany, A. (in press). An economic production and remanufactur-
    ing model with learning effects. *International Journal of Production Economics*
    (doi:10.1016/j.ijpe.2009.04.019).
Jaber, M.Y., Goyal, S., and Imran, M., 2008. Economic production quantity model for items
    with imperfect quality subject to learning effects. *International Journal of Production
    Economics* 115(1): 143–150.
Jaber, M.Y., and Guiffrida, A., 2004. Learning curves for process generating defects requiring
    reworks. *European Journal of Production Research* 159(3): 663–672.
Jaber, M.Y., and Guiffrida, A., 2007. Observations on the economic order (manufacture)
    quantity model with learning and forgetting. *International Transactions in Operational
    Research* 14(2): 91–104.
Jaber, M.Y., and Guiffrida, A., 2008. Learning curves for imperfect production processes
    with reworks and process restoration interruptions. *European Journal of Operational
    Research* 189(1): 93–104.
Jaber, M.Y., and Kher, H., 2004. Variant versus invariant time to total forgetting: The learn–
    forget curve model revisited. *Computers & Industrial Engineering* 46(4): 697–705.
Jaber, M.Y., Kher, H.V., and Davis, D., 2003. Countering forgetting through training and
    deployment. *International Journal of Production Economics* 85(1): 33–46.
Jaber, M.Y., and Sikström, S., 2004a. A note on "An empirical comparison of forgetting mod-
    els". *IEEE Transactions on Engineering Management* 51(2): 233–234.
Jaber, M.Y., and Sikström, S., 2004b. A numerical comparison of three potential learning and
    forgetting models. *International Journal of Production Economics* 92(3): 281–294.
Janiak, A., and Rudek, R., 2008. A new approach to the learning effect: Beyond the learning
    curve restrictions. *Computers & Operations Research* 35(11): 3727–3736.
Jovanovic, B., 1979. Job matching and the theory of turnover. *Journal of Political Economy*
    87(5): 972–990.
Kanfer, R., 1990. Motivation and individual differences in learning: An integration of devel-
    opmental, differential and cognitive perspectives. *Learning and Individual Differences*
    2(2): 221–239.
Kannan, V., and Palocsay, S., 1999. Cellular versus process layouts: An analytic investigation
    of the impact of learning on shop performance. *Omega* 27(5): 583–592.
Karaoz, M., and Albeni, M., 2005. Dynamic technological learning trends in Turkish manufac-
    turing industries. *Technological Forecasting-Social Change* 72(7): 866–885.
Klenow, P., 1998. Learning curves and the cyclical behavior of manufacturing industries.
    Review of Economic Dynamics 1(2): 531–550.
Knecht, G., 1974. Costing, technological growth and generalized learning curves. *Operations
    Research Quarterly* 25(3): 487–491.
Kopcso, D., and Nemitz, W., 1983. Learning curves and lot sizing for independent and depen-
    dent demand. *Journal of Operations Management* 4(1): 73–83.
Kortge, G., 1993. Link sales training and product life cycles. *Industrial Marketing Management*
    22(3): 239–245.
Kortge, G., Okonkwo, P., Burley, J., and Kortge, J., 1994. Linking experience, product life
    cycle, and learning curves: Calculating the perceived value price range. *Industrial
    Marketing Management* 23(3): 221–228.
Koulamas, C., 1992. Quality improvement through product redesign and the learning curve.
    Omega 20(2): 161–168.
Kuijer, P.P.F.M., Visser, B., and Kemper, H.C.G., 1999. Job rotation as a factor in reducing
    physical workload at a refuse collecting department. *Ergonomics* 42(9): 1167–1178.
Lam, K., Lee, D., and Hu, T., 2001. Understanding the effect of the learning-forgetting
    phenomenon to duration of projects construction. *International Journal of Project
    Management* 19(7): 411–420.
Levy, F., 1965. Adaptation in the production process. *Management Science* 11(6): 136–154.

Li, G., and Rajagopalan, S., 1997. The impact of quality on learning. *Journal of Operations Management* 15(3): 181–191.

Li, G., and Rajagopalan, S., 1998. A learning curve model with knowledge depreciation. *European Journal of Operational Research* 105(1): 143–154.

Lieberman, M., 1984. The learning curve and pricing in the chemical processing industries. *The RAND Journal of Economics* 15(2): 213–228.

Mazur, J.E., and Hastie, R., 1978. Learning as accumulation: A reexamination of the learning curve. *Psychological Bulletin* 85(6): 1256–1274.

Mazzola, J., Neebe, A., and Rump, C., 1998. Multiproduct production planning in the presence of work-force learning. *European Journal of Operational Research* 106(2–3): 336–356.

McIntyre, E., 1977. Cost-volume-profit analysis adjusted for learning. *Management Science* 24(2): 149–160.

Mosheiov, G., 2001. Scheduling problems with learning effect. *European Journal of Operational Research* 132(3): 687–693.

Mosheiov, G., and Sidney, J., 2003. Scheduling with general job-dependent learning curves. *European Journal of Operational Research* 147(16): 665–670.

Muth, E.J., and Spremann, K., 1983. Learning effects in economic lot sizing. *Management Science* 29(2): 102–108.

Naim, M.M., and Towill, D.R., 1990. An engineering approach to LSE modeling of experience curves in the electricity supply industry. *International Journal of Forecasting* 6(4): 549–556.

Natter, M., Mild, A., Feurstein, M., Dorffner, G., and Taudes, A., 2001. The effect of incentive schemes and organizational arrangements on the new product development process. *Management Science* 47(8): 1029–1045.

Nembhard, D.A., and Osothsilp, N., 2002. Task complexity effects on between-individual learning/forgetting variability. *International Journal of Industrial Ergonomics* 29(5): 297–306.

Nembhard, D.A., and Uzumeri, M.V., 2000a. An individual-based description of learning within an organization. *IEEE Transactions on Engineering Management* 47(3): 370–378.

Nembhard, D.A., and Uzumeri, M.V., 2000b. Experiential learning and forgetting for manual and cognitive tasks. *International Journal of Industrial Ergonomics* 25(3): 315–326.

Ortega, J., 2001. Job rotation as a learning mechanism. *Management Science* 47(10): 1361–1370.

Pananiswaml, S., and Bishop, R.C., 1991. Behavioral implications of the learning curve for production capacity analysis. *International Journal of Production Economics* 24(1–2): 157–163.

Park, S., Lee, J., and Kim, T., 2003. Learning by doing and spillovers: An empirical study on the TFT-LCD industry. Engineering Management Conference IEMC, 2003.

Pinedo, M., 2008. *Scheduling, theory, algorithms and systems.* New York: Springer.

Plaza, M., and Rohlf, K., 2008. Learning and performance in ERP implementation projects: A learning-curve model for analyzing and managing consulting costs. *International Journal of Production Economics* 115(1): 72–85.

Plaza, M., Ngwenyama, O., and Rohlf, K., 2010. A comparative analysis of learning curves: Implications for new technology implementation management. *European Journal of Operational Research* 200(2): 518–528.

Pramongkit, P., Shawyun, T., and Sirinaovakul, B., 2000. Analysis of technological learning for the Thai manufacturing industry. *Technovation* 20(4): 189–195.

Pramongkit, P., Shawyun, T., and Sirinaovakul, B., 2002. Analysis of technological learning for the Thai manufacturing industry. *Technological Forecasting and Social Change* 69(1): 89–101.

Pratsini, E., 2000. The capacitated dynamic lot size problem with variable technology. *Computers & Industrial Engineering* 38(4): 493–504.

Rachamadugu, R., and Tan, C., 1997. Policies for lot sizing with setup learning. *International Journal of Production Economics* 48(2): 157–165.

Rea, P., and Kerzner, H., 1997. *Strategic planning: A practical guide*. New York: John Wiley.

Rink, D.R., and Swan, J.E., 1979. Product life cycle research: A literature review. *Journal of Business Research* 7(3): 219–242.

Salameh, M.K., Abdul-Malak, M.U., and Jaber, M.Y., 1993. Mathematical modeling of the effect of human learning in the finite production inventory model. *Applied Mathematics Modeling* 17(11): 613–615.

Salameh, M., and Jaber, M.Y., 2000. Economic production quantity model for items with imperfect quality. *International Journal of Production Economics* 4(1):59–64.

Saraswat, S.P., and Gorgone, J.T., 1990. Organizational learning curve in software installation: An empirical investigation. *Information and Management* 19(1): 53–59.

Serel, D.A., Dada, M., Moskowitz, H., and Plante, R.D., 2003. Investing in quality under autonomous and induced learning. *IIE Transactions* 35(6): 545–555.

Shafer, S., Nembhard, D.A., and Uzumeri, M.V., 2001. The effects of worker learning, forgetting, and heterogeneity on assembly line productivity. *Management Science* 47(12): 1639–1653.

Smunt, T.L., 1999. Log-linear and non-log-linear learning curve models for production research and cost estimation. *International Journal of Production Research* 37(17): 3901–3911.

Smunt, T.L., and Watts, C.A., 2003. Improving operations planning with learning curves: overcoming the pitfalls of "messy" shop floor data. *Journal of Operations Management* 21(1): 93–107.

Spence, A., 1981. The learning curve and competition. *The Bell Journal of Economics* 12(1): 49–70.

Sturm, R., 1999. Cost and quality trends under managed care: Is there a learning curve in behavioral health carve-out plans? *Journal of Health Economics* 18(5): 593–604.

Tapiero, C., 1987. Production learning and quality control. *IIE Transactions* 19(4): 362–370.

Tarpey, T., and Kinateder, K.K., 2003. Clustering functional data. *Journal of Classification* 20(1): 93–114.

Teng, J., and Thompson, G., 1996. Optimal strategies for general price-quality decision models of new products with learning production costs. *European Journal of Operational Research* 93(3): 476–489.

Teplitz, C.J., 1991. *The learning curve deskbook: A reference guide to theory, calculations and applications*. New York: Quorum Books.

Terwiesch, C., and Bohn, R., 2001. Learning and process improvement during production ramp-up. *International Journal of Production Economics* 70(1): 1–19.

Towill, D.R., 1985. Management systems applications of learning curves and progress functions. *Engineering Costs and Production Economics* 9(4): 369–383.

Towill, D.R., 1990. Forecasting learning curves. *International Journal of Forecasting* 6(1): 25–38.

Uzumeri, M., and Nembhard, D., 1998. A population of learners: A new way to measure organizational learning. *Journal of Operations Management* 16(5): 515–528.

Vigil, D.P., and Sarper, H., 1994. Estimating the effects of parameter variability on learning curve model predictions. *International Journal of Production Economics* 34(2): 187–200.

Vits, J., and Gelders, L., 2002. Performance improvement theory. *International Journal of Production Economics* 77(3): 285–298.

Wahab, M.I.M., and Jaber, M.Y., 2010. Economic order quantity model for items with imperfect quality, different holding costs, and learning effects: A note. *Computers & Industrial Engineering* 58(1): 186–190.

Waterworth, C.J., 2000. Relearning the learning curve: A review of the derivation and applications of learning-curve theory. *Project Management Journal* 31(1): 24–31.

Wickens, C.D., Gordon, S.E., and Liu,Y., 1998. *An introduction to human factors engineering.* New York: Addison Wesley Longman.

Womer, N., 1979. Learning curves, production rate, and program costs. *Management Science* 25(4): 312–319.

Wright, T.P., 1936. Factors affecting the cost of airplanes. *Journal of the Aeronautical Sciences* 3(2): 122–128.

Yang, L., Wang, Y., and Pai, S., 2009. On-line SPC with consideration of learning curve. *Computers & Industrial Engineering* 57(3): 1089–1095.

Yelle, L.E., 1976. Estimating learning curves for potential products. *Industrial Marketing Management* 5(2–3): 147–154.

Yelle, L.E., 1979. The learning curve: Historical review and comprehensive survey. *Decision Sciences* 10(2): 302–328.

Yelle, L.E., 1980. Industrial life cycles and learning curves: Interaction of marketing and production. *Industrial Marketing Management* 9(2): 311–318.

Yelle, L.E., 1983. Adding life cycles to learning curves. *Long Range Planning* 16(6): 82–87.

Zamiska, J., Jaber, M.Y., and Kher, H., 2007. Worker deployment in dual resource constrained systems with a task-type factor. *European Journal of Operational Research* 177(3): 1507–1519.

Zangwill, W.I., and Kantor, P.K., 1998. Toward a theory of continuous improvement and the learning curve. *Management Science* 44(7): 910–920.

Zangwill, W.I., and Kantor, P.K., 2000. The learning curve: A new perspective. *International Transactions in Operational Research* 7(6): 595–607.

2 Inside the Learning Curve: Opening the Black Box of the Learning Curve

Michael A. Lapré

CONTENTS

INTRODUCTION

The learning curve phenomenon is well known. As organizations gain operating experience, organizational performance improves, although at a decreasing rate. Scholars have frequently used the power curve to model this relationship in manufacturing contexts. In these models, the logarithm of unit cost decreases linearly as a function of the logarithm of cumulative number of units produced (Yelle 1979). The decrease in cost (i.e., improvement) is attributed to organizational learning, hence the name "learning curve." Scholars have extended the power curve by incorporating forgetting (recent experience matters more than older experience) and learning from others (transfer of experience). For an overview of forgetting and learning from others, see Argote (1999). This chapter focuses on learning from own experience.

    The disappointing implication of the typical use of the power curve is that management can only accelerate learning from own experience by producing more. There are several limitations to this traditional view of the learning curve. First, the

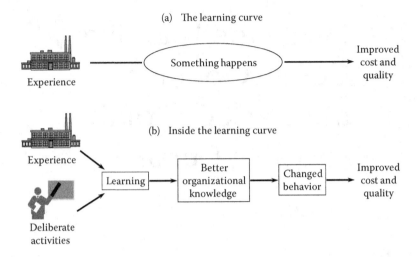

(a) The learning curve

Experience — Something happens → Improved cost and quality

(b) Inside the learning curve

Experience / Deliberate activities → Learning → Better organizational knowledge → Changed behavior → Improved cost and quality

FIGURE 2.1 Two views of the learning curve. (Adapted from Bohn, R.E., *Sloan Management Review* 36(1), 1994.)

rate of improvement—the learning rate—is typically treated as some exogenously given constant. However, there is ample evidence that learning rates vary widely across industries, within industries across organizations, and within organizations across organizational units (Dutton and Thomas 1984; Hayes and Clark 1985; Lapré and Van Wassenhove 2001). Hence, the learning rate should be treated as an endogenous variable. In other words, management is actually responsible for managing the rate of improvement. Second, experience—typically measured by cumulative production volume—is not the only source for learning. Organizations can engage in deliberate learning activities such as quality improvement projects. Third, and most importantly, the traditional view treats the learning in the learning curve as some "black box" (see Figure 2.1a). Yet, there is an actual learning process inside the learning curve. Learning results from experience and deliberate activities. It can yield better organizational knowledge, and better organizational knowledge can persuade organizational members to modify behavior. Changed behavior, in turn, can improve organizational performance (Bohn 1994) (see Figure 2.1b). None of these steps are trivial. Scholars have merely scratched the surface in terms of studying these steps. No single empirical study has incorporated all of the steps.

This chapter reviews empirical findings in the literature in terms of (i) different sources for learning, and (ii) partial assessments of the steps that make up the actual learning process inside the learning curve. The chapter concludes by identifying opportunities for future research that should provide insights for organizations to better manage their learning curves.

EXPERIENCE AS A SOURCE FOR LEARNING

As far as learning from own experience goes, scholars have started to investigate three interesting themes: First, what constitutes own experience and how should it be measured? Is cumulative volume the most relevant proxy, or would a variant of cumulative volume be a more accurate measure of experience? Second, to what extent can organizations gain a competitive learning curve advantage from specialization? Third, what factors contribute to the rather significant variation in learning rates from own experience?

NATURE OF EXPERIENCE

The learning curve literature has focused on three common experience variables: (1) cumulative volume, (2) calendar time, and (3) maximum volume. As mentioned before, cumulative production volume is the typical experience variable (Yelle 1979; Argote 1999). Repetition allows organizations to gain experience and to fine tune operations. Cumulative volume captures the notion of "learning by doing." Some scholars have used calendar time elapsed since the start of operation (Levin 2000; Field and Sinha 2005). When time for reflection is more important for learning, calendar time captures the notion of "learning by thinking." Mishina (1999) proposed the third experience variable: maximum output produced to date, or maximum proven capacity to date. When a plant is scaling up production, the production system faces significant challenges and new situations. Factory personnel need to figure out how to solve such challenges during scale-up. Mishina's experience variable captures the notion of "learning by new experiences" or "learning by stretching" (Mishina 1999; Lapré et al. 2000). Only two studies have compared learning curve estimations with all three measures of experience. Interestingly, both studies concluded that maximum volume was the best measure. Mishina (1999) found that the learning curve estimation for bomber airplanes suffered from autocorrelation with cumulative volume and calendar time, but not with maximum volume. In tire-cord manufacturing, Lapré et al. (2000) found that maximum volume explained the learning curve much better than cumulative volume and calendar time. Future research should assess whether these findings generalize beyond airplane and tire-cord manufacturing.

Not all experience is necessarily equally effective in improving organizational performance. First, organizations could learn significantly from their own failures (Cannon and Edmondson 2005). Whenever a production unit produces defective units, such defects provide opportunities to learn from and improve the production system. Li and Rajagopalan (1997) found that the cumulative number of defective units is statistically more significant than the cumulative number of good units in explaining learning curve effects. For manufacturing contexts that require defective units to be reworked, Jaber and Guiffrida (2004) propose a model that incorporates rework time. Depending on the evolution of rework time, a learning curve may continue to improve, plateau, or deteriorate. Future learning curve research is needed to empirically investigate the role of defects and rework.

Second, experience can accumulate at the individual, team, and organizational levels. Recently, scholars have investigated the impact of team experience. In addition to organizational experience measured with the usual cumulative volume variable, organizations with stable teams could potentially reach higher performance levels. In stable teams, team members learn how to better coordinate work with one another, because team members learn (1) who is best at performing which role, and (2) to trust one another. Scholars have found that team experience is a significant driver for learning curves in health care (Reagans et al. 2005) and software development (Huckman et al. 2009).

Sinclair et al. (2000) have questioned the role that experience plays in achieving better organizational performance. Their study suggests that cumulative volume provides an indication of future volume. Future expected volume conditioned future expected returns from research and development (R&D) and, by extension, the choice of R&D projects. Research and development projects—not cumulative volume—were the real source of cost reduction. Future research should continue to investigate the question of "under what conditions does a certain type of experience trigger actual learning?"

EXPERIENCE AND SPECIALIZATION

In 1974, Skinner introduced the notion of a "focused factory." Focused factories that specialize in executing fewer tasks outperform factories that perform a wider set of tasks. Task homogeneity, coupled with a higher frequency of repetition, allows a factory to learn more quickly from its experience. Several learning curve studies have investigated the benefits of specialization. In the U.S. hotel industry, Ingram and Baum (1997) found that hotel chains operating in a limited geographic region (geographic specialists) benefited more from their own experience than hotel chains operating nationwide (geographic generalists). In a study of incidents and accidents in the U.S. airline industry, Haunschild and Sullivan (2002) found that specialist airlines benefited more from analyzing heterogeneous causes than generalist airlines. Heterogeneous causes allow for deeper analysis. The authors concluded that focus helped specialist airlines to analyze heterogeneous causes. Interestingly, however, it might not be optimal to focus as much as possible. An experimental study showed that some degree of variation yields faster learning rates. Schilling et al. (2003) found that learning rates for related experience are greater than for specialized or unrelated experience. An analysis of an offshore software services operation confirmed the potential benefits of related variation. According to Narayanan et al. (2006), exposure to a greater variety of tasks improves long-term productivity. However, investments in learning new tasks can impede short-term productivity. A learning curve study on customer dissatisfaction compared specialist and generalist airlines (Lapré and Tsikriktsis 2006). Average specialist airlines did not learn faster than average generalist airlines. However, the best specialist airline did learn faster than the best generalist airline. So, focus provides an opportunity for faster learning, but there are no guarantees for superior performance. A promising area for future work would be to investigate under what conditions does a particular level of specialization result in faster learning from a certain type of experience?

VARIATION IN LEARNING RATES

It has been well documented that organizations show tremendous variation in learning rates. Dutton and Thomas (1984), for example, graphed a distribution of learning rates from a sample of over 100 studies. Dutton and Thomas (2004, 238) expressed the learning rate as a progress ratio: "When cumulative volume doubles, the cost per unit declines to p% of original cost." "P" is called the progress ratio. Progress ratios ranged from 55% to 108%. Understanding the dynamics that cause learning curve heterogeneity has been an important area for learning curve research. In a study of adoption of minimally invasive cardiac surgery in sixteen hospitals, Pisano et al. (2001) found significant learning curve heterogeneity. The authors used case data from two hospitals to explore differences that might have contributed to variation in learning rates. The two hospitals differed markedly in terms of: (i) their use of formal procedures for new technology adoption, (ii) cross-functional communication, (iii) team and process stability, (iv) team debrief activities, and (v) surgeon coaching behavior. In a follow-up study, Edmondson et al. (2003) found that learning curve heterogeneity is greater for aspects of performance that rely on tacit knowledge (as opposed to codified knowledge). Wiersma (2007) investigated how four conditions affected learning rates across twenty-seven regions in the Royal Dutch Mail Company. A higher degree of temporary employees, a higher level of free capacity, a higher degree of product heterogeneity, and less conflicting concerns about other performance measures all had a favorable impact on the learning rate. It will be worthwhile for future research to further quantify conditions that vary across organizations and include such quantitative data in learning curve analyses.

DELIBERATE ACTIVITIES AS A SOURCE FOR LEARNING

Organizations do not have to limit themselves to learning from experience. They can also engage in a more pro-active approach to managing learning curves. Levy (1965) introduced the distinction between "autonomous learning" from experience and "induced learning" from deliberate activities designed to improve production processes. Examples of deliberate activities include both pre-production planning before a process starts, as well as industrial engineering after a process starts. Levy found that prior experience and training explain differences in the estimated learning rates for individual workers. This was a landmark study even though the explanatory variables prior experience and training did not evolve over time. Adler and Clark (1991) made the next step by incorporating longitudinal variables for deliberate activities in productivity learning curves: cumulative engineering activity and cumulative training activity. In one production department, engineering activity enhanced productivity while training activity disrupted productivity. In a second production department, the exact opposite occurred. Thus, deliberate learning activities can both help and hurt. The authors provided some case-based explanations for these surprising findings. For example, if producibility concerns trigger engineering activity, engineering activity enhances productivity. On the other hand, if product performance concerns trigger engineering changes, such changes could be disruptive. Hatch and Mowery (1998) studied the impact of cumulative engineering in yield learning curves in semiconductor manufacturing. Yield learning curves for processes in the early stages of manufacturing were driven by cumulative engineering as opposed to cumulative volume. In more mature processes, cumulative engineering and cumulative volume were both sources for learning to improve yields.

However, the introduction of new processes disrupted the ongoing learning activities of existing processes. Hatch and Dyer (2004) further examined yield learning curves and showed that, in addition to cumulative engineering, human capital variables such as screening tests and statistical process control training significantly enhanced yield improvements. Human capital variables differed across processes, but were constant over time within a process.

Lapré et al. (2000) provide a systemic explanation for the seemingly unpredictable effect of deliberate learning activities observed by Adler and Clark (1991). In a tire-cord manufacturing plant, the authors studied quality improvement projects as deliberate learning activities. They found that the cumulative number of quality improvement projects that generated both know-why and know-how accelerated waste reduction, whereas the cumulative number of quality improvement projects that generated know-why without know-how slowed down waste reduction. A production line run as a learning laboratory, called a "model line," consistently produced the learning-rate-enhancing mix of know-why and know-how. Replication of this model line concept in other plants in the same firm fell short of expectations (Lapré and Van Wassenhove 2001). These replications neglected conditions of management buy-in and knowledge diversity to solve interdepartmental problems.

Two studies provide further evidence that deliberate activities should be included in learning curves in addition to autonomous learning-by-doing variables. Ittner et al. (2001) showed that both cumulative quality engineering and cumulative design engineering significantly reduced defect rates. Quality engineering includes prevention activities such as quality planning, developing and maintaining the quality planning and control systems, quality improvement activities, and internal quality improvement facilitation and consulting. Design engineering covers product design engineering expenses incurred for prevention activities. Arthur and Huntley (2005) studied cost reduction ideas submitted by staff in an auto-parts manufacturing plant's gainsharing program. The authors found that the cumulative number of implemented employee suggestions (their measure of deliberate learning) significantly reduced costs. More research is needed to understand under what conditions does a certain type of deliberate activities enhance knowledge creation, adoption, and organizational performance.

STEPS INSIDE THE LEARNING CURVE

Very few studies have addressed the steps inside the learning curve as indicated in Figure 2.1. Only two studies have incorporated a step from Figure 2.1 with longitudinal variables in learning curve estimations (Lapré et al. 2000; Arthur and Huntley 2005). This section also reviews related research—cross-sectional studies without a link to longitudinal organizational performance.

Mukherjee et al. (1998) studied 62 quality-improvement projects undertaken in a tire-cord plant over a decade. A cross-sectional analysis of these projects showed significant variation on two dimensions of the learning process: conceptual learning and operational learning. Conceptual learning consists of using science and statistical experimentation to develop a deeper understanding of cause-and-effect relationships—in other words, the development of know-why. Operational learning consists of modifying action variables and obtaining follow-up of experiments—in other words, the development of know-how. Both conceptual and operational learning enhanced changed behavior measured by modifications in standard operating procedures and statistical process control rules. Lapré et al. (2000) used the dimensions of conceptual and operational learning to split a sample of quality improvement projects into four categories according to high or low conceptual learning and high or low operational learning. For each of the four categories, the authors constructed longitudinal variables capturing the cumulative number of projects completed to date. The four cumulative project variables were incorporated in a learning curve estimation for the factory's waste rate—the percentage of products that had to be scrapped because of irreparable defects. Only two cumulative project variables had a statistically significant impact on waste evolution. Projects with high conceptual learning and low operational learning were disruptive. These "non-validated theories" were often advanced by experts from central R&D, yet the insights obtained at central R&D were not developed sufficiently for a full-scale manufacturing environment. Projects with high conceptual learning and high operational learning, on the other hand, accelerated waste reduction. These "operationally validated theories" provided solutions that worked, backed by scientific principles explaining why these solutions worked. This study explicitly incorporates the "better organizational knowledge" step in Figure 2.1 in a longitudinal learning curve estimation. The research site lacked historical data on all modifications to standard operating procedures and statistical process control rules. If such data had been available, the "changed behavior" variable could also have been included. The Arthur and Huntley (2005) study mentioned in the previous section used a cumulative number of implemented employee suggestions, which does capture "changed behavior" in Figure 2.1.

None of the steps depicted in Figure 2.1 are self-evident. Tucker et al. (2002), for example, investigated problem-solving behavior by front-line workers. Faced with a problem, nurses typically engage in "first-order problem solving"—fixing a problem without doing anything to prevent a similar problem from occurring in the future. Rarely do nurses engage in "second-order problem solving"—conducting root-cause analysis to change underlying causes. In 92% of the 120 problems observed by the authors, nurses ignored possible root causes. Such a focus on first-order problem solving prevents actual learning. Despite the opportunity for root-cause analysis, no attempt has been made to create better organizational knowledge. Tucker et al. (2002) identified several factors contributing to a continued emphasis on first-order problem solving at the expense of learning. First, front-line personnel feel good about themselves by patching a problem, demonstrating independence and competence. Second, a high workload focuses front-line workers on completing pressing tasks now, rather than thinking about improvement for the future. Third, the lower status of nurses compared to physicians might prevent nurses "to intrude upon a physician's time." The authors conclude that it is necessary to create an organizational environment that is psychologically safe (Edmondson 1999). Tucker et al. (2007) investigated the importance of psychological safety (a supportive organizational context) in deliberate learning activities. The authors conducted a cross-sectional study of organizational learning in 23 neonatal intensive care units (NICUs). The authors studied learn-how, which concerns "understanding why a practice works, as well as how to carry it out" Tucker et al. (2007, 898). Like Mukherjee et al. (1998), the authors link "learning" to "changed behavior" in Figure 2.1. Learn-how in improvement projects enhanced project implementation success. Tucker et al. (2007) found that psychological safety was an antecedent of learn-how. Furthermore, the level of published evidence for a practice (a measure of knowledge) also enhanced implementation success. In a follow-up study, Nembhard et al. (2009) investigated the impact of learn-how on organizational performance. The authors used data on 1061 infant patients from the same 23 NICUs. Learn-how significantly reduced patient mortality rates (measured at the organizational level). Moreover, interdisciplinary collaboration was found to mediate the relationship between learn-how and organizational performance.

A survey of 188 six sigma projects in a manufacturing firm provides further cross-sectional evidence for Figure 2.1. Choo et al. (2007) found that "learning behaviors" enhanced "knowledge created," which in turn enhanced "project performance." Moreover, the authors found that "use of a structured method" was an antecedent of "learning behaviors," whereas "psychological safety" was an antecedent of "knowledge created."

AVENUES FOR FUTURE RESEARCH

So far, this chapter has identified the following avenues for future research:

- What are the conditions that determine which measure of experience best captures the learning curve phenomenon?
- When does what type of experience trigger actual learning?
- Quantify conditions that vary across organizations and include such quantitative data in learning curve analyses to explain learning curve heterogeneity.
- When does how much specialization result in faster learning from what type of experience?
- Under what conditions do which deliberate activities enhance knowledge creation, adoption, and organizational performance?

A major reason for learning curve heterogeneity is the nature of the existing knowledge base in organizations. As mentioned earlier, Edmondson et al. (2003) found more heterogeneity for aspects of performance that rely on tacit knowledge. Many organizations have incomplete knowledge of their production systems (Jaikumar and Bohn 1992). Examples of settings characterized by incomplete knowledge include kitchens in commercial food firms (Chew et al. 1990), semiconductor manufacturing (Hatch and Mowery 1998), tire-cord manufacturing (Lapré and Van Wassenhove 2001), cardiac surgery (Edmondson et al. 2003), and electromechanical plants (Field and Sinha 2005). In the face of incomplete knowledge, organizations rely more on art as opposed to science. As a result, organizations depend on workers to figure out how to control processes, create better knowledge, and improve performance. "Transforming operators into quasi-engineers requires investments in human capital but pays big dividends in learning performance" (Hatch and Dyer 2004, 1173) How can we assess whether an organization is making progress in moving from an art to a science? Jaikumar and Bohn (1992), and Bohn (1994, 1995, 2005) introduced the "stages of knowledge" to gage progress of knowledge creation.

STAGES OF KNOWLEDGE

Bohn (1994, 62) defined technological knowledge as "understanding the effects of the input variables on the output. Mathematically, the process output, Y, is an unknown function f of the inputs, x: $Y = f(x)$; x is always a vector (of indeterminate dimension)." Inputs include raw materials, control variables, and environmental variables. Jaikumar and Bohn (1992) and Bohn (1994) developed "stages of knowledge" detailing how much an organization knows about $Y = f(x)$. In 1995, Bohn refined the concept, recognizing that stages of knowledge need to be measured along two separate dimensions: causal knowledge and control knowledge. Table 2.1 depicts the stages of causal knowledge and control knowledge. (See also Bohn 2005.)

Causal knowledge assesses how much an organization knows about the relationship between an input x_i and output y. At stage 1 "ignorance," the organization is unaware that x_i might affect y. At stage 2 "awareness," the organization is aware that x_i and y are related, but the direction of causality is unknown. At stage 3 "direction," the organization knows that x_i affects y. At stage 4 "magnitude," the organization can quantify the impact of a small change in x_i on y. At stage 5 "scientific model," the organization has a functional specification with parameters describing the relationship between x_i and y. At stage 6 "interactions," the organization has extended stage 5 knowledge to include interactions with all other input variables (never obtained in practice).

Control knowledge, on the other hand, assesses an organization's ability to keep an input variable x_i at its desired target level. At stage 1 "ignorance," the organization is unaware of x_i. At stage 2 "awareness," the organization is aware of the existence of x_i. At stage 3 "measure," the organization is able to measure x_i routinely. At stage 4 "control of the mean," the organization can control x_i at the mean level, but there is significant variation in the level of x_i. At stage 5 "control of the variance," the

---

TABLE 2.1
Stages of Causal Knowledge and Control Knowledge

Causal Knowledge Know How x_i affects y	Control Knowledge Know How to Control x_i
1. Ignorance	1. Ignorance
2. Awareness	2. Awareness
3. Direction	3. Measure
4. Magnitude	4. Control of the mean
5. Scientific model	5. Control of the variance
6. Interactions	6. Reliability

Note: Author's notes taken during a presentation by Bohn (1995).

---

organization can control the variance of x_i. At stage 6 "reliability," the organization can always keep x_i at its target level (never obtained in practice).

The two dimensions of causal and control knowledge closely mirror the dimensions of the learning process identified by Mukherjee et al. (1998). Conceptual learning should allow an organization to climb the stages of causal knowledge; whereas operational learning should allow an organization to climb the stages of control knowledge. Lapré et al. (2000) demonstrated the significance of incorporating the two learning dimensions into a learning curve estimation. It would be a major contribution to include longitudinal progress on the stages of knowledge in a learning curve estimation. At what stages of causal knowledge and control knowledge can an organization expect to make more than merely incremental improvements? Do breakthrough improvements require balanced climbing of the stages knowledge; that is, should causal knowledge and control knowledge progress at the same pace? Primary variables are variables that directly impact output. Secondary variables are variables that directly impact primary variables. What is the impact of climbing the stages of knowledge for primary variables versus secondary variables? One challenge in addressing such questions is in finding a research site where progress along the stages of knowledge can be captured. However, some organizations are aware of progress along stages of knowledge. Ittner et al. (2001), for example, used a research site that measured four stages of quality-based learning: (1) "aware of need," (2) "process characterized and sources of variation identified," (3) "critical process parameters understood," and (4) "knowledge institutionalized." In a longitudinal field study of an electromechanical motor assembly plant, Field and Sinha (2005) found that actions to control the mean (control knowledge stage 4) do indeed precede actions to control the variance (control knowledge stage 5). Scholars have yet to quantitatively measure progress on the stages of knowledge and include such measures in learning curve estimations.

LEARNING TO IMPROVE MULTIPLE DIMENSIONS OF ORGANIZATIONAL PERFORMANCE

Learning curve scholars have focused their attention on single dimensions of organizational performance. However, organizations might have to perform on more than one dimension. The operations strategy literature has advanced cost, quality, delivery, and flexibility as typical candidates for competitive priorities. Typically, not all competitive priorities are equally important at all times. For example, an entrepreneurial firm might successfully compete on quality, whereas a mature firm might find it necessary to have low cost. Hence, the importance of different competitive priorities might change over time (Corbett and Van Wassenhove 1993).

Initially, competitive priorities were thought of as fundamental trade-offs. Higher quality implies higher cost, while cost reductions imply worse quality. Ferdows and De Meyer (1990) challenged this inherent trade-off view. Their sand-cone model proposes that capabilities are built cumulatively: first invest in improving quality, then delivery and flexibility, ending with cost reduction. Lapré and Scudder (2004) found empirical evidence for the sand-cone model in the U.S. airline industry. Airlines that ended up in a sustainable superior quality-cost position made larger initial improvements in quality compared with cost, although trade-offs do occur when operating close to asset frontiers. Gino et al. (2006) propose that learning to improve one dimension may come at the expense of learning on another dimension. In a sample of 16 hospitals, the authors found evidence of a learning trade-off between efficiency and application innovation.

Future learning curve research should address learning along multiple performance measures. Do different types of experiences enhance different dimensions of performance? How do organizations learn to improve internal performance (such as cost) and external performance (such as customer loyalty) at the same time? What is the relationship between internally oriented operating experience and externally oriented competitive experience?

In the past two decades, learning curve scholars have made important contributions to understanding what processes are "behind the learning curve." Much work, however, remains to be done. Hopefully, this chapter inspires others to further our understanding of organizational learning curves.

ACKNOWLEDGMENTS

Ingrid Nembhard and Bradley Staats provided very helpful comments. Support for this chapter was provided by the Dean's Fund for Faculty Research, Owen Graduate School of Management at Vanderbilt University.

REFERENCES

Adler, P.S., and Clark, K.B., 1991. Behind the learning curve: A sketch of the learning process. *Management Science* 37(3): 267–281.

Argote, L., 1999. *Organizational learning: Creating, retaining and transferring knowledge.* Norwell: Kluwer Academic.

Arthur, J.B., and Huntley, C.L., 2005. Ramping up the organizational learning curve: Assessing the impact of deliberate learning on organizational performance under gainsharing. *Academy of Management Journal* 48(6): 1159–1170.

Bohn, R.E., 1994. Measuring and managing technological knowledge. *Sloan Management Review* 36(1): 61–73.

Bohn, R.E., 1995. Measuring technological knowledge in the hard disk-drive industry. Presentation at the Fall meeting of the Institute of Management Sciences and Operations Research (INFORMS), New Orleans.

Bohn, R.E., 2005. From art to science in manufacturing: The evolution of technological knowledge. *Foundations and Trends in Technology, Information and Operations Management* 1(2): 129–212.

Cannon, M.D., and Edmondson, A.C., 2005. Failing to learn and learning to fail (intelligently): How great organizations put failure to work to innovate and improve. *Long Range Planning* 38(3): 299–319.

Chew, W.B., Bresnahan, T.F., and Clark, K.B., 1990. Measurement, coordination, and learning in a multiplant network. In *Measures for Manufacturing Excellence*, ed. R.S. Kaplan, 129–162. Boston: Harvard Business School Press.

Choo, A.S., Linderman, K.W., and Schroeder, R.G., 2007. Method and psychological effects on learning behaviors and knowledge creation in quality improvement projects. *Management Science* 53(3): 437–450.

Corbett, C., and Van Wassenhove, L., 1993. Trade-offs? What trade-offs? Competence and competitiveness in manufacturing strategy. *California Management Review* 35(4): 107–122.

Dutton, J.M., and Thomas, A., 1984. Treating progress functions as a managerial opportunity. *Academy of Management Review* 9(2): 235–247.

Edmondson, A.C., 1999. Psychological safety and learning behavior in work teams. *Administrative Science Quarterly* 44(2): 350–383.

Edmondson, A.C., Winslow, A.B., Bohmer, R.M.J., and Pisano, G.P., 2003. Learning how and learning what: Effects of tacit and codified knowledge on performance improvement following technology adoption. *Decision Sciences* 34(2): 197–223.

Ferdows, K., and De Meyer, A., 1990. Lasting improvements in manufacturing performance: In search of a new theory. *Journal of Operations Management* 9(2): 168–184.

Field, J.M., and Sinha, K.K., 2005. Applying process knowledge for yield variation reduction: A longitudinal field study. *Decision Sciences* 36(1): 159–186.

Gino, F., Bohmer, R.M.J., Edmondson, A.C., Pisano, G.P., and Winslow, A.B., 2006. *Learning tradeoffs in organizations: Measuring multiple dimensions of improvement to investigate learning-curve heterogeneity*. Working Paper 05–047. Boston: Harvard Business School.

Hatch, N.W., and Dyer, J.H., 2004. Human capital and learning as a source of sustainable competitive advantage. *Strategic Management Journal* 25(12): 1155–1178.

Hatch, N.W., and Mowery, D.C., 1998. Process innovation and learning by doing in semiconductor manufacturing. *Management Science* 44(11): 1461–1477.

Haunschild, P.R., and Sullivan, B.N., 2002. Learning from complexity: Effects of prior accidents and incidents on airlines' learning. *Administrative Science Quarterly* 47(4): 609–643.

Hayes, R.H., and Clark, K.B., 1985. Exploring the sources of productivity differences at the factory level. In *The uneasy alliance: Managing the productivity-technology dilemma*, eds. K.B. Clark, R.H. Hayes, and C. Lorenz, 151–188. Boston: Harvard Business School Press.

Huckman, R.S., Staats, B.R., and Upton, D.M., 2009. Team familiarity, role experience, and performance: Evidence form Indian Software Services. *Management Science* 55(1): 85–100.

Ingram, P., and Baum, J.A.C., 1997. Opportunity and constraint: Organizations' learning from the operating and competitive experience of industries. *Strategic Management Journal* 18(S1): 75–98.

Ittner, C.D., Nagar, V., and Rajan, M.V., 2001. An empirical examination of dynamic quality-based learning models. *Management Science* 47(4): 563–578.

Jaber, M.Y., and Guiffrida, A.L., 2004. Learning curves for processes generating defects requiring reworks. *European Journal of Operational Research* 159(3): 663–672.

Jaikumar, R., and Bohn, R.E., 1992. A dynamic approach to operations management: An alternative to static optimization. *International Journal of Production Economics* 27(3): 265–282.

Lapré, M.A., Mukherjee, A.S., and Van Wassenhove, L.N., 2000. Behind the learning curve: Linking learning activities to waste reduction. *Management Science* 46(5): 597–611.

Lapré, M.A., and Scudder, G.D., 2004. Performance improvement paths in the U.S. airline industry: Linking trade-offs to asset frontiers. *Production and Operations Management* 13(2): 123–134.

Lapré, M.A., and Tsikriktsis, N., 2006. Organizational learning curves for customer dissatisfaction: Heterogeneity across airlines. *Management Science* 52(3): 352–366.

Lapré, M.A., and Van Wassenhove, L.N., 2001. Creating and transferring knowledge for productivity improvement in factories. *Management Science* 47(10): 1311–1325.

Levin, D.Z., 2000. Organizational learning and transfer of knowledge: An investigation of quality improvement. *Organization Science* 11(6): 630–647.

Levy, F., 1965. Adaptation in the production process. *Management Science* 11(6): B136–B154.

Li, G., and Rajagopalan, S., 1997. The impact of quality on learning. *Journal of Operations Management* 15(3): 181–191.

Mukherjee, A.S., Lapré, M.A., and Van Wassenhove, L.N., 1998. Knowledge driven quality improvement. *Management Science* 44(11): S35–S49.

Mishina, K., 1999. Learning by new experiences: Revisiting the flying fortress learning curve. In *Learning by doing in markets, firms, and countries*, eds. N.R. Lamoreaux, D.M.G. Raff, and P. Temin, 145–179. Chicago: The University of Chicago Press.

Narayanan, S., Balasubramanian, S., and Swaminathan, J.M., 2006. *Individual learning and productivity in a software maintenance environment: An empirical analysis*. Working Paper, University of North Carolina at Chapel Hill. http://isb.edu/faculty/upload/Doc4120072316.pdf (accessed February 9, 2009).

Nembhard, I.M., Tucker, A.L., Bohmer, R.M.J., Carpenter, J.H., and Horbar, J.D., 2009. *Learn-how to improve collaboration and performance*. Working Paper 08–002. Boston: Harvard Business School.

Pisano, G.P., Bohmer, R.M.J., and Edmondson, A.C., 2001. Organizational differences in rates of learning: Evidence from the adoption of minimally invasive cardiac surgery. *Management Science* 47(6): 752–768.

Reagans, R., Argote, L., and Brooks, D., 2005. Individual experience and experience working together: Predicting learning rates from knowing who knows what and knowing how to work together. *Management Science* 51(6): 869–881.

Schilling, M.A., Vidal, P., Ployhart, R.E., and Marangoni, A., 2003. Learning by doing something else: Variation, relatedness, and the learning curve. *Management Science* 49(1): 39–56.

Sinclair, G., Klepper, S., and Cohen, W., 2000. What's experience got to do with it? Sources of cost reduction in a large specialty chemicals producer. *Management Science* 46(1): 28–45.

Skinner, W., 1974. The focused factory. *Harvard Business Review* 52(3): 113–121.

Tucker, A.L., Edmondson, A.C., and Spear, S., 2002. When problem solving prevents organizational learning. *Journal of Organizational Change Management* 15(2): 122–137.

Tucker, A.L., Nembhard, I.M., and Edmondson, A.C., 2007. Implementing new practices: An empirical study of organizational learning in hospital intensive care units. *Management Science* 53(6): 894–907.

Wiersma, E., 2007. Conditions that shape the learning curve: Factors that increase the ability and opportunity to learn. *Management Science* 53(12): 1903–1915.

Yelle, L.E., 1979. The learning curve: Historical review and comprehensive survey. *Decision Sciences* 10(2): 302–328.

3 Learning and Thinking Systems

J. Deane Waldman and Steven A. Yourstone

CONTENTS

WHAT IS LEARNING AND HOW DO WE DO IT?

A dictionary definition of learning is: "To gain or acquire knowledge of or skill in (something) by study, experience or being taught." Humans are thinking systems (Waldman 2007, 2010) and when learning, thinking systems are unique in three ways: (1) they continuously learn; (2) they can structure their learning; and (3) they can, and often do, have multiple goals they wish to achieve from their learning.

TYPES OF LEARNING

There are three different ways to learn: (1) rote learning; (2) practiced variations; and (3) innovation. The first two are important. Indeed, the first is a prerequisite to the second. Only the third can improve on what we currently have or do now.

Rote Learning

Say that someone teaches you something new, like serving a tennis ball. First, you only occasionally make contact with the ball. Then, with practice, you begin to hit it every time and learn the right angle of the racquet and the proper amount of force to apply to make the ball land on the opposite side of the net. Continued practice enables you to place the ball within the service area with each serve. Eventually, with arduous practice, you can make the ball land in different parts of the service area, with a variety of speeds and even with a kick or curved bounce after landing. This is a result of rote practice, or what athletes call "muscle memory."

Rote practice allows you to perfect your ability to do some thing by doing it over and over with the same results—*assuming that the conditions and the desired outcomes* are the same every time. However, what if the conditions are not always the same? What if the desired outcome changes?

Practiced Variations

A second type of learning is practiced variations—this is the type of learning that airline pilots benefit from through their use of simulators. Pilots first practice the rote actions of starting an airplane, taking-off, flying, and landing until they can do this the same way every time.

After mastering rote learning, the pilot practices variations. What if conditions are icy? Practice taking-off and landing in such conditions in a simulator. What if you are flying through a hurricane? Simply practice flying in a simulated hurricane. A more recent simulation that pilots and attendants must practice involves a hijacking scenario. What if the hydraulics fail and you cannot lower the landing gear? It seems best that the pilot should not have to learn how to deal with this—as a first experience—with us on-board the plane.

Flight planners think up every situation that they can imagine and have pilots practice those variations. As passengers, we can board with confidence that the pilot has practiced every conceivable situation. However, what if the plane encounters a new, and therefore never before simulated, situation? The highest and most difficult form of learning involves innovation.

Innovation

Innovation strictly means to "introduce something new; make a change in something established." We prefer to use the word innovation more broadly in order to include the invention phase—the step of creating (and not just introducing) something original. Sir William Bragg, 1915 Nobel Laureate in Physics said: "The most important thing in science is not so much to obtain new facts as to discover new ways of thinking about them." Bruner's (1983, 183) definition of creativity captures a similar essence: "Figuring out how to use what you already know to go beyond what you currently think."

The airplane pilot in a simulator does not innovate: the programmer of the simulator did. In the 1989 Tour de France, American bike racer Greg LeMond used a radical innovation, aerodynamic handlebars, in the final stage to win a three-week long race by the narrowest margin in history—eight seconds!

For over 100 years, investigators have studied how innovation occurs, how it is absorbed into the base of general knowledge, and how it is adopted (Rogers 1983). A necessary precursor is a person's mindset: open; not certain of his/her correctness; and willing to consider silly, obviously wrong, and/or unpopular ideas. A second prerequisite is an environment and culture that encourages new ideas and risk taking.

Learning Requires "E&F"

Structuring a learning process often tends to focus only on the technical: who teaches whom, when, how, and what. An effective learning process must include "E&F": evidence and (embedded) feedback. To determine the effect of the learning experience—didactic lecture; mediated experience; heuristic learning—and to learn how to improve on this the next time (in other words, to learn from learning), one must have evidence of the effect and must feed that information back to those who make decisions and first implement the learning experience.

Most teachers get fed back very little evidence of what they really accomplish. The most important learning that occurs in schools is not defined by student test scores: it is the development of a productive and self-actualizing member of society. This is not measured and therefore cannot be fed back to teachers as evidence of what they have achieved.

Good manufacturing systems, such as Toyota production systems, structure the process as a continuous learning experiment with both interim and terminal results and utilize embedded feedback throughout the process (Spear and Bowen 1999).

Figure 3.1a represents a classic manufacturing value chain. By simply changing the labels, this value chain would be appropriate for a service endeavor. An example of a value chain in health care (HC) is shown in Figure 3.1b, along with one of its subsystems. These linear value chains ignore what has been learned in systems dynamics. To have a stable and yet continuously improving process, whether building a car or providing a service like health care, proper feedback is essential [dashed lines in (c)]. What is fed back is evidence of the effects—intended and unintended; adverse and beneficial; positive and negative; interim and final.

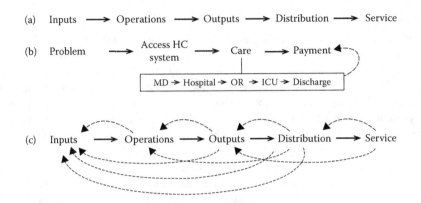

FIGURE 3.1   Value chains.

LEARNING BY SYSTEM

Many things seem to come in threes, including systems: (1) machine; (2) complex adaptive; and (3) what we call "thinking systems" (Waldman 2007). Any machine, from a plow to a super-computer, does only what it is programmed to do. Machine systems have no emotions or specific purpose(s) except as given by the designer. Even with proper maintenance, machine systems wear out or become obsolete because they neither learn nor adapt.

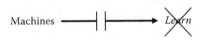

Much has been written about complex adaptive systems (CAS). Forests, beehives, cities—all systems with biologic elements—are CAS (Beinhocker 1997; McDaniel 1997). While machine outputs are predictable, CAS can produce unexpected or emergent results (Johnson 2001); that is, they learn. The process by which a CAS learns is random rather than directed to a specific predetermined outcome. A CAS learns for a single purpose: to improve its chances of survival.

Consider a biologic system such as a flock of birds. Over the course of millennia, birds learned to fly in a V-shaped formation, constantly rotating the lead position. The bird up front does the most work breaking the wind while the others slipstream behind. Those behind the lead bird rest until they take their turn at the front. In this manner, flocks of birds can travel further and faster, in order to survive until their destination.

No ancient imaginative bird genius set up a series of experiments to determine the best flying formation. Birds did not pre-specify what outcome they wanted.

Evolution drove them to co-evolve, to self-organize in different patterns, and to learn how best to fly with species survival emerging as the outcome.

Thinking Systems

Humans in groups (or even alone) are thinking systems and, as such, they continuously learn (Waldman 2007). Thinking systems have two unique capabilities lacking in other systems: (i) to structure their own learning, and (ii) to have multiple, and even contradictory, goals for their learning.

Regardless of what your organization does—whether it is commerce, manufacturing, finance, or a service such as health care—you are a thinking system. Failure to harness your special capabilities is throwing away the potential to create a competitive advantage, whether sustainable or transient.

Unplanned versus Structured Learning

Humans can learn just by doing, without planning either how to learn or what to learn. Simply by driving to work every day, we learn traffic patterns and the quickest routes versus the shortest distances. Most of what we learn is unplanned.

Employer's do not intend for us to learn not to take a strong public stand on controversial issues. Our manager does not intend for us to learn that there is only one right way to do things. That is, however, what many workers learn, though the lessons are unplanned and unintended.

Learning can and should be structured for both process and outcomes. In medical school, there is a necessary sequence of learning steps, each step planned to build on previously acquired knowledge. You must learn anatomy before you study surgery; you need to learn how nerves operate before you can consider how to treat nervous conditions.

The method of learning, the order of things learned, and testing of what has been learned—these things can all be planned. The old Bell telephone repairman had to learn about electricity and switches before he was allowed to climb up the pole to fix the line (Mobley 1982).

Mediated Learning Experience

The learning process can and should be mediated, as shown by the important work of Reuven Feuerstein. A teacher may recite and the student may be taught—a purely passive exercise where the student is like a computer hard drive onto which the teacher writes memory. A real educator helps the student to learn by showing where to access information and how to use it.

In an MLE, the mediator places himself/herself between the external or internal stimuli and the learner and transmits the data in a particular way for a specific purpose. The mediator seeks to impart meaning, rather than just information, by adding context (Weick 1993). The mediator might say: "You need to know this in order to watch out for that." The context makes learning easier, more useful, and permanent.

"Social interaction is the key to cognitive development and to learning" (Fickeisen 1991, 42). Anything that disconnects us from one another decreases the effectiveness of the learning experience. A mediated learning experience is the exact opposite of the comedy movie image of a big lecture hall with a video of the professor droning on in front of 150 tape recorders on the desks instead of students.

The mediator is not bound by anything, certainly not traditional didactic technique. The effective mediator chooses mode, location, technique, and even the person who should mediate, all with one objective: to help the student to learn. The maintenance man in the following story probably never heard of MLE. He is just a natural mediator.

A private school in Washington, DC was recently faced with a unique problem. A number of 12-year-old girls were beginning to use lipstick and would put it on in the bathroom. After they put on their lipstick, they would press their lips on the mirror leaving dozens of little lip prints.

Every night the maintenance man would remove the prints and the next day the girls would put them back.

Finally the principal decided that something had to be done. She called all the girls to the bathroom and met them there with the maintenance man. She explained that all these lip prints were causing extra work because he had to clean the mirrors every night.

To demonstrate how difficult it was to clean the mirrors, the principal asked the maintenance man to show the girls how much effort was required. He took out a long-handled squeegee, dipped it in the toilet, and cleaned the mirrors.

Since then there have been no lip prints.

The ultimate goal of MLE learning is to achieve what is called "transcendence": the ability to make connections between the specific and the general. "Transcendence is the heart of mediation. It involves moving beyond the immediate needs of whatever is going on in the current situation, or task, or what you're thinking about to develop the potential to apply it elsewhere in slightly different ways" (Fickeisen 1991, 45). A person capable of transcendence can create connections never conceived of before, and thus innovate.

There are many different ways to structure the learning process; MLE is just one. The keys are: (i) to recognize that the process can and should be structured; (ii) to be sure of the desired general result from the learning experience; and (iii) to accept that the results of the process will emerge, rather than be precisely predictable (this is not a machine system.)

Learning with Purposes (Plural)

Machines do not learn. CAS learn by random action with a single goal—to survive—the end point being procreation. Thinking systems like humans—and their organizations—typically want to survive, but thinking systems always have multiple purposes (Maslow 1943). Some may be even more important than survival. Just recall the New York City firefighters on September 11, 2001.

Think back to the flock of birds—how and why they learned. As a complex adaptive system, they learned to fly in a V-shaped formation through trial and error in order to survive as a species. Now consider the peloton of bicycle racers in the Tour de France—it looks just like a flock of migratory birds. The cyclist at the front of the pack works very hard for a brief period of time and then allows another to take his place while he recovers within the large mass of riders. In this way, the peloton covers large distances very quickly, for hours maintaining average speeds of 40–48 km per hour. How did they learn and why?

Months before the actual Tour de France, each bicycle team begins to experiment. They test which team members are better at climbing than at sprinting; which bike frame is more efficient on which stage; the best order of riders in line; and optimal aerodynamic position, and even clothing. All these activities represent structured learning aimed at multiple pre-specified goals, not just survival. Only humans can structure their own learning and have multiple, pre-determined goals in mind. Thinking systems learn differently than CAS, and machine systems do not learn at all. (Keep in mind that many modern companies still use the machine model as the foundation for making managerial decisions despite its incapacity for learning.)

Why We Learn

In *The Fifth Discipline*, Peter Senge (1990) writes that we all learn because we love to learn. World-renowned organizational behavior expert, Edgar Schein, took an opposing view when he compared learning to brainwashing: "You can't talk people out of their learning anxieties; they're the basis for resistance to change, none of us would ever try something new unless we experienced the second form of anxiety, survival anxiety – the horrible realization that in order to make it, you're going to have to change" (Coutu 2002).

Whether we learn because we enjoy the activity or we learn simply to survive, we always learn because it is in our nature as thinking systems. It is a unique characteristic of humans that we can learn with intent; we can innovate, and we have specific goals in mind. No other "system" shares these traits with us.

LEARNING NEEDS SYSTEMS THINKING

Thinking systems such as your organization need systems thinking (Waldman, Yourstone, and Smith 2007) because of the fundamental nature of a thinking system: it continuously learns; it can structure its learning; and it has multiple goals.

You cannot program a thinking system; you must help it to learn. Doing this requires the concepts of systems thinking, from its archetypes to causal loops, and even the butterfly effect (where consequences are disproportionate to the strength of the initial action) (Ackoff 1999; Bertalanffy 1975; Kauffman 1980, 1995; McDaniel 1997; Sterman 2002).

If an organization, especially a service one, seeks to acquire a sustainable competitive advantage, it must make use of the inherent advantages of a thinking system and facilitate that system's learning. The organization that treats its workforce according to the machine model will fail. The one that accepts its thinking system nature and enables learning will succeed.

LEARNING AND OUTCOMES

Every process and organization, whether manufacturing or service, determines its success by the quantification of outputs or outcomes. A focus on outcomes is central to learning. "Outcomes" are the answer to the question "what do we learn from?"

TYPES OF OUTCOMES

When outcome data are analyzed, they become evidence. Outcome evidence can come in different forms, which may not be mutually exclusive, such as the following:

- Intended versus unintended
- Adverse versus beneficial
- Positive or negative
- Interim or final
- Surrogate or actual
- Survey
- Relative versus absolute.

Intended versus Unintended

The outcomes from a manufacturing process can be carefully defined in advance, with specifications clearly delineated, timetables laid out, and with a predictable delivery time. Service "success" is more amorphous and is determined by the end user. It cannot be precisely quantified and certainly not guaranteed. Whether you are dry cleaning a dress, writing a will, or providing health care, the quality of success is ultimately in the eyes of the beholder—that is, the customer or consumer.

A favorite management aphorism is "if you cannot measure it, it does not exist or it did not happen." Whether a quantitative metric such as cost per unit or a semi-quantitative one such as customer satisfaction, some measurement is necessary in order to assess performance. Anyone who has studied management knows that you get what you measure, along with its corollary—what you measure will improve. Unfortunately, few act as though they know this. When we should be measuring the actual final outcomes—both intended and unintended—there is instead a tendency to limit our metrics to interim metrics, surrogate measures, and either absolute or relative measures, but not both.

Adverse versus Beneficial

Adverse or beneficial (Figure 3.2) outcomes can refer to either the customers and/or the organization. Whether the value judgment is adverse or beneficial, good or bad, it must be fed back into the system.

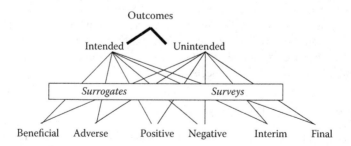

FIGURE 3.2 Types of outcomes.

Intended outcomes are not all beneficial and unintended ones are not invariably adverse. Minoxidil was a drug developed to lower high blood pressure. It had an variable effect on blood pressure but an "adverse" side effect: patients given this drug intravenously began to grow hair everywhere—on their cheeks, their palms, their back, and so on. Minoxidil became the popular medication Rogaine, which, when applied topically, is widely purchased for its "adverse side effect" of promoting hair growth.

The invention of antibiotics started as a failed botany experiment. The heart medication digoxin was an unexpected by-product of an epidemiologic study of older women in England who lived longer than expected. They drank tea made from plants in a garden that contained foxglove, which contains digoxin—a heart-strengthening medication. They were medicating themselves without knowing. 3M Corporation sought to create stronger, residue-free glue, they failed—and eventually produced one of their best-selling products: Post-it Notes.

These were all fortuitous discoveries—beneficial unintended outcomes—that would have been missed by people focused exclusively on the intended outcomes. Conversely, there are adverse unintended outcomes such as a surprise allergic reaction to a drug or late complications such as thyroid tumors decades after irradiation of tonsils in childhood.

Positive or Negative

Positive or negative outcomes would be better described as being those that are either aligned or not aligned with organizational goals. Positive or negative tend to be confused with good and bad, which refer to beneficial or adverse, as above. To be certain that outcomes are aligned (positive), organizational goals must be known, clear, and consistent. When, as is all too common, the publicly stated goals are different from the actual goals, it becomes impossible for a worker or manager to learn from analyzing organizational outcomes.

If the sign on the front door reads "customers come first"; "at the forefront of medicine"; or "quality is our number one priority," and yet the budget remains the ultimate arbiter of all decisions, how do you know what is a positive outcome? If you cannot determine whether the outcomes are positive or negative, what does the evidence mean? What will be fed back? And how (or what) can people learn?

Interim versus Final

There is a tendency, particularly in service activities, for people to track interim results and use surrogate measures in place of the desired final outcomes. This impairs learning and is detrimental to organizational or institutional success.

Interim outcome measures include metrics such as: number of coatings on an automobile chassis or a camera lens; spin rate of a drill bit; cost of materials or personnel needed to create a product or deliver a service; number of rings before a telephone call is answered; and the length of time spent on a heart–lung machine.

Interim results or intermediate outcomes are frequently confused with final outputs. Collins and Porras (1997) related the CEO's opening remarks for an annual stockholders' meeting of a company that made hand-held drills. He started out by shocking his audience, saying, "gentlemen, no one wants our drills." (The company supplied over 90% of the U.S. market for these products.) He then added, "they want

---

TABLE 3.1

Surrogate Measure versus Desired Outcome

Common Surrogate	Desired Outcome
Number of coatings on a camera lens[a]	Sharp pictures
Spin rate of drill rotor	Efficient drilling of hole
Inflation speed of auto air bag	Reduced accident fatalities
Turnover of employees	Retention of employees
Compliance with regulations[a]	Error- or injury-free activity
Deaths (health care)	Long, healthy life

[a] Both surrogate metric and interim measurement.

---

holes." The drill's spin rate (Table 3.1), its torque pounds, or even its cost are intermediate measures that describe features of a necessary tool used in the process of building things. What is built is the final outcome.

Surrogate Measures

Surrogate comes from the Latin *surrogare*, meaning "to stand in place of another." A surrogate measure is used as a substitute for the thing you really want to measure. We use them all the time. When you look at a thermometer, you are using the number as a surrogate for the air temperature. What you really want is the weather report so that you can decide what to wear.

Use of a surrogate measure generally means that the desired metric is difficult, costly, impossible to obtain, or takes too long. Generally, those are excuses rather than good reasons to measure something other than what you want. A surrogate measure is only as good as how closely it represents that for which it is a substitute. Unfortunately, most surrogates are poor stand-ins for what the consumer wants.

Surrogate measures are actually the standard in many service industries, including health care. Hospitals measure mortality rate, using it as a surrogate for survival without considering that there are gradations of living but not of death. Error-free medicine is not the same as high-quality health care. Being sued does not correlate with, and therefore is a poor surrogate for, practicing low-quality medicine (Brennan et al. 1996; Charles 1992; Levinson et al. 1997; Ogbrun et al. 1988; Taragin et al. 1992; Waldman and Spector 2003; Ward 1991).

Regulations are an especially pernicious surrogate measure. Regulations assume that: (a) they specify the right way to do something or to avoid an adverse outcome; (b) they protect us as consumers; and (c) following them will produce the desired outcome. Each assertion is widely held and untrue.

Regulations specify how a legislative subcommittee or professional organization believes that something should be done. That way often has no relation to the best practice. More important is that there is virtually never any hard evidence to prove that following regulations produces the desired outcome.

The Health Insurance Portability and Accountability Act (HIPAA) of 1998 was passed to protect the confidentiality of medical information. Although it now dominates the display, transmission, and storage of all health information, there is no evidence that the HIPAA protects anyone. There was and is no cost/benefit analysis. Worst of all, there is no data proving that there was a problem in the first place.

Surveys

We are all bombarded daily with surveys: online, web based, hard copy (regular mail). Do you prefer Coke or Pepsi? Do you agree with or oppose current foreign policy? How was your experience in our emergency room? Surveys are always being carried out.

Surveys are commonly employed tools that are used to learn. They are both interim as well as surrogate measures. Whether the instrument is a market survey, a benchmark study, a polling analysis, or a workforce survey, they all suffer from the same two problems: bad science and improper usage.

Surveys are often designed to prove the designer's bias or to satisfy the commissioning organization rather than to produce objective, hard evidence to prove or disprove the null hypothesis. Surveys almost always suffer from statistical anemia. A "high" response rate is 18%–24% from the population of interest (Baker 1999). When the survey is unsolicited as in "cold calls," the response rate is typically 3%–5%. Imagine that you are a scientist and that 76%–97% of your experiments showed no results. Would you then make strong statements with solid recommendations? I hope not, but surveyors do so every day.

Surveys are rarely an objective science, but they are often perceived and used as such. They indicate how a select subgroup within a population behave or think. The "benchmark" describes the majority of this subgroup, *not* the best or most effective way to do something. While you are trying to learn the right answer, a survey gives you the most common answer. "Fifty million Frenchmen can't be wrong," wrote Cole Porter in 1929. Look up the *Life Magazine* advertisement proclaiming that smoking must be good for you: after all, 100,000 doctors smoke! Sometimes "crowds" can be wise (Surowiecki 2004) but in surveys, the herd mentality prevails and the lowest common denominator dictates.

The latest survey fad in service occupations—benchmarking best practices—tends to encourage mediocrity. As long as you are doing better than average, you are doing okay. This is the opposite of what highly successful companies say: "good enough never is" (Collins and Porras 1997). Most important, you do not learn the "right way" by benchmarking: you learn the most common way.

Many companies survey their own workforce to determine satisfaction and to try to predict and prevent turnover. Such surveys can be harmful to an organization. When an employee is surveyed three times in five years and, based on the survey results, nothing changes, the employee—a thinking system—learns. She or he learns that what they think or say does not matter and that the organization does not care about them.

Relative versus Absolute

Surveys—benchmark, satisfaction, and polling—are not *the* answer but are part of *an* answer, one-half of what you need in order to learn from outcome evidence.

FIGURE 3.3 My student's scored first place!

To comprehend and improve a process, you need both relative and absolute outcome metrics.

The basic problem with surveys and especially benchmarks becomes easy to understand when you answer this question: what are you trying to achieve? If you have the best customer satisfaction in your area on a market survey, but customers are buying a different product because everyone's service is poor, being best does not translate to organizational success. You have to meet both an absolute standard and the relative standard of competitive success.

As Thoreau once wrote, analogies help us to understand. Imagine this one: you are driving your car toward a four-way stop. Approaching the intersection, on your left and quite close to you is a bicyclist who might hit you. On your right, further away from you, is a big truck that also might crash into your car. Which one should you worry about: the nearer or the more dangerous? The answer is both. Just as you need to consider both the more immediate and the more dangerous threats, so you also need both relative and absolute measurements in order to make good decisions.

LEARNING CURVE THEORY

A learning curve is a graphic representation of the relationship between experience and outcomes. The more you do, the better your results will be, through learning. This effect has been observed in numerous activities ranging from aircraft manufacture to the production of violins (Argote 1990; Baloff 1971; Wright 1936; Yelle 1979) to service endeavors—particularly health care (Begg et al. 1998; Birkmeyer et al. 1999; Clark 1996; Ellis et al. 1997; Glasgow 1996; Hannan et al. 1992, 1997, 1998; Hillner 1998; Hosenpud 1994; Jollis et al. 1994, 1997; Kastrati et al. 1998; Klein et al. 1997; Konvolinka et al.

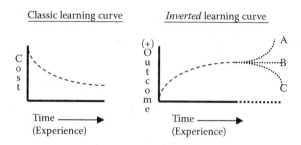

FIGURE 3.4   Learning curves.

1995; McGrath et al. 1998; Phibbs et al. 1996; Showstack et al. 1987; Simunovic et al. 1999; Sollano et al. 1999; Sowden et al. 1995; Thiemann et al. 1999; Zelen et al. 1991).

The most widely known learning curve (called "classic" in Figure 3.4) relates unit production cost (vertical axis) to time for experience (horizontal axis). As experience is gained, people learn to do things more efficiently, which reduces cost. For many activities, but particularly services, it is more useful to measure positive outcomes—such as call response time, satisfaction score, or restoration of function (health care)—rather than tracking only negative ones such as cost. Thus, in Figure 3.4, we have inverted the horizontal axis and compared acquisition of positive outcomes to time. The learning curve is an asymptotic curve, as described by Theorem 2 (Table 3.2).

The fate of the learning curve is unknown. Does it eventually become horizontal (unchanging) as in line B in Figure 3.4? Is there a point where outcome improvement

TABLE 3.2

Three Classical Learning Curve Theorems and Their Applications

Classic Theorems	In General Business	In Service Business	In Health Care
1. Unit production time [and therefore cost] decreases with each iteration	Both cost and quality improve with experience	The one with the most volume is likely to be the best and is certainly *perceived* as the best	The 100th patient is at much less risk than the first[a]
2. Unit production time improvement lessens over time	There is a finite limit to incremental improvement	Consumers implicitly reject the finite limit concept	Outcomes in healthcare can never be perfect[a]
3. Unit production time [and quality] improvement follows a predictable pattern	Learning, especially structured learning, can improve results	Risk-averse businesses cannot improve	Care pathways can improve patient and payer outcomes through learning[a]
		Consumers expect "perfect" service	*All doctors graduated in the top 10% of their class*

[a] Waldman, Yourstone, and Smith (2003).

starts to fall off with more experience (line C)? Might the curve accelerate after a certain amount of time (line A)? There are no data.

Three primary theorems more precisely describe the interaction of experience to outcome (Table 3.2). The first theorem states that unit production time decreases with each repetition. As workers gain experience, they can do more over the same time frame:they become more efficient. That is why the inverted learning curve (Figure 3.4) is upward. Since time translates to labor cost, the more experience a worker acquires, the faster he/she will produce and the cheaper it is to produce. For some products and most services, the gain from experience also translates into improved quality, all through learning. In manufacturing, and particularly in service endeavors, the one with the most volume is perceived as being best and often this perception is valid.

The first theorem has a disquieting translation to health care. Since learning makes people quicker and better at their jobs, it is to be expected that care providers perform "better" after they have acquired experience. In other words, the first patient is at much greater risk than the hundredth. Someone has to be the first or there will never be a second, much less a hundredth. Everyone agrees and everyone wants the first patient to be someone else.

The second theorem states that the improvement in unit production time due to experience (learning) lessens over time. This is why the inverted learning curve (Figure 3.4) is not a straight line but an asymptote. There are certain natural limits and after we have learned a great deal, we approach that limit. As mentioned previously, the ultimate fate of this curve is unknown. Both theorists and pragmatists know that the curve approaches horizontal but many consumers tend to reject this limit and believe they can always do better.

The second theorem means that outcomes can never be perfect. No matter how much you study and practice, in a large population, there will be some bad outcomes, including fatalities. When a vaccination has a $1:1,000,000$ risk, this is a very low risk but not zero. When you give the shot to a million people, one person will be harmed. No amount of practice and learning can eliminate this fact. This injured patient was not harmed by mistake or incompetence: statistical reality did it.

The third learning curve theorem tells us that unit production (or service) improvement—in time and quality—follows a predictable pattern. This means that learning is specific to the type of activity: it takes a longer and different learning curve to do brain surgery than to hit a tennis ball inside the service area. Structuring the learning experience can facilitate the acquisition of a learning curve. Furthermore, learning means risk (even if only to one's ego) and risk-averse activities—such as business, commerce, finance, manufacturing, and medicine—constrain their ability to improve by their hesitance to accept risk.

APPLYING LEARNING CURVE THEORY TO SERVICE OCCUPATIONS

Learning curve theory can be useful in virtually all activities. However, there is a spectrum of difficulty in its application, ranging from the easiest (producing a fungible commodity) to the hardest (health care where every product is unique and willful). Service industries have a greater challenge than manufacturing when seeking to apply learning curve theory.

Structured scientific testing is the best way to produce evidence for learning. First, you must choose the best metrics—those metrics that describe all possible final outcomes in the appropriate time frame. If you are measuring the effectiveness of diabetes prevention or of our education system, the final desired outcomes are decades in the future.

If you want to learn the best rigidity for a bicycle, put various frames in torque vises, and twist and test them. If you want to find the safest automobile frames, put crash dummies in the seats and run the cars into walls.

How do you test for the best consulting service? How do you (ethically) test drugs on people, or animals? Testing—a necessary element in structured learning—is particularly problematic in service industries. And while the ethical, legal, and technical issues are significant, the major constraint is often the corporate culture.

LEARNING AND CULTURE

There is possibly no prerequisite more critical to organizational success than how the corporate culture handles learning (Baker 2001; Collins and Porras 1997). Success, defined by whatever measure, depends on improving efficiency and, particularly in service industries, effectiveness. Improvement, in turn, requires learning by both the individual and the organization (Ferlie and Shortell 2001).

An organization that worships the status quo; that demands rigid adherence to rules and regulations; and that punishes the odd-ball free thinker—that organization does not learn and thus cannot improve (Ashmos et al. 1998; McDaniel and Driebe 2001; Owen 1991). The world changes but the organization with an inflexible culture cannot adapt because it does not reward learning. A culture that encourages innovation and risk taking can learn, improve, and succeed (Pfeffer 1994; Senge 1990).

LEARNING AND RISK

Doing something previously untried or simply doing something known, yet new, carries risks: financial, legal, specific (as in a complication related to health care), and organizational. However, the most important risk—often the greatest impediment to learning—may be the least obvious: danger to the ego.

To learn requires the foregoing of one's ego in order to publicly declare that one does not know something. In most modern activities, particularly in health care, the person who is sure—the one who will take charge—is respected and rewarded. In a debate (or a presidential election), who wins: the person who immediately snaps back a definitive answer, or the person who hesitates in their response? How many teachers are willing to say to their students "I do not know"? Answer: only the good ones.

Not only does learning require the ego-risk of saying that one does not know out loud, one must also do it internally. This is called unlearning, and requires that we say to ourselves that what we think we know may not be correct. The more learned and accomplished a person is, the harder it is to question the knowledge and experience that got them to where they are in the first place.

Medical students and interns, proud of their hard-won and newly acquired medical knowledge, look at me as though I am speaking Swahili when I say that what they have just learned may turn out to be completely wrong in ten years or so. How, they cry, can we depend on anything if medical truth is subject to change? If you cannot question accepted wisdom, then you cannot unlearn and cannot learn anything new.

> There is no learning without risk.
> There is no improvement without learning.
> There is no learning without unlearning.

INCENTIVES AND LEARNING

We are each more than a simple *homo economicus* (Mullainathan and Thaler 2006), and are greater than rational economic units. So you might ask: why did they do it—both the September 11 terrorists and the NYC firefighters? Because they had incentives that were more important to them than living.

Everyone knows that incentives determine behavior. Unfortunately, few recognize how commonly the incentives provided are perverse. In his seminal paper of the same title, Stephen Kerr (1973) described the "Folly of rewarding A while hoping for B." We encourage (or incentivize) "A" and then wonder why we never get what we wanted, but get "B" instead.

A key to organizational success is the recognition that all organizations are thinking systems and that the people within the organization are also thinking systems. All thinking systems have their own goals and for the organization to succeed, it must devise an incentive structure that aligns the employees' goals with the organization's desired outcomes. This may be difficult in the manufacturing world but can be even harder in service industries.

LEARNING AND RETENTION

The remarks that follow may seem obvious, but they are critically important and all too often ignored.

Creating structured learning algorithms and acquisition of learning curves takes *time*. The more complex, judgment intense, and dangerous the behavior, the more time will be required to learn it. An employee might acquire job mastery of rote behaviors and simpler activities in as little as six months, i.e., the electric line worker (Mobley 1982). Contrast this to the ICU nurse who must judge whether the heart rhythm shown on the monitor is life threatening or trivial. She will (hopefully) bring to bear the years of experience, mistakes, learning curves, and better judgment than she had when she was younger (and less experienced).

Time for learning means worker retention. If there is a high employee turnover, neither they nor the organization can learn. This is true for manufacturing endeavors but more so for service activities as the latter always require more judgment. In manufacturing, there is (or should be) a right way to do something; however, in the future there may be an even better way, acquired through innovation and testing (Spear and Bowen 1999). In service businesses, the right way satisfies the customer. Customers are different—each has his or her personal definition of "satisfaction." To get the best results, employees must constantly use judgment, which takes longer to acquire than rote learning.

We emphasize that organizations must track retention, not turnover. They are not inverses (Waldman and Arora 2004). Turnover is non-cumulative as well as a surrogate measure. Managers need to track desired outcomes, while learning to produce superior outcomes, with the direct, positive metric—retention—necessary to produce those outcomes (Waldman et al. 2004).

At medical center X, nurse turnover was at the national median (14%) and therefore, the hospital thought it was doing reasonably well. This same institution had net five-year nurse retention of 17%. This means that 83% of all the nurses hired in 1995 had left the hospital by 2000. How long does it take for a nurse to acquire those critical learning curves to have excellent (i.e., clinically safe) judgment? Even the ones who might have acquired these skills in as little as three to five years would have mostly left. Retention (not low turnover) is what you need in order to have a workforce and an organization with established learning curves and good judgment.

CONCLUSIONS

Learning is the only means to improve outcomes. Outcomes are the evidence that should be fed back to complete the causal loops and to produce continuous improvement. Given that the evidence (of outcomes) is more diffuse and harder to quantify in service industries compared with manufacturing products, managers and planners in service industries are faced with a great challenge.

Humans in groups are thinking systems. Thinking systems continuously learn. They can structure their own learning process and pre-define outcomes that they desire. Outcomes from thinking systems emerge unpredictably rather than follow inexorably. It is the same management model that was so successful in the nineteenth century—the machine system with its predictably standardized outcomes—that shackles organizations in the twenty-first century.

In all modern endeavors, but especially in service occupations, effective learning and the creation of continuous improvement require:

1. An appropriate mental model for the organization (thinking system)
2. The proper application of learning curve theorems and structured learning
3. Appropriate interim and final outcomes metrics
4. Embedded evidence creation and feedback systems
5. Aligned incentives systems
6. The inclusion of systems thinking into the learning process
7. The retention of workers
8. A culture that encourages risk taking

REFERENCES

Ackoff, R.L., 1999. *Ackoff's best—His classic writings on management*. New York: Wiley & Sons.

Argote, L., and Epple, D., February 1990. Learning curves in manufacturing. *Science* 247: 920–924.

Ashmos, D.P., Duchon, D., and McDaniel, R.R., 1998. Participation in strategic decision making: The role of organizational predisposition and issue interpretation. *Decision Sciences* 29(1): 25–51.

Baker, E., 2001. Learning from the Bristol inquiry. *Cardiology in the Young* 11: 585–587.

Baker, T., 1999. *Doing well by doing good*. Washington, DC: Economic Policy Institute.

Baloff, N., 1971. Extension of the learning curve—Some empirical results. *Operational Research Quarterly* 22(4): 329–340.

Begg, C.B., Cramer, L.D., Hoskins, W.J., and Brennan, M.F., 1998. Impact of hospital volume on operative mortality for cancer surgery. *JAMA* 280: 1747–1751.

Beinhocker, E.D., 1997. "Strategy at the edge of chaos." *The McKinsey Quarterly* Winter #1, 24–40.

Bertalanffy, L., 1975. *Perspectives on general systems theory: Scientific-philosophical studies*. New York: Braziller.

Birkmeyer, J.D., Finlayson, S.R., Tosteson, A.N., et al., 1999. Effect of hospital volume on in-hospital mortality with pancreaticoduodenectomy. *Surgery* 125: 250–256.

Brennan, T.A., Sox, C.M., and Burstin, H.R., 1996. Relation between negligent adverse events and the outcomes of medical-malpractice litigation. *N Engl J Med* 335: 1963–1967.

Bruner, E.M., ed. 1983. *Text, play, and story: The construction and reconstruction of self and society: 1983 Proceedings of the American Ethnological Society*. Prospect Heights, Ill, 1988: Waveland Press.

Charles S.C., Gibbons R.D., Frisch P.R., et al., 1992. Predicting Risk for Medical Malpractice Claims Using Quality-of-Care Characteristics. *West J Med* 157:433–439.

Clark, R.E., 1996. Outcome as a function of annual coronary artery bypass graft volume. *Annals of Thoracic Surgery* 6(1): 21–26.

Collins, J.C., and Porras, J.I., 1997. *Built to last*. New York: Harper Business.

Coutu, D.L., 2002. "The anxiety of learning." [HBR interview with Edgar Schein]. *Harvard Business Review,* 100–106.

Ellis, S.G., Weintraub, W., Holmes, D., et al., 1997. Relation of operator volume and experience to procedural outcome of percutaneous coronary revascularization at hospitals with high interventional volumes. *Circulation* 95: 2479–2484.

Ferlie, E.B., and Shortell, S.M., 2001. Improving the quality of health care in the United Kingdom and the United States: A framework for change. *Milbank Quarterly* 79(2): 281–315.

Fickeisen, D.H., Winter 1991. "Learning how to learn: An interview with Kathy Greenberg." *The Learning Revolution (IC#27) by the Context Institute*, 42. www.context.org/ICLIB/IC27/Greenbrg.htm (accessed December 2004).

Glasgow, R.E., and Mulvihill, S.J., 1996. Hospital volume influences outcome in patients undergoing pancreatic resection for cancer. *West Journal of Medicine* 165: 294–300.

Hannan, E.L., Kilburn, H., O'Donnell, J.F., et al., 1992. A longitudinal analysis of the relationship between in-hospital mortality in New York State and the volume of abdominal aortic aneurysm surgeries performed. *Health Studies Research* 27: 517–542.

Hannan, E.L., Racz, M., Kavey, R.-E., et al., 1998. Pediatric cardiac surgery: The effect of hospital and surgeon volume on in-hospital mortality. *Pediatrics* 101(6): 963–969.

Hannan, E.L., Racz, M., Ryan, T.J., et al., 1997. Coronary angioplasty volume-outcome relationships for hospitals and cardiologists. *JAMA* 277: 892–898.

Hillner, B.E., and Smith, T.J., 1998. Hospital volume and patient outcomes in major cancer surgery: A catalyst for quality assessment and concentration of cancer service. *JAMA* 280: 1783–1784.

Hosenpud, J.D., Breen, T.J., Edwards, E.B., et al., 1994. Effect of transplant center volume on cardiac transplant outcome: A report of the United Network for organ sharing scientific registry. *JAMA* 271(23): 1844–1849.

Johnson, S., 2001. *Emergence*. New York: Simon & Schuster.

Jollis, J.G., Peterson, E.D., DeLong, E.R., et al., 1994. The relation between the volume of coronary angioplasty procedures at hospitals treating medicare beneficiaries and short-term mortality. *New England Journal of Medicine* 331: 1625–1629.

Jollis, J.G., Peterson, E.D., Nelson, C.L., et al., 1997. Relationship between physician and hospital coronary angioplasty volume and outcome in elderly patients. *Circulation* 95: 2485–2491.

Kastrati, A., Neumann, F.-J., and Schömig, A., 1998. Operator volume and outcome of patients undergoing coronary stent placement. *Journal of the American College of Cardiology* 32(4): 970–976.

Kauffman, S.A., 1995. *At home in the universe*. New York: Oxford University Press.

Kauffman, D., 1980. *Systems one: An introduction to systems thinking*. Minneapolis, MN: SA Carlton.

Kerr, S., 1975. On the folly of rewarding A while hoping for B. *Academy of Management Journal* 18: 769–783.

Klein, L.W., Schaer, G.L., Calvin, J.E., et al., 1997. Does low individual operator coronary interventional procedural volume correlate with worse institutional procedural outcome? *Journal of the American College of Cardiology* 30(4): 870–877.

Konvolinka, C.W., Copes, W.S., and Sacco, W.J., 1995. Institution and per-surgeon volume versus survival outcome in Pennsylvania's trauma centers. *American Journal of Surgery* 170: 333–340.

Levinson, W., Roter, D.L., Mullooly, J.P., Dull, V.T., and Frankel, R.M., 1997. Physician–patient communication: The relationship with malpractice claims among primary care physicians and surgeons. *JAMA* 277: 553–559.

Maslow, A.H., 1943. A theory of human motivation. *Psychological Review* 50: 370–396.

McDaniel, R.R., 1997. Strategic leadership: A view from quantum and chaos theories. *Health Care Management Review* 22(1): 21–37.

McDaniel, R.R., and Driebe, D.J., 2001. Complexity science and health care management. *Advances in Health Care Management* 2: 11–36.

McGrath, P.D., Wennberg, D.E., Malenka, D.J., et al., 1998. Operator volume and outcome in 12,998 percutaneous coronary interventions. *JACC* 31(3): 570–576.

Mobley, W.H., 1982. *Employee turnover: Causes, consequences and control*. Reading, MA: Addison-Wesley.

Mullainathan, S., and Thaler, R.H., 2000. "*Behavioral economics.*" NBER Working Paper Series #7948. 2000. http://introduction.behaviouralfinance.net/MuTh.pdf (accessed November 2006).

Ogbrun, P.L., Julian, T.M., Brooker, D.C., et al., 1988. Perinatal medical negligence closed claims from the St. Paul Company, 1980–1982. *Journal of Reproductive Medicine* 33: 608–611.

Owen H. Winter, 1991. Learning as transformation. *The Learning Revolution (IC#27) by the Context Institute*, 42. www.context.org/ICLIB/IC27/Owen.htm (accessed December 2004).

Pfeffer, J., 1994. *Competitive advantage through people*. Cambridge, MA: Harvard Business School Press.

Phibbs, C.S., Bronstein, J.M., Buxton, E., and Phibbs, R.H., 1996. The effects of patient volume and level of care at the hospital of birth on neonatal mortality. *JAMA* 276: 1054–1059.

Rogers, E.M., 1983. *Diffusion of innovation*. New York: The Free Press.

Senge, P.M., 1990. *The fifth discipline—The art and practice of the learning organization*. New York: Currency Doubleday.

Showstack, J.A., Rosenfeld, K.E., Garnick, D.W., et al., 1987. Association of volume with outcome of coronary artery bypass graft surgery: Scheduled versus nonscheduled operations. *JAMA 257*: 785–789.

Simunovic, M., To, T., Theriault, M., and Langer, B., 1999. Relation between hospital surgical volume and outcome for pancreatic resection for neoplasm in a publicly funded health care system. *CMAJ* 160: 643–648.

Sollano, J.A., Gelijns, A.C., Moskowitz, A.J., et al., 1999. Volume-outcome relationships in cardiovascular operations. *Journal of Thoracic & Cardiovascular Surgery* 117(3): 419–430.

Sowden, A.J., Deeks, J.J., and Sheldon, T.A., 1995. Volume and outcome in coronary artery bypass graft surgery: true association or artifact? *British Medical Journal* 311(6998): 151–155.

Spear, S., and Bowen, H.K., 1999. Decoding the DNA of the Toyota production system. *Harvard Business Review* September/October, 97–106.

Sterman, J.D., 2002. "Systems dynamics modeling: Tools for learning in a complex world." *IEEEE Engineering Management Review* First Quarter, 42–52.

Surowiecki, J., 2004. *The wisdom of crowds*. New York: Anchor Books.

Taragin, M.I., Wilczek, A.P., Karns, M.E., Trout, R., Carson, J.L., 1992. Physician demographics and the risk of medical malpractice. *Amer J Med* 93: 537–542.

Thiemann, D.R., Coresh, J., Oetgen, W.J., and Powe, N.R., 1999. The association between hospital volume and survival after acute myocardial infarction in elderly patients. *New England Journal of Medicine* 340: 1640–1648.

Waldman, J.D., 2007. Thinking systems need systems thinking. *Systems Research & Behavioral Science* 24: 1–15.

Waldman, J.D., 2010. *Uproot healthcare*. Indianapolis, IN: Trafford Publishing.

Waldman, J.D., and Arora, S., 2004. Measuring retention rather than turnover—A different and complementary HR calculus. *Human Resource Planning* 27(3): 6–9.

Waldman, J.D., and Cohn, K., September 2007. *Mend the gap: In the business of health*, eds. K.H. Cohn and D. Hough, New York: Praeger Perspectives.

Waldman, J.D., and Spector, R.A., 2003. Malpractice claims analysis yields widely applicable principles. *Pediatric Cardiology* 24(2): 109–117.

Waldman, J.D., Yourstone, S.A., and Smith, H.L., 2003. Learning curves in healthcare. *Health Care Management Review* 28(1): 43–56.

Waldman, J.D., Yourstone, S.A., and Smith, H.L., 2007. Learning—The means to improve medical outcomes. *Health Services Management Research 2007;* 20: 227–237.

Waldman, J.D., Smith, H.L., Kelly, F., and Arora, S., 2004. The shocking cost of turnover in health care. *Health Care Management Review* 29(1): 2–7.

Ward, C.J., 1991. Analysis of 500 obstetric and gynecologic malpractice claims: Causes and prevention. *Am J Obstet Gynecol* 165: 298–306.

Weick, K.E., 1993. The collapse of sensemaking in organizations: The Mann Gulch Disaster. *Administrative Science Quarterly* 38: 628–652.

Wright, T.P., 1936. Factors affecting the cost of airplanes. *Journal of Aeronautical Sciences* 3: 122–128.

Yelle, L.E., 1979. The learning curve—Historical review and comprehensive survey. *Decision Sciences*, 302–328.

Zelen, J., Bilfinger, T.V., and Anagnostopoulos, C.E., 1991. Coronary artery bypass grafting: The relationship of surgical volume, hospital location, and outcome. *New York State Journal of Medicine* 91(7): 290–292.

4 From Men and Machines to the Organizational Learning Curve

Guido Fioretti

CONTENTS

LEARNING CURVES AS EMERGENCE OF ROUTINES

Learning curves were first discovered in the aerospace industry where a large number of items must be assembled together in order to build an airplane (Wright 1936). This is possibly not a chance occurence, for it has been observed that the slope of organizational learning curves is generally more pronounced in assembling operations (where organizational learning has a prominent role) than in machining operations, where individual learning has a prominent role (Hirsch 1952, 1956). This insight suggests that organizational learning curves may stem from the coordination of large sets of men and machines.

According to this point of view, organizational learning curves reflect a distributed development of patterns of behavior, leading to the emergence of routines (Levitt and March 1988; Weick 1991) that are meant as recurrent but flexible patterns of operations (Pentland 1992; Pentland and Rueter 1994; Dubuisson 1998). Routines arise spontaneously in both structured environments (Cicourel 1990) and informal communities of practice (Brown and Duguid 1991; Wenger 1998, 2000) out of the repetition of successful coordination schemes. During their development, the sequencing of operations is improved, which implies that the required task is accomplished earlier.

On the contrary, learning curves disappear if routines are destroyed. For instance, it is known that if production is suspended and subsequently restarted—e.g., because of a prolonged strike—production time is generally longer than it was before the interruption (Baloff 1970; Birkler et al. 1993; Argote 1999; Sikström and Jaber 2002).

A prototypical example of this vision of organizational learning is Hutchins' (1990, 1991, 1995) detailed story of the slow emergence of a manual calculation routine among the crew of a large ship suffering a breakdown that had disabled an important piece of navigational equipment while entering a harbor. In this story, a set of men had to learn how to use certain mechanical tools in order to read the coordinates of the ship by observing reference points on the ground. Some operations had to be carried out necessarily before others; in other cases, the sequencing of operations was a matter of convenience. Some members of the crew had unique abilities to use particular tools; other tools could be easily used by anyone.

Through trial and error, the crew learned which tools should be used by particular members of the crew, along with a sequencing of the operations that each member had to carry out. In other words, they were discovering increasingly faster routines until one was found that satisfied their particular needs. Or, in the parlance of learning curves, production time decreased while the crew was exploring the possibilities of available tools, until a plateau was reached. Or, maintaining the parlance of organizational science, an organization formed out of a set of men and machines.

In general, one may distinguish the following two aspects in the formation of routines:

1. Given sequences of operations must be routed on a set of organizational units.
2. A set of operations must be arranged into feasible sequences at the same time as they are routed on a set of organizational units.

Pure routing problems (1) arise when customers, managers, or other actors require an organization to carry out a certain sequence of operations. Sequencing problems (2) arise when customers, managers, or other actors require an organization to carry out a certain set of operations, no matter in what sequence. Sequencing involves routing, so it cannot be found in pure form unless each operation can only be carried out by one single organizational unit.

Obviously, reality is made of a mixture of routing and sequencing problems. However, it makes sense to analyze the features of extreme cases in order to understand their reality. We shall see, in this chapter, that routing problems (1) are sufficient to generate learning curves, but more interesting possibilities arise if sequencing problems (2) are considered.

In order to disentangle the impacts of routing and sequencing on learning curves, sequencing will be considered under conditions where routing problems do not arise. In particular, learning curves arising out of pure routing [case (1) above] will be analyzed by means of an agent-based simulator in the "Learning Curves: Routing" section, whereas in the "Learning Curves: Sequencing" section a theoretical model will be used in order to understand some features of learning curves arising out of pure sequencing problems [case (2) above].

Both in the simulations and in the theoretical model an organization is conceived as a graph. Nodes are its organizational units, edges are flows of semi-manufactured goods, and routines are recurring paths between units. Organizational learning means striving toward optimal paths of operations through units, which is reflected in the decreasing throughput time. In the "Final Discussion" section, a possible relation between features of these organizational units and the slope of the learning curve will be discussed.

LEARNING CURVES: ROUTING

In this section, learning curves arising out of routing problems will be investigated by means of the *java Enterprise Simulator** (henceforth *jES*). This is an agent-based platform for the modelization of firms where orders that are composed of sequences of elementary operations are routed on a set of organizational units. Organizational units are compounds of men and machines capable of carrying out a subset of operations each—such as the crew of the ship entering a harbor in the above example.

Henceforth, the following assumptions will be made:

- Orders of given length are random sequences of operations drawn from a uniform distribution defined over the set of possible operations.
- Orders are routed on organizational units with the criterion that, if two or more units are able to carry out the required operation, the unit with the shortest waiting list is chosen.
- The outcomes of accomplished orders are stored in one end unit, which may represent an inventory of finished products.

The *jES* is able to generate learning curves due to pure routing problems. In fact, customers, managers, or other decision makers generate orders that must be routed on available organizational units in order for them to be accomplished. Since units route orders wherever waiting lists are shortest, the firm as a whole learns how to minimize throughput time.

Despite this simple setting, organizational learning still occurs—even if the single units neither overview the whole process nor remember their own decisions. In fact, although these units behave according to a rule so simple that it does not allow for individual learning, the organization learns how to route orders in order to decrease the throughput time.

Let us first consider simulations where each organizational unit is able to carry out one single elementary operation. This may be the case, for instance, where workers are operating quite simple machines.

Figure 4.1a reports the ratio between the actual throughput time and the minimum feasible throughput time. This ratio describes a learning curve. Orders were composed of 10 elementary operations, drawn at each step from a set of 10 possible operations by means of a uniform distribution. These orders were routed over 20 units performing one operation each; thus, each operation could be carried out by 2 units. Although outcomes generally changed with the random seed, the learning curve illustrated in Figure 4.1a is quite typical.

Figure 4.1b shows the outcome provided by the simulator when orders are routed randomly, with all parameters and the random seed as in Figure 4.1a. Remarkably, Figure 4.1b gives the impression that organizational learning is still taking place, albeit to a much smaller extent than in the case where orders were routed to the unit with the shortest waiting list. However, this actually occurs because, once sufficiently diverse orders have accumulated, random routing is quite an efficient

---

* Freely available at http://web.econ.unito.it/terna/jes under the GNU public license.

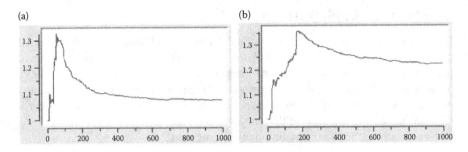

FIGURE 4.1 The ratio of actual throughput time to minimum feasible throughput time. Random orders composed of 10 elementary operations, drawn from a set of 10 different operations, and routed over 20 organizational units capable of 1 operation each. (a) Orders were routed to the unit with the shortest waiting list. (b) Orders were routed randomly. (a) and (b) were generated with the same random seed.

strategy. It is at least questionable as to whether this effect may be labeled as "organizational learning."

Figure 4.2 illustrates four learning curves obtained on organizations endowed with 10, 20, 30, and 40 units, respectively. Figure 4.2b is the same as Figure 4.1a. All curves have been obtained with the same sequence of random orders composed of 10 elementary operations.

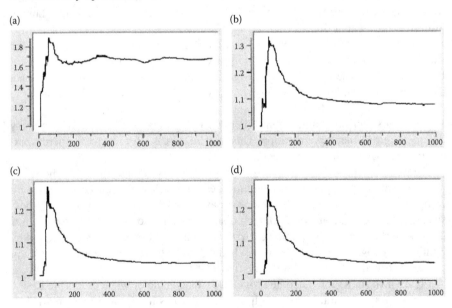

FIGURE 4.2 The ratio of actual throughput time to the minimum feasible throughput time. Random orders composed of 10 elementary operations, drawn from a set of 10 different operations, routed to the unit with the shortest waiting list. (a) refers to 10 units, each devoted to 1 operation. (b) refers to 20 units, 2 for each operation. (c) refers to 30 units, 3 for each operation. (d) refers to 40 units, 4 for each operation. The data in (a), (b), (c), and (d) were generated with the same random seed.

The clearest pattern among learning curves that originate from pure routing is that they decrease with the number of organizational units. Figure 4.2 makes it clear that organizational learning is greater when more units are available. In fact, from (a) to (b) and (c) the plateau is ever smaller.

However, Figure 4.2 also points to the fact that beyond a threshold where many units stay idle, adding organizational units does not improve the organization's throughput time. This can be seen by comparing cases (c) and (d), which are nearly indistinguishable from each other.

Figure 4.3 illustrates a similar comparison when the set of possible operations entails 20 items; that is, orders are still composed of 10 operations, but these 10 operations are drawn from a set of 20. Thus, in this case, there is a higher variety of orders.

Figures 4.3a and 4.3b refer to 20 and 40 organizational units, respectively. So far it regards the number of organizational units, they correspond to Figures 4.2b and 4.2d, respectively.

A comparison between Figures 4.2 and 4.3 highlights that, so far in regard to learning curves arising out of the pure routing of random orders, only the number of organizational units matter. In fact, Figure 4.3a is nearly identical to Figure 4.2b: both were obtained with 20 units; likewise, Figure 4.3b is nearly identical to Figure 4.2d: both were obtained with 40 units. The variety of orders has no impact.

Since adding organizational units improves the performance of organizational learning, one may speculate that endowing organizational units with the ability to perform several operations may produce a similar effect. Figure 4.4 shows that this is not the case.

Figure 4.4 compares Figure 4.2a, reproduced as Figure 4.4a, with a simulation where each unit was capable of two elementary operations, all else being equal. The outcome of this simulation is illustrated in Figure 4.4b.

Instead of aiding the task of routing, having more flexible organizational units may have made things slightly worse. In fact, in Figure 4.4b the learning curve never starts to descend, and the plateau is higher than in Figure 4.4a. If the same comparison is made when 20 units are available, no appreciable difference appears when the units are more flexible. Thus, it appears that increasing the flexibility of organizational units has, at best, no effect, and, at worst, has a negative impact on organizational learning.

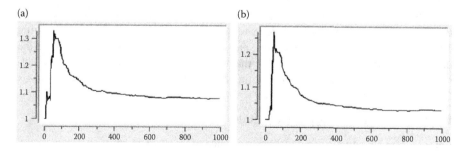

FIGURE 4.3 The ratio of actual throughput time to minimum feasible throughput time. Random orders composed of 10 elementary operations, drawn from a set of 20 different operations, routed to the unit with the shortest waiting list. (a) Refers to 20 units, each carrying out a different operation. (b) Refers to 40 units, 2 for each of the 20 operations. The data of (a) and (b) were generated with the same random seed.

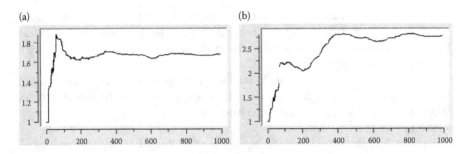

FIGURE 4.4 The ratio of actual throughput time to minimum feasible throughput time. Random orders composed of 10 elementary operations, drawn from a set of 10 different operations, routed on 10 units with the criteria of the shortest waiting list. In (a), each unit was devoted to a different operation. In (b), each unit was capable of carrying out two operations. The data in (a) and (b) were generated with the same random seed.

This is possibly because if two or more flexible units carry out the same operation, some other operation may find no unit able to process it. This does not occur if 20 units are available, but it may become a problem if—as in Figure 4.4—only 10 units are available.

The shape of learning curves originating from pure routing depends very strongly on the random seed, as well as on the number of elementary operations of which an order is composed. The variety obtained by chaining the random seed or the length of orders is very large, and no clear pattern can be distinguished by changing these two parameters. Some curves descend smoothly after an initial peak, as in Figure 4.1a; others exhibit several peaks, each followed by a descent; and a few have a very low initial peak followed by a weak descent. There are also a few instances where the descending phase never starts, but this occurs only when the peak is extremely low or non-existent.

On the whole, one may conclude that, with problems of pure routing, organizational learning necessarily sets in if random orders generate queuing problems. The number of available units is the only relevant parameter for this kind of learning curve.

However, it is remarkable that if the simulator is fed with deterministic orders— that is, a series of operations that repeat themselves with no variation—no learning curve appears. Real routing problems, and organizational learning, begin when the orders are unpredictable. We shall see that this is a crucial insight in order to understand the arousal and the slope of learning curves.

LEARNING CURVES: SEQUENCING

Let us consider a situation in which the sequence of given elementary operations is not specified. The sequence of operations is not chosen randomly and obeyed by the units, but is instead chosen by the units themselves. In a sense, organizational units randomize orders rather than receive them, as shown in the "Learning Curves: Routing" section.

This can occur if managers, customers, or other actors require that a set of operations be carried out, but, except for a few constraints, they do not care about the exact sequence. Orders are sets of operations rather than required sequences of operations. In this case, finding out optimal sequences is the task that the organization must learn.

In general, sequencing involves routing. In fact, once operations are arranged in a sequence, they must be routed onto organizational units. However, by considering organizations where each operation can only be carried out by one single unit, we can ignore routing and focus on pure sequencing. We shall do so in order to deal with a problem that is simple and opposite to that of the "Learning Curves: Routing" section.

However, even if in this simple setting, managers leave organizational units with the freedom to arrange a set of operations in any possible sequence, the organizational units, in general, do not combine the required operations in any conceivable way. Technical, legal, or administrative constraints generally force organizational units to work on subsets of all possible sequences. Thus, units must classify the sequences of operations that they receive, distinguishing those on which they can carry out their operation from those on which they cannot.

Henceforth, all sequences with an operation at a position such that it can be processed by a particular unit will be called the set of *feasible sequences* for that unit. Organizational units must be endowed with categories in order classify the sequences proposed by other units as feasible or unfeasible. Only feasible sequences are accepted and scheduled for processing.

Categories may be coarse or sharp, depending on the ability of a unit to perform its operation at various stages or at different stages; for example, more flexible machines may not care whether a certain operation has been performed before they carry out their own operation. Coarseness of categories will often be referred to as the flexibility of an organizational unit.

Sequences of operations are strings of integers, each representing one operation. Let us represent categories by means of strings made up of the integers used to represent operations, plus "don't care" characters #s. A category classifies all strings as having either the same number at the same positions, or whatever number where the category has a number. Figure 4.5 illustrates an example.

The following assumptions allow a simple analytic treatment of organizational learning that arises out of sequencing problems:

- Each operation can be carried out by one, and only one, unit. Consequently, this is a problem of pure sequencing.
- Each organizational unit has one instance of all categories represented in the organization.

$$
\begin{bmatrix} \# \\ 5 \\ \# \\ 0 \end{bmatrix} \quad \begin{bmatrix} 7 \\ 5 \\ 3 \\ 0 \end{bmatrix} \begin{bmatrix} 0 \\ 5 \\ 1 \\ 0 \end{bmatrix} \begin{bmatrix} 3 \\ 5 \\ 0 \\ 0 \end{bmatrix} \begin{bmatrix} 2 \\ 5 \\ 5 \\ 0 \end{bmatrix}
$$

FIGURE 4.5 A category (left) and four strings that it can classify (right).

- Any sequence is feasible for at least one category. Thus, each organizational unit can accept strings from any other.
- A category has the same probability to process any feasible sequence. Contrary to classifier systems, no strength is passed on.

Let $N \in \mathcal{N}$ denote the number of possible elementary operations. Let $L \in \mathcal{N}$ denote the length of sequences of operations, as well as the length of categories. Let $H \in \mathcal{N}$ denote the number of different categories available in the organization. It is obviously $H \leq (N+1)^L$, the number of dispositions with the repetition of $N+1$ elements (the N operations plus the #) of class L.

Let $K \in \mathcal{N}$ denote the number of different sequences produced in the organization. It is $H \leq N^L$, the number of dispositions with the repetition of N elements of class L.

Since categories exist in order to make classifications, it must be $H < K$. Thus, H can never reach its upper bound.

Let p denote the probability that a link exists between any two organizational units. This parameter captures the range of possibilities for changing the production routines. In fact, the more possibilities there are for connecting the stages of production, the more possibilities there are for creating new routines.

Links are not established without a purpose. Links represent flows of semi-finished products through organizational units until an end unit is reached, where the product is finished. For instance, in the simulations of the "Learning Curves: Routing" section there was a single end unit, representing an inventory of finished products. Thus, searching for better routines means finding shorter paths to the end unit.

This search is generally not random. In general, it reflects an organization's ability to adopt better procedures and discard the inefficient ones. Let r denote the probability of eliminating the unproductive links departing from an organizational unit. Thus, $r = 1$ corresponds to a perfect decision procedure, whereas $r = 0$ corresponds to a random walk.

Shrager et al. (1988) and Huberman et al. (2001) derived a learning curve as a function of these parameters. In its turn, the following model links p and r to observable magnitudes such as H and K (Fioretti 2007a, 2007b).

Let us consider one single attempt to connect two organizational units. Let us assume that the probability that a category accepts a sequence from another unit is $1/(K-1)$, as would be the case if the number of units was equal to the number of sequences and each category could classify any sequence. This is obviously not the case, for, in general, there are (a) many more sequences than elementary operations or units, and (b) a category does not classify all sequences, only the feasible ones. However, (a) and (b) push in opposite directions so, as a first approximation, one may hope that they nearly cancel each other out.

Since each unit has been assumed to own all categories, the probability of establishing at least one link between any two organizational units is:

$$p = \frac{H}{K-1}, \tag{4.1}$$

where $K - H \geq 1$.

Parameter r represents the probability that the search for better arrangements is effective. In the limit of infinite attempts to establish connections to other units, sooner or later the end unit is reached. On the contrary, if novel connections are no longer tried, the end unit may never be reached because the same connection to one and the same unit is endlessly repeated.

Let us calculate the probability of getting stuck connecting to a particular unit again and again. Parameter r will be its complement to one.

Equation 4.1 expresses the probability that at least one link is established between any two units during a procedure where H trials are made. The probability that this happens twice if the whole procedure is carried out twice is $(H/(K-1))^2$, and so on. If these probabilities are summed, it is safe to divide the sum by the coefficient $K-1$ in order to ensure that the sum will be less than unity. In the end, the probability of repeating a connection endlessly is the sum of:

$$\frac{H}{(K-1)^2} \sum_{i=1}^{\infty} \left(\frac{H}{K-1} \right)^i,$$

which amounts to $H/(K-1)(K-H-1)$.

Consequently, the probability of choosing the right path is $1 - H/(K-1)(K-H-1)$, which can be written as:

$$r = \frac{(K-1)^2 - KH}{(K-1)(K-H-1)}, \tag{4.2}$$

with $K-H>1$.

Note that Equation 4.2 is quite a good estimate of r when routines are under construction, but it is definitely wrong once good routines have been established. In fact, it captures the extent to which organizational units try novel paths. This is appropriate when describing the beginning of organizational learning, but loses its relevance once a good routine has already been found. Equation 4.2 makes sense in a theoretical model that estimates possible learning curves, but it would not fit a simulation logic as in the "Learning Curves: Routing" section.

The implications of Equations 4.1 and 4.2 become clear when one plots them for various values of H and K. For obvious reasons, interesting values appear when the difference $K-H$ is not too small with respect to the absolute values of H and K.

Equation 4.1 is defined for $K-H \geq 1$ but is trivial if $K-H=1$. Equation 4.2 is defined for $K-H>1$. Thus, the smallest possible value of $K-H$ is 2. Correspondingly, the range of values of H and K should be close to the origin. Figure 4.6 illustrates p and r for $H \in [1, 10]$ and $K \in [3, 12]$.

The higher the parameter p, the more attempts are made at improving on the current arrangement of production. Equivalently, the higher the value of p, the steeper the learning curve would be.

Thus, Figure 4.6 shows that the greater the number of operation sequences and categories, the more possibilities there are for improvement. In short, the more there is to learn, the more that can be learned.

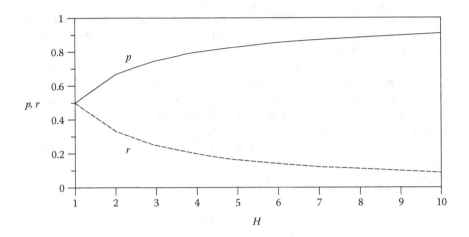

FIGURE 4.6   Parameters p (solid line) and r (dashed line) for $K - H = 2$.

However, learning may not proceed if the search for better arrangements of production becomes stuck in a vicious circle. The parameter r captures the likelihood of this possibility.

Figure 4.6 shows that the greater the number of sequences and categories, the more likely it is that no improvement will take place at all. To be more concise: the more there is to learn, the more likely it is that nothing will be learned at all.

Thus, Figure 4.6 illustrates a trade-off between the possibility of improving the arrangement of an organization and the danger of getting lost in an endless search. In fact, the more possibilities that exist for improvement, the more difficult it is to realize them.

Let us consider an organization with fewer categories, which often implies more generic categories. This means that workers have a wider knowledge so they can do more diverse jobs, or that machines are more flexible so they can process a wider range of semi-finished goods, or both.

Let us choose $K - H = 3$. Figures 4.7 and 4.8 show the ensuing effect on p and r, respectively.

Even with so small a change in the number of categories, the differences are impressive. The possibilities for improvement—captured by the parameter p—have slightly decreased. On the contrary, the likelihood that better arrangements are found—captured by the parameter r—have increased dramatically. Furthermore, the greater H, the more pronounced are these effects.

Figures 4.7 and 4.8 suggest that, by employing a few general categories, a large gain in effectiveness can be attained at the expense of a small loss on the possibilities for improvement. An organization of open-minded generalists and flexible machines may lose a fraction of the learning possibilities afforded by specialization, but will not get stuck in meaningless routines that lead nowhere.

FINAL DISCUSSION

Learning curves would be a valuable tool for business planning if they were predictable. The trouble is that this is generally not the case. The slope of the learning curve

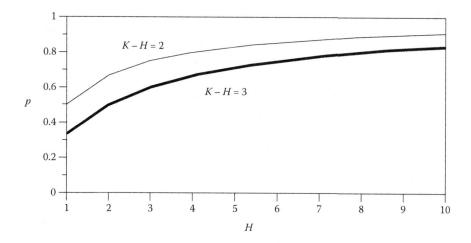

FIGURE 4.7   Parameter p when $K - H = 2$ (thin line) and $K - H = 3$ (bold line).

is something of a guess, and it may even happen that no decreasing pattern sets in. Given that there is always a small but positive probability that the learning curve will not descend at all, it is hard for managers to rely on it in the evaluations of future costs.

It is obvious that it is necessary to understand the reasons behind why learning curves arise in order to be able to predict whether they will arise or not. This chapter moved from the idea that organizational learning is grounded on the formation of routines and attempted to highlight some features on which the shape of learning curves depends.

In the "Learning Curves: Routing" section, we found that orders must be produced randomly in order for learning curves to exist, and that organizational units must be sufficiently many for learning curves to be effective. In the "Learning Curves: Sequencing" section, we found that there must be a sufficient amount of things to do for learning curves to set in (large H and large K), and that organizational units must

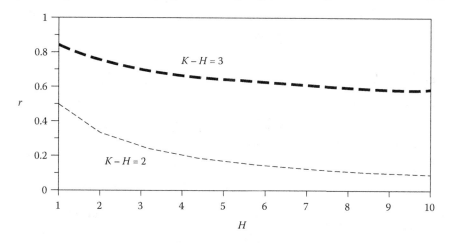

FIGURE 4.8   Parameter r when $K - H = 2$ (thin line) and $K - H = 3$ (bold line).

be sufficiently flexible to enable the formation of routines (large difference between K and H). It is possible that these findings point to a common pair of principles for organizational learning to take place; namely, that (i) there is a sufficient number of novel possibilities for conceiving novel routines; and (ii) organizational units are sufficiently high in number and sufficiently flexible to implement novel routines.

Point (i) is exemplified by the case of technical innovations, quite often taking place at the same time that organizational units are striving to develop better routines (Baloff 1966). A few cases where innovation was totally absent (Baloff 1971; Dutton and Thomas 1984; Reis 1991) highlighted that without continuous stimulation and the injection of novelties, the learning curve reaches a plateau (Hirschmann 1964; Levy 1965; Baloff and McKersie 1966; Adler and Clark 1991). On the contrary, a changing and stimulating environment is beneficial to both production time (Shafer et al. 2001; Macher and Mowery 2003; Schilling et al. 2003) and qualitative improvements (Levin 2000).

Point (ii) is exemplified by the fact that, among industrial plants, learning curves are most pronounced where assembling operations are involved (Hirsch 1952, 1956). Assembling operations require a large number of units that must be flexible enough to interact with one another in multiple configurations—a circumstance that facilitates the emergence and modification of routines. On the contrary, plants based on conveyor belts are not the typical settings where organizational learning takes place.

More detailed simulations are in order. It is necessary to integrate all factors giving rise to organizational learning curves and to investigate their consequences beyond the stylized models amenable to mathematical formalization, and this is only possible by means of numerical simulations.

The application of concepts derived from numerical simulations to real cases poses yet another kind of problem, for organizational units are, in general, not just machines but compounds of men and machines. The features of machines can be easily measured; those of human beings often cannot. Human beings exert a large influence on learning curves, as testified by the fact that the slope of the learning curve may differ across identical plants of the same firm (Billon 1966; Yelle 1979; Dutton and Thomas 1984; Dutton et al. 1984), or even across shifts in the same plant (Argote and Epple 1990; Epple et al. 1991; Argote 1999). These episodes suggest that there are some limits to the extent to which learning curves can be managed and predicted.

REFERENCES

Adler, P.S., and Clark, K.B., 1991. Behind the learning curve: A sketch of the learning process. *Management Science* 37(3): 267–281.

Argote, L., 1999. *Organizational learning: Creating, retaining and transferring knowledge.* Norwell: Kluwer Academic Publishers.

Argote, L., and Epple, D., 1990. Learning curves in manufacturing. *Science* 247(4945): 920–924.

Baloff, N., 1966. The learning curve – Some controversial issues. *The Journal of Industrial Economics* 14(3): 275–282.

Baloff, N., 1970. Start-up management. *IEEE Transactions on Engineering Management* 17(4): 132–141.

Baloff, N., 1971. Extension of the learning curve – Some empirical results. *Operational Research Quarterly* 2(4): 329–340.

Baloff, N., and McKersie, R., 1966. Motivating start-ups. *The Journal of Business* 39(4): 473–484.

Billon, S.A., 1966. Industrial learning curves and forecasting. *Management International Review* 1(6): 65–79.

Birkler, J., Large, J., Smith, G., and Timson, F., 1993. *Reconstituting a production capability: Past experience, restart criteria, and suggested policies.* Technical Report MR–273, RAND Corporation.

Brown, J.S., and Duguid, P., 1991. Organizational learning and communities-of-practice: Toward a unified view of working, learning, and innovation. *Organization Science* 2(1): 40–57.

Cicourel, A.V., 1990. The integration of distributed knowledge in collaborative medical diagnosis. In *Intellectual teamwork: Social and technological foundations of cooperative work*, eds. J. Galegher, R.E. Kraut, and C. Egido, 221–243. Hillsdale: Robert Erlsbaum Associates.

Dubuisson, S., 1998. Regard d'un sociologue sur la notion de routine dans la théorie évolutionniste. *Sociologie du Travail* 40(4): 491–502.

Dutton, J.M., and Thomas, A., 1984. Treating progress functions as a managerial opportunity. *The Academy of Management Review* 9(2): 235–247.

Dutton, J.M., Thomas, A., and Butler, J.E., 1984. The history of progress functions as a managerial technology. *Business History Review* 58(2): 204–233.

Epple, D., Argote, L., and Devadas, R., 1996. Organizational learning curves: A method for investigating intra-plant transfer of knowledge acquired through learning by doing. In *Organizational learning*, eds. M.D. Cohen and L.S. Sproull, 83–100. Thousand Oaks: Sage Publications.

Fioretti, G., 2007a. A connectionist model of the organizational learning curve. *Computational and Mathematical Organization Theory* 13(1): 1–16.

Fioretti, G., 2007b. The organizational learning curve. *European Journal of Operational Research* 177(3): 1375–1384.

Hirsch, W.Z., 1952. Manufacturing progress functions. *The Review of Economics and Statistics* 34(2): 143–155.

Hirsch, W.Z., 1956. Firm progress ratios. *Econometrica* 24(2): 136–143.

Hirschmann, W.B., 1964. Profit from the learning curve. *Harvard Business Review* 42(1): 125–139.

Huberman, B.A., 2001. The dynamics of organizational learning. *Computational and Mathematical Organization Theory* 7(2): 145–153.

Hutchins, E., 1990. The technology of team navigation. In *Intellectual teamwork: Social and technological foundations of cooperative work*, eds. J. Galegher, R.E. Kraut, and C. Egido, 191–220. Hillsdale: Robert Erlsbaum Associates.

Hutchins, E., 1991. Organizing work by adaptation. *Organization Science* 2(1): 14–39.

Hutchins, E., 1995. *Cognition in the wild*. Cambridge: The MIT Press.

Levin, D.Z., 2000. Organizational learning and the transfer of knowledge: An investigation of quality improvement. *Organization Science* 11(6): 630–647.

Levitt, B., and March, J.G., 1988. Organizational learning. *Annual Review of Sociology* 14: 319–340.

Levy, F.K., 1965. Adaptation in the production process. *Management Science* 11(6): 136–154.

Macher, J.T., and Mowery, D.C., 2003. "Managing" learning by doing: An empirical study in semiconductor manufacturing. *Journal of Product Innovation Management* 20(5): 391–410.

Pentland, B.T., 1992. Organizing moves in software support hot lines. *Administrative Science Quarterly* 37(4): 527–548.

Pentland, B.T., and Rueter, H.H., 1994. Organizational routines as grammars of action. *Administrative Science Quarterly* 39(3): 484–510.

Reis, D.A., 1991. Learning curves in food services. *The Journal of the Operational Research Society* 42(8): 623–629.

Schilling, M.A., Vidal, P., Ployhart, R.E., and Marangoni, A., 2003. Learning by doing something else: Variation, relatedness, and the learning curve. *Management Science* 49(1): 39–56.

Shafer, S.M., Nembhard, D.A., and Uzumeri, M.V., 2001. The effects of worker learning, forgetting, and heterogeneity on assembly line productivity. *Management Science* 47(12): 1639–1653.

Shrager, J., Hogg, T., and Huberman, B.A., 1988. A graph-dynamic model of the power law of practice and the problem-solving fan-effect. *Science* 242(4877): 414–416.

Sikström, S., and Jaber, M.Y., 2002. The power integration diffusion model for production breaks. *Journal of Experimental Psychology: Applied* 8(2): 118–126.

Weick, K.E., 1991. The non-traditional quality of organizational learning. *Organization Science* 2(1): 116–124.

Wenger, E., 1998. *Communities of practice.* Cambridge: Cambridge University Press.

Wenger, E., 2000. Communities of practice and social learning systems. *Organization* 7(2): 225–246.

Wright, T.P., 1936. Factors affecting the cost of airplanes. *Journal of the Aeronautical Sciences* 3(2): 122–128.

Yelle, L.E., 1979. The learning curve: Historical review and comprehensive survey. *Decision Sciences* 10(2): 302–328.

5 Management at the Flat End of the Learning Curve: An Overview of Interaction Value Analysis

Walid F. Nasrallah

CONTENTS

INTRODUCTION

There are always new things to learn in most organizations. In order to function in a certain role within an organization, an individual must constantly find out new things about tasks to be performed, resources needed in order to do the work, and methods of performing these tasks under different circumstances. More significantly, one must also learn how to navigate one's way around the organization to find information, authorization, access to resources, and any other useful thing that the organization can provide via the mediation of different gatekeepers.

Learning about who in the organization can be useful in what situation is an example of a network search. You find out about someone's (broadly defined) capabilities on reaching the node in the organization's network of relationships that represents that person. To reach this point, you must first learn about the other nodes representing the people who have relationships that can bridge the gap between you and the person you are seeking. Searching through a network to find increasingly valuable nodes can lead to an exponential decay in the incremental benefits (as a function of the amount of searching); that is, what we know as the standard learning curve. Indeed, it could be argued that the source of the learning curve effect is a confluence of several such processes (Adler and Clark 1991).

This chapter considers how best to find and use the information that needs to be learned before one is competent enough to fulfill a certain role within an organization. This could be as simple a task as finding out who within the organization can do what, and how well, in order to help one fulfill one's duties. As long as changes take place both inside and outside the organization, this learning process can become a continuous one, because there are always changes occurring, which in turn create new things to learn about. It is also possible to envision a static situation, which may last for a long time or a short time, where everyone has to discover the fixed value that can be derived from interacting with everyone else: task-specific skills, capabilities, organizational privileges, legal and contractual powers, and so forth. The amount of output that an organization generates during this learning process follows a curve that goes up quickly at first, then more slowly, until it levels out. Through examining this static form of the organizational learning problem, we can evaluate different ways of conducting the search that leads to the flat end of the learning curve (the steady state).

Even if the situation changes in future, what we learn about how to reach the steady state of the learning curve the first time will be useful if those changes force us to ascend another new learning curve in order to reach its new steady state, or flat end.

MODELS OF OPTIMAL NETWORKS IN ORGANIZATIONS

Huberman and Hogg (1995) wrote about a model of a community where various levels and niches of expertise exist, making it expedient to ask for "hints" from different members when someone encounters a need to do something new. One assumption of this model was that meta-knowledge—that is, knowledge of where to go in order to ask certain questions—is freely available.

This could be through experience (i.e., prior learning by the members of the community) or through some sort of written record or directory that someone maintains and shares.

This model was named "communities of practice" because it envisioned a diffuse group of people with sparse interactions, such as members of a public interest group or hobbyists on an informal network. The reason for this requirement is that each member who receives requests for "hints" must be assumed to have enough time to satisfactorily answer these requests. If the community was replaced by an organization, or the hobby replaced by a day job, then the model becomes more complex. In Huberman and Hogg's model, the requesting party will either acquire the information needed or will not, depending on whether or not the information source is in possession of it. This led the authors to conclude that community members would do best to keep a "shortlist" of the people most likely to have useful information, and to rotate requests through different names on that list in order to give different people a chance to offer more information.

The model that will be described in this chapter builds on the Huberman and Hogg model, but will allow for requests to be unsuccessful, even if the information is available. People working within organizations are often very busy (Nasrallah 2006)—not only with the requirements of their jobs, but with the demands of responding to prior requests from other sources. This means that many requests may need to wait until previous requests have been fulfilled, and this delay can last beyond the period of original need for the information requested. With this simple change in the model assumptions, the request can now be for something other than simple information. The same model could equally well represent a request for anything work-related that some organization member needs to spend time to generate.

One commonality with the Huberman and Hogg (1995) model is that the meta-knowledge is still assumed to be there. The members of the organization being modeled must have already gone through a learning process to find out not only who can offer what, but also who is likely to be so busy that the request is likely to fail. The model is only useful if this learning has happened and if changes have not overtaken the meta-knowledge gained from this learning. Job turnover must be low enough that the person relied on for, say, access to a certain system, or advice on certain features of a technology, is still there when needed next. External circumstances must be similar enough that someone with the expertise to deal with a certain government bureaucracy or to sway a certain external stakeholder is also still as effective in that job as the last time. In this relative lull, interesting conclusions can be gleaned from assuming that interactions may fail due to either being requested from someone who is too busy to respond in time, or due to being requested from someone who has already yielded what they can from similar interactions in the very recent past.

What follows will be a step-by-step description of the model and its results, based on the four journal papers that built up to the full model (Nasrallah et al. 1998, 2003; Nasrallah and Levitt 2001; Nasrallah 2006).

WHO DOES THE SEARCH?

The *raison d'être* of any model is to answer some research question. The interaction value analysis (IVA) model has provided an answer to the following question: "How much value can management add to an organization?". We therefore start the description of the model with the part that pertains to the role of management.

If should be said from the outset that management fulfils many roles (Mintzberg 1989) and no single model can capture all of the value that is added through these various contributions. However, we also know that developments in education, contract law, technology, popular culture, and even human rights have reduced the need for many functions that used to be performed by managers. It is possible to envision, and sometimes even observe, an organization without managers (Davidow 1992). However, one role of managers that will never go away is the role of "traffic policeman," by which I mean the role of preventing people from going in different directions that, taken individually, seem perfectly consistent with getting where the organization needs them to go, but which could result in a gridlock or collision that neither party can unilaterally prevent. Only a central entity, namely management, can simultaneously see the learning that has taken place among all organization members and deduce that leaving each member to behave according to his or her local knowledge may or may not best serve the goals of the organization.

However, if the search for knowledge is conducted only by individual organization members, and the results of this search were to yield behavior (i.e., the selection of favorite interaction partners for various needs) that cannot be improved on by a central actor, then one can conclude that management would not be adding value in that particular way. As long as the organization provides certain minimal criteria for its own existence, such as incentives and a communication infrastructure, the role of management should be reduced or else the organization would be less efficient than a comparable organization that had reduced the role of management and encouraged individual initiative.

The converse case is when individual decisions about the selection of favorite interaction partners for various needs *did* result in some sort of gridlock. This might occur when popular individuals are swamped with requests and cannot properly respond to them all. In that case, an organization that made sure that it centrally coordinated its learning—or, more specifically, its reaction to this learning, would be more likely to perform well. More management control is needed to accomplish this coordination, and investing in additional (managerial) personnel, systems, or procedures will pay off in greater efficiency.

SIMPLIFYING

IVA is an extremely simplified representation of what it takes to get work done, and hence to add value to an organization.

INTERACTION VALUE

Coase (1988) and Williamson (1979) argued that organizations add economic value by allowing repeated transactions between individuals while incurring part of the transaction cost only once. Individuals who joint the organization get some knowledge about about each other and set constraints on each other's future behavior, incurring a fixed cost. This cost is then amortized over all future transactions that rely on this knowledge or these constraints. Adopting this point of view allows us to measure the effectiveness of the organization by summing up the value of all the transactions that occur between the members of the organization, which we now call "interactions" since they happen within the organization. IVA dispenses with the content of the interaction and simply keeps count of how many interactions are successfully completed. The assumption is that the different values of the various interactions will average out; that is, that the variance in these values will have no significant effects. The effect of higher or lower variance on the model results is an open research question, as are the effects of relaxing the other simplifying assumptions described in this section.

Varying Usefulness

Since all interactions are homogenized to add the same (normalized) value when they succeed, every individual or team that initiates an interaction is as capable of adding value as every other. In addition, if the partners who are sought for this interaction are ranked in order of decreasing usefulness for the purposes of this interaction, the value of selecting the most useful partner is also assumed to be the same.

Since both large and small organizations might be represented by small or large models, model granularity has to be taken out of the equation affecting success (Nasrallah and Levitt 2001). This is accomplished by reducing the value of an interaction by a known proportion when the respondent's suitability for the interaction is reduced by one rank. In other words, if the second most useful partner contributes 80% of what the most useful partner contributed, then the *third* most useful partner contributes 80% of that (i.e., 64% of the highest value) and so on.

This constant rate of usefulness loss is the first of the IVA model's six variables and it is referred to as the *differentiation* level within the organization (see the "Variables" section for a fuller explanation.)

Why would one not want to always seek the most useful partner? The primary reason is that this partner might be very busy as he/she is sought out by everyone. If partners are less popular, then they are less likely to be busy, and if they are not busy and already members of the same organization, then they should be under an obligation to engage in the interaction. However, if conditions are such that popular partners exist and are busy, then different factors come into play, such as how likely the partner is to prioritize your request over others, and how much management intervention is applied to change this prioritization.

How Many Categories Are Searched For?

Another thing to think about when designing the IVA model is the number of different types of needs that a typical job will entail. Every different sort of resource (e.g., access to materials, budget approval, past customer feedback) leads to a different ordering of the possible set of interaction partners based on how much they can help access that particular resource. Several simplifying approximations were made in this regard, again based on a general assumption that any trends revealed by comparing averages are likely to be significant even when all the detailed data are considered.

Categories per Task

Although a typical organization might have thousands of members controlling hundreds of resources, if the average task needs six resources, then studying an organization with six resources can tell a lot about all organizations.

Homogeneity of Resource Value

Each resource might contribute a different proportion of the end task's value, and most resources only add value in the presence of other "catalytic" resources. This distinction was also omitted from the IVA model in order to concentrate on more generalized effects of the number of resources needed for a typical or average task. If there are three resources needed, then getting half the value of one of them reduces the task's value by one-half of one-third (i.e., by one-sixth).

Popularity

If we are looking for the value added by management in the context of mitigating resource gridlock, then there must be some contention for resources in order for gridlock to be a problem in the first place. If there are only two cars passing an intersection per minute in each direction, then no one would need stop signs or traffic police. Similarly, in an organization where everyone requires different resources from everyone else, everyone can help themselves without fearing an adverse effect on other members. Nasrallah et al. (1998) derived the relationship between the number of different criteria for ranking in interaction partner and the highest number of seekers likely to want an interaction with the same partner. This can be called the "popularity" of the partner, since a partner who is preferred by 100 interaction seekers is more "popular" than one who is preferred by 50 interaction seekers. In a population of P individuals, absolute homogeneity means that all follow one ranking criterion, whereas absolute heterogeneity means that there are P independent ranking criteria, one per person. If the number of ranking criteria is N, then $N = 1$ implies that the most popular individual in that population is preferred by all P of them (including him or herself.), If $N = P$ then the number preferring the most popular individual is much smaller. In general:

$$\text{Popularity} = \frac{\log N}{\log(\log N)}. \tag{5.1}$$

To put this into perspective, if we imagine that five billion people each followed independent methods of choosing a preferred interaction partner for any reason (conversation, entertainment, leadership), it would then take only seven people selecting the same friend/performer/politician to make that person the planet's most popular friend/performer/politician. If, on the other hand, the number of equally weighted and statistically independent criteria was six, then we would have $\log(6)/\log(\log(6)) = 3$ criteria pointing to the same individual. Since the criteria have equal followership among the 5 billion, $3/5 \times 5$ billion would select the same most popular individual.

There has been a lot of recent research into the properties of networks, such as social links between people and hypertext links between web pages (Barabási 2003). Popular individuals, or popular websites, exist even when the average person or website has very few links to others. One way in which this comes about in nature is when popularity itself is a reason for establishing a new link. This common metaphor for describing selective attachment is that "the rich get richer." There are many possible avenues of future research into whether these types of networks characterize the links between colleagues who need to interact in order to do their jobs. Even on a purely theoretical front, there is much to be learned about the value and saturability of central versus distributed learning in networks that are, or are not, for example:

1. Small world networks, where every node is linked via a relatively short path to all, or most, other nodes.
2. Clustered networks, where nodes linked to the same neighbor are more likely to be linked to one another too.
3. Scale-free networks, where the incidence of a certain number of nodes with a high number of links implies that there must be at least one node with some multiple of this already large number of links.

Having said all that, we can get back to the very simple IVA model that focuses, as a first step, on the average organization member and whether that member's own work is dependent on a popular individual or not. In real life, I can accomplish all of my goals without ever meeting the pope or the president of the United States. Similarly, organization members, on average, are only likely to need to interact with someone too busy to help them if these popular individuals are a sizable contingent in the organization's population.

As implemented by Nasrallah (2006), the IVA model considers random networks with one of two values of N: 6 or 3. There are distinct independent criteria of selection, corresponding to types of resources needed on a typical task, and yielding a population where either one-third or one-half of the organization members favor the same interaction partner. It is as if there were three or six roles that each of the thousands of organization members take on at any point, and in that role they need someone else in one of the remaining roles to help them add value to the organization.

The number 3 or the number 6 is the *diversity* of the organization, as will be further explained in the "Variables" section below.

WHY DOES AN INTERACTION FAIL?

Not all interaction attempts add value. This is true both in real life and in IVA. The IVA model only becomes non-obvious when it combines the effects of different ways in which an interaction attempt can fail.

SEEKING AN INTERACTION AT THE WRONG TIME

Huberman and Hogg (1995) proposed that interactions within a community of far-flung individuals with a common interest serve to transfer a fresh insight. This implied that a certain amount of time must elapse before the same individuals had generated new insights that they could share. This amount of time is, of course, not a constant but varies for each interaction. Huberman and Hogg (1995) used a probability distribution to investigate the average success rate over a large number of requests for a given average value of this time. IVA uses the same approach but gives it a different interpretation (Nasrallah et al. 2003). An interaction still fails if the same interaction partner had been used too recently, but this is because, within an organization, most value-adding tasks involve multiple interactions. If value accrues to the organization from each individual interaction, then the incremental value of subsequent interactions with the same partner will go down.

As the learning process settles on an optimal frequency of interactions with each partner, the effect of going to an individual partner too frequently is balanced against the value lost by going to other partners who add less value per interaction, but whose contributions are valuable components of the whole task. The component of the aggregate value equation that accounts for loss from going too often to the same partner comes from the expression for the outcome of a race between two random (Poisson-distributed) processes. There is one random process producing the interaction request, and another random process representing everything else that needs to be in place in order for the interaction to succeed. If the random process producing the request beats the random process producing readiness, then the interaction fails: it went out too soon! (The probability that the two are equal is zero because we model the processes with a Poisson distribution, which is derived from the limit as time increments go to zero of the time between consecutive events that are equally likely to happen in any time interval.)

If we define $S1_{ij}$ as the success of an interaction requested by party i of party j, then this depends on two things: (1) the proportion of i's requests directed to j; and (2) the ratio of the two process rates: ρ, the average rate at which requests are made, and ω, the average rate at which interactions become ready to add value:

$$S1_{ij} = \frac{\omega}{\omega + p_{ij}\rho} = \frac{1}{1 + p_{ij}\rho/\omega}. \tag{5.2}$$

The ratio ρ/ω represents the *interdependence* of the organization's work (see the "Variables" section below).

GOING TO A COMPETENT PERSON WHO IS TOO BUSY

Another reason why an interaction might fail is if the respondent is so busy that, when he or she finally gives a response, the requestor no longer needs interaction. Again, the "deadline" for the interaction can vary randomly, and for a lack of more specific information we can assume that this is equally likely to occur at any given future moment, if it has not occurred already. Again, we have a Poisson distribution, and again a race between the process that gives rise to the requestor's deadline and the process that generates the responder's response. This process can be a Poisson process too, but only if the respondent starts working on his or her response as soon as the request is received. This was true in the Huberman and Hogg (1995) model of a community of experts who seldom interact, and would be true in an organization of full-time employees only if the respondent was not very popular (that is, busy), or if the organization as a whole was not very busy. Otherwise, there will be a queue of requests ahead of each incoming request and we would need to use a different kind of random distribution. Again, for lack of additional information, we will make the simplest possible assumption—namely, that each item in the queue is as likely to be processed at any point in time as it is at any other point in time. This leads us to what queuing theorists call the "M/M/1 queue," and what probability texts call the "Erlang distribution." The equation for the outcome of a race between a Poisson process and an Erlang process is more complex than just a ratio of rates, and involves the incomplete gamma function, as worked out by Ancker and Gafarian (1962). There are three rates involved:

1. The rate at which an interaction request becomes stale; that is, no longer capable of benefiting from a response (let us call this β)
2. The rate at which requests are received (λ)
3. The rate at which an average interaction request is processed (μ)

Since multiplying all the rates by the same constant does not change the probability of one process beating the other, it is the ratios of these rates that are used as parameters in IVA:

- *Load* is the ratio of the request arrival rate to the request processing rate (λ/μ). An organization with low load has less work to do for its customers, so people in the organization have fewer reasons for issuing interaction requests relative to the large number of people and other resources (e.g., computers) that can respond to these requests. This excess capacity might be a result of having had to deal with higher loads in the past, or it might be due to prior preparation for higher loads in future.
- *Urgency* is the ratio of request staleness rate to request processing rate (β/μ). Under high-urgency conditions, the value that the organization provides to its clients drops off rapidly with the passage of time, whereas low-urgency markets might be equally able to benefit from the organization's efforts a few days later.

It is important at this point to distinguish between the rate at which the average members generate requests, which is used to calculate load, and the rate at which a popular member *receives* requests. The former is the same as the rate at which the *average* member receives requests. To calculate the failure rate due to request staleness, popular members matter more, so we need to multiply the average arrival rate by a factor that reflects the recipient's popularity. Recalling that p_{ij} represents the proportion of interaction requests made by member i that are sent to member j, this factor is $\sum_{i=1}^{\text{Diversity}} p_{ij}$—namely, the sum over all senders i of request proportions that i selects to send to any particular recipient j. This will be higher than 1 for popular members and lower than 1 for less popular members, since by definition $\sum_{j=1}^{\text{Diversity}} p_{ij} = 1$ for any given i. (Note: Diversity, as defined earlier, is the number of distinct roles salient for the average task. It is the distinct roles that have distinct p_{ij} values, hence the summation over the diversity parameter).

The last thing that is needed to adapt the Ancker and Gafarian (1962) equation to the IVA model is to consider how many prior requests are likely to be ahead of any given request.

SOMEONE WHO HAS TIME, JUST NOT FOR YOU

The most straightforward queuing models assume "first-in-first-out" (FIFO) discipline. Every request is treated equally, and the only requests ahead of a given request are those that arrived before it. It is possible to imagine organizations where this is indeed the way that requests are treated. It is also not too difficult to think of reasons why some organizations might include members who do not act like that. Some requests are simply given more priority than others.

Out of the near-infinite variety of patterns of precedence when it comes to responding to a request, IVA focuses on three broad cases that have very different effects on behavior. These are:

1. The FIFO case, which represents a "disciplined" organizational climate where people uphold the principle of waiting one's turn and expect to be treated similarly.
2. A priority scheme where the request whose response has the greatest potential to add value is picked ahead of other requests. We say that an organization that follows this sort of system has a "fraternal" climate, since resources are preferentially given to those in the greatest need. (Of course "need" here refers to a thing required for professional rather than personal purposes.)
3. A priority scheme where the greatest priority is given to requests from the member whose future responses are most valuable to the current respondent. We could call this a "cynical" climate (a change from the term "capitalist" climate used in Nasrallah 2006).

Note that words like "disciplined," "fraternal," and "cynical" are merely labels for the general behavior patterns, not an implication that any other characteristics associated with these words need to be attached to the organizations in question. The use of the word "climate," on the other hand, is intended to suggest something similar to what management scientists refer to as "corporate climate" or sometimes incorrectly as "culture." To see how this is the case, consider the simplifying assumptions of the model; namely, that all members act in the same way; and the strategic nature of the optimization process, where every member acts with the foreknowledge of how all the other members will also react to this course of action. The combination of a principle that describes what constitutes acceptable behavior, and an expectation that others will also follow that principle is a good definition of what it feels like to operate in a certain organizational climate.

SUMMARY OF THE MODEL

VARIABLES

Each of the variables introduced in the "Simplifying" section above has something to say about the nature of the organization, its work, or its external environment. Putting an interpretation on each model variable is a subjective process and can be said, if one happens to be in a charitable mood, to be more of an art than a science.

Only validation against actual measurements can lend credence to the nominative process that ties a numeric parameter in a simple mathematical model to an observable description of an organization or its environment (see the "Validation Study" section below.)

As discussed above:

Diversity is the number of different sorts of resources needed for the organization's work. This represents things like division of labor, system complexity, and technical advancement level. IVA distinguishes between low diversity with 3, and medium diversity with 6. Higher levels of diversity always lead to zero value added by management intervention into individual choices, but this is only because of the assumption of relative homogeneity in the connectivity of different members made in the current version of the model.

Differentiation is the amount of value added by the interaction that adds the most value, divided by the amount of value added by a successful interaction with the partner who adds the least value. This corresponds to the range of qualifications or educational level of organization members. A chain gang or a Navy SEAL unit will both have low differentiation because the members have similar levels of preparation. A medical team consisting of a doctor, a medical technician, and a porter/driver has high differentiation. The numerical value used in generating the result reported here was varied from 2 for low differentiation, to 30 for high differentiation, with 10 used as the value for medium.

Note: Using the highest-to-lowest rather than the first-to-second ratio as the definition of differentiation allows us to model the same organization at different diversity levels without changing the differentiation value. At a higher diversity of 6, there are four intermediate levels between the highest and the lowest, but there is only one intermediate level under a low diversity of 3. Going from highest to second highest successful interaction under a diversity of 3 means losing a portion of the value equal to the square root of differentiation, but going from the highest to the second highest under a diversity of 6 loses only a portion of the value equal to the fifth root of the differentiation.

Interdependence is the degree to which interactions with the different types of parties (i.e., those who control the different types of resources) complement one another in order to add value. Task decomposability is another way of referring to the same property. The values used to generate the results were 1 for low interdependence—meaning that, on average, one more interaction with someone else was required between successful interactions with a particular partner—to 9, indicating a requirement for 10 interaction attempts with others. The value chosen for medium was 3.

Load is the ratio of interaction request generation rate to interaction processing rate. It corresponds to how busy an organization is, relative to its capacity to do work. Load is the most likely factor to vary over time in any given organization in real life, since adding capacity by hiring or through technology is a slow process, but market demand for a product can spike or decay more quickly. Low load was defined as a ratio of 15%, meaning that 15% of the time between requests was needed to process the request. Medium load was defined as 75%; and 125% was used for high loads, meaning that, on average, five requests would appear in the time it takes to process four.

Urgency is another ratio of rates of random processes, and it measures how quickly things need to be done in order to effectively add value. Some industries are inherently higher in urgency, due to perishable materials or life-threatening outcomes. Others only experience high urgency sporadically, such as when a deadline approaches or progress needs to be accelerated following a delay or failure. Low urgency was defined as 5%, meaning that 1 out of 20 requests were likely to be no longer needed by the time it took to process the request by someone who was not otherwise occupied. Medium urgency was defined as 50%, and high as 90%.

Climate can be a contentious term in organization science (Denison 1996), but it can be loosely defined as the way people tend to act when there is no hard rule that specifies their behavior. IVA models climate as the "queuing discipline"— that is, the method used to select among requests when several are waiting to be processed. FIFO selection, corresponding to a climate where there is no expectation of preference between different types of members, is given the name "disciplined." The other two types of climate defined in IVA, "cynical" and "fraternal," are meant to correspond to two extremes of systematic preferential treatment. The "cynical" organization has a climate of preference for those who have more power to punish or reward when the time comes to request things back from them, but of course this expectation is illusory because at that time the powerful will defer to those that they perceive to be most useful to them, not to those who had sought approval from them in the past. (An interesting future study might include the effects of past loyalty, but doing so would probably entail a full simulation approach instead of a mathematical solution of game-theory equilibria.) Finally, the "fraternal" climate means a bias, and expectation of bias, in favor of the most needy, defined as those for whom the interaction represents a higher proportion of their expected contribution to the organization.

This list is summarized in Table 5.1.

CALCULATION

The final model consists of a matrix of preference rankings, namely, three permutations of 1, 2, and 3 for the three-party version, and six permutations of 1 through 6 for the high-diversity six-party version. The learning that takes place begins with the discovery of which parties are best for which sort of interaction (the entries in the matrix). Next, it is necessary to predict which other members will reach similar conclusions and have similar preferences, and hence make certain parties busier and possibly less useful. The "flat end" of the learning curve is reached only when every party has reacted to all the expected reactions of other parties, and equilibrium (in the game-theoretic sense) has been reached.

TABLE 5.1

Summary of IVA Parameters

Name	Meaning in Model	Meaning in Organization
Diversity	Number of independent, equally weighted criteria used for evaluating potential interaction partners	Number of different skill types or disciplines needed for the organization's work
Differentiation	Ratio of highest value added to lowest value added from an interaction evaluated by the same criterion	Spread in skill levels for a given skill type used in the organization
Interdependence	Proportion of organization's disciplines that must be consulted in the course of any value-adding activity	Degree of coordination between concurrent activities necessary for success
Load	Ratio of rate at which interactions are proposed to rate at which responses are received	Resources available to meet demand for information
Urgency	Ratio of rate at which interaction requests cease to be relevant to rate at which responses are received	Incidence of deadlines and other situations where delay results in failure
Climate	Queuing service discipline: FIFO verus different priority schemes	Shared and perceived values extant in the organization

When management intervention is high, the cooperative (Von Neumann) equilibrium determines where the learning curve becomes flat. This is because management can force members to go along with the most globally efficient allocation of their interaction requests, possibly reducing the proportion of their requests sent to the party they find the most useful, and sending more requests to the second or third most useful (to a greater extent than they already need to spread requests around in order to account for interdependence.)

The *laissez faire* approach leads to a learning curve that ends at the non-cooperative (Nash) equilibrium because each member is only concerned with maximizing their own productivity, not that of the whole organization. Every other member knows this to be the case and acts accordingly.

The equations for finding these two equilibrium points for each of the three climate regimes were derived by Nasrallah (2006). The solutions were found numerically for 486 combinations of the six variables above. Finding an analytic solution to these equations remains an open research question.

MODEL OUTPUT

The main result of this whole exercise is that the total value added by the organization under the two different learning curve equilibria might be the same, or it might be different. The non-cooperative equilibrium cannot be higher than the cooperative. When the two are equal, we can conclude that *laissez faire* management, in all its popular modern names and forms, is sufficient for coordination, leading to a learning curve that reaches as high as it can. This means that it is better to avoid the extra expenditure and frustrations of heavy-handed management in those instances.

(One can also argue that the learning curve might take longer to reach equilibrium when one point of learning has to process all the information for all organization members. Deriving the exact shape of the learning curve under the group-learning assumptions of IVA is also an open research question.)

Otherwise, the higher the ratio of cooperative to non-cooperative equilibrium value, the more management needs to do to both discover and enforce the higher-value behavior.

What the IVA model must do, therefore, is this: find the value added by the whole organization under cooperative (Von Neumann) equilibrium, and divide that by the value added by the whole organization under non-cooperative (Nash) equilibrium. For clarity, we can express this as a percentage improvement by subtracting 1. The table of these percentage improvements for each of 486 combinations or the six parameters is the output of the model.

These values are tabulated in Figures 5.1 through 5.3, with darker shading for the higher values of management usefulness. Figure 5.1 shows the values under low load; Figure 5.2 shows medium load; and Figure 5.3 shows high load.

INTERPRETATION OF RESULTS

The low load plot in Figure 5.1 shows hardly any value from management playing the role of regulator, as might be expected since there is very little activity to regulate under low load. Organizations typically form to carry out processes that require more than 15% of the members' processing capacity, but it is possible for many organizations to go through periods of low load, during which the value added by layers of management is hard to gauge. For example, a standing army is typically lampooned as the epitome of bureaucratic overkill because the load on it outside of wartime does not call for the large hierarchy that is typically in place. In business, organizations that no longer have the load present during their creation typically shed management layers or go out of business.

Figures 5.2 and 5.3 show very similar patterns, with the higher load in Figure 5.3 leading to higher values of the usefulness of management. We can focus on Figure 5.3 to note the patterns. Observations include:

1. In general, a "cynical" climate needs management to centralize learning more than a "disciplined" climate, and a "fraternal" climate needs it less than disciplined.
2. In general, low urgency coupled with high or medium load gives management an advantage over high urgency, possibly because high-urgency work needs the quick response of decentralized learning.
3. In general, high differentiation gives management an advantage.
4. High interdependence gives management a big advantage under a "cynical" climate, and a small advantage under a "disciplined" climate. Under a "fraternal" climate, the effect of interdependence is mixed: under low diversity and high or medium differentiation, decentralized learning has an advantage as interdependence goes up.

Diversity	Interdep	Urgency	Disciplined climate			Cynical climate			Fraternal climate		
			High differentiation	Medium differentiation	Low differentiation	High differentiation	Medium differentiation	Low differentiation	High differentiation	Medium differentiation	Low differentiation
Low diversity	High interdep	High urgency	0.0%	0.0%	0.0%	0.1%	0.1%	0.0%	0.1%	0.0%	0.0%
		Medium urgency	0.0%	0.0%	0.0%	0.1%	0.1%	0.0%	0.1%	0.0%	0.0%
		Low urgency	0.0%	0.0%	0.0%	0.0%	0.0%	0.0%	0.0%	0.0%	0.0%
	Medium interdep	High urgency	0.0%	0.0%	0.0%	0.1%	0.0%	0.0%	0.0%	0.0%	0.0%
		Medium urgency	0.0%	0.0%	0.0%	0.0%	0.0%	0.0%	0.0%	0.0%	0.0%
		Low urgency	0.0%	0.0%	0.0%	0.0%	0.0%	0.0%	0.0%	0.0%	0.0%
	Low interdep	High urgency	0.0%	0.0%	0.0%	0.0%	0.0%	0.0%	0.0%	0.0%	0.0%
		Medium urgency	0.0%	0.0%	0.0%	0.0%	0.0%	0.0%	0.0%	0.0%	0.0%
		Low urgency	0.0%	0.0%	0.0%	0.0%	0.0%	0.0%	0.0%	0.0%	0.1%
Medium diversity	High interdep	High urgency	0.0%	0.0%	0.0%	0.1%	0.1%	0.0%	0.0%	0.0%	0.0%
		Medium urgency	0.0%	0.0%	0.0%	0.1%	0.0%	0.0%	0.0%	0.0%	0.0%
		Low urgency	0.0%	0.0%	0.0%	0.0%	0.0%	0.0%	0.0%	0.0%	0.0%
	Medium interdep	High urgency	0.0%	0.0%	0.0%	0.0%	0.0%	0.0%	0.0%	0.0%	0.0%
		Medium urgency	0.0%	0.0%	0.0%	0.0%	0.0%	0.0%	0.0%	0.0%	0.0%
		Low urgency	0.0%	0.0%	0.0%	0.0%	0.0%	0.0%	0.0%	0.0%	0.0%
	Low interdep	High urgency	0.0%	0.0%	0.0%	0.0%	0.0%	0.0%	0.0%	0.0%	0.0%
		Medium urgency	0.0%	0.0%	0.0%	0.0%	0.0%	0.0%	0.0%	0.0%	0.0%
		Low urgency	0.0%	0.0%	0.0%	0.0%	0.0%	0.0%	0.0%	0.0%	0.0%

FIGURE 5.1 Value added by centralized learning under low load.

Diversity	Interdep	Urgency	Disciplined climate			Cynical climate			Fraternal climate		
			High differentiation	Medium differentiation	Low differentiation	High differentiation	Medium differentiation	Low differentiation	High differentiation	Medium differentiation	Low differentiation
Low diversity	High interdep	High urgency	1.3%	0.7%	0.1%	3.1%	2.4%	1.3%	0.9%	0.6%	0.5%
		Medium urgency	1.9%	1.0%	0.1%	4.7%	3.7%	2.0%	1.3%	0.9%	0.8%
		Low urgency	1.7%	0.9%	0.1%	4.3%	3.4%	1.9%	1.4%	1.0%	0.8%
	Medium interdep	High urgency	0.6%	0.3%	0.0%	1.6%	1.3%	0.7%	0.8%	0.5%	0.3%
		Medium urgency	1.1%	0.6%	0.1%	2.4%	2.0%	1.1%	1.3%	0.8%	0.4%
		Low urgency	1.0%	0.5%	0.1%	2.1%	1.8%	1.0%	1.4%	0.8%	0.4%
	Low interdep	High urgency	0.1%	0.3%	0.0%	0.1%	0.9%	0.6%	0.6%	0.5%	0.2%
		Medium urgency	0.3%	0.5%	0.1%	0.3%	1.4%	0.9%	1.3%	0.9%	0.3%
		Low urgency	0.4%	0.5%	0.1%	0.3%	1.2%	0.8%	1.4%	0.9%	0.3%
Medium diversity	High interdep	High urgency	0.5%	0.3%	0.0%	1.8%	1.3%	0.9%	1.2%	1.0%	0.9%
		Medium urgency	0.8%	0.4%	0.0%	2.5%	1.8%	1.2%	1.6%	1.4%	1.2%
		Low urgency	0.8%	0.4%	0.0%	2.3%	1.7%	1.1%	1.5%	1.3%	1.1%
	Medium interdep	High urgency	0.3%	0.1%	0.0%	1.0%	0.7%	0.5%	0.7%	0.6%	0.5%
		Medium urgency	0.5%	0.2%	0.0%	1.6%	1.1%	0.7%	1.0%	0.8%	0.7%
		Low urgency	0.5%	0.2%	0.0%	1.5%	1.0%	0.6%	1.0%	0.8%	0.6%
	Low interdep	High urgency	0.2%	0.1%	0.0%	0.9%	0.7%	0.4%	0.5%	0.5%	0.4%
		Medium urgency	0.5%	0.3%	0.0%	1.4%	1.0%	0.5%	0.8%	0.8%	0.5%
		Low urgency	0.5%	0.3%	0.0%	1.4%	1.0%	0.5%	0.8%	0.7%	0.5%

FIGURE 5.2   Value added by centralized learning under medium load.

Diversity	Interdep	Urgency	Disciplined climate			Cynical climate			Fraternal climate		
			High differentiation	Medium differentiation	Low differentiation	High differentiation	Medium differentiation	Low differentiation	High differentiation	Medium differentiation	Low differentiation
Low diversity	High interdep	High urgency	4.4%	2.3%	0.2%	10.5%	7.8%	3.8%	1.8%	1.2%	1.4%
		Medium urgency	8.1%	4.5%	0.5%	25.0%	20.2%	10.7%	2.9%	2.3%	3.3%
		Low urgency	8.3%	4.7%	0.5%	27.4%	22.6%	12.6%	3.1%	2.5%	3.8%
	Medium interdep	High urgency	2.1%	1.1%	0.1%	5.0%	4.2%	2.2%	1.8%	1.2%	0.8%
		Medium urgency	4.9%	2.6%	0.3%	12.5%	10.5%	5.9%	3.5%	2.5%	2.1%
		Low urgency	5.2%	2.8%	0.3%	13.9%	11.7%	6.7%	3.9%	2.8%	2.3%
	Low interdep	High urgency	0.6%	0.8%	0.1%	0.5%	3.0%	1.9%	1.8%	1.2%	0.6%
		Medium urgency	1.9%	2.2%	0.3%	4.1%	7.2%	4.8%	4.4%	2.9%	1.6%
		Low urgency	2.3%	2.5%	0.3%	4.8%	7.8%	5.4%	4.9%	3.3%	1.8%
Medium diversity	High interdep	High urgency	1.8%	0.8%	0.1%	5.0%	3.7%	2.5%	3.1%	2.7%	2.4%
		Medium urgency	2.9%	1.4%	0.2%	9.8%	7.6%	5.4%	5.6%	5.1%	4.9%
		Low urgency	2.9%	1.5%	0.2%	10.6%	8.3%	5.9%	5.8%	5.3%	5.4%
	Medium interdep	High urgency	0.9%	0.4%	0.0%	3.2%	2.3%	1.5%	2.0%	1.7%	1.5%
		Medium urgency	2.1%	1.0%	0.1%	7.2%	5.2%	3.3%	4.3%	3.7%	3.2%
		Low urgency	2.3%	1.1%	0.1%	7.8%	5.6%	3.5%	4.6%	3.9%	3.5%
	Low interdep	High urgency	0.8%	0.4%	0.0%	2.8%	2.1%	1.2%	1.5%	1.5%	1.2%
		Medium urgency	2.1%	1.1%	0.1%	6.8%	5.0%	2.7%	3.9%	3.5%	2.7%
		Low urgency	2.3%	1.3%	0.1%	7.5%	5.4%	2.9%	4.3%	3.8%	2.9%

FIGURE 5.3 Value added by centralized learning under high load.

VALIDATION STUDY

Since the IVA model is fairly recent, only one validation study has been performed on it (Nasrallah and Al-Qawasmeh 2009).

The study involved interviewing the managers of 23 companies in the kingdom of Jordan using a structured questionnaire. The questionnaire assessed the operating environment of the company as well as the extent to which different mechanisms were used to specify how employees should act. These mechanisms were combined in a somewhat arbitrary fashion into one "management control" measure, accounting for the effects of:

1. High formalization, which means that rules and procedures are written down and enforced
2. High centralization, which means that a greater proportion of operating decisions are made by higher managers
3. Bureaucratic structure, which means that one permanent manager oversees a person's work rather than delegating some authority to a temporary team or project leader
4. Micro-management, which means that mangers ask for reports and give instructions beyond what is required by the formal structure

The interview responses about the work of the organization were mapped to the six parameters of the IVA model and the tables in Figures 5.1 to 5.3 were used to deduce management usefulness. The companies that had an amount of management control (the sum of the four factors in the list above) proportional to the management usefulness indicated by IVA were expected to be doing better financially than those that did not match practice to need. Financial performance was derived from public stock market listings of the companies. It was necessary to exclude 3 of the 23 companies because they were losing money, which indicates things being wrong with the company other than over- or under-management. Another eight company managers indicated that they faced an unpredictable competitive environment, which indicates that their learning curves were not at the flat part yet, thus making the IVA model useless.

The remaining 12 companies showed a statistically significant correlation between positive growth in the "return on assets" (ROA) measure over three years and the degree of fit between management practice and IVA prescription, as shown in Figure 5.4.

While this finding is promising, much more remains to be done in order to improve the mapping of the model to real and measurable variables encountered in management practice. It is nevertheless safe to conclude that two distinct organizational learning curves can and do coexist, and that the higher-cost learning curve that involves centralizing all the learning is only beneficial in certain limited situations. The cheaper decentralized learning model is likely to be the key to profitability and growth in other situations For those organizations using the mode of learning that is not appropriate for their organizational climate, workload, degree of urgency, work interdependence, skill diversity and differentiation, it will be profitable to change. This can be done either through consultants or internally, by shifting

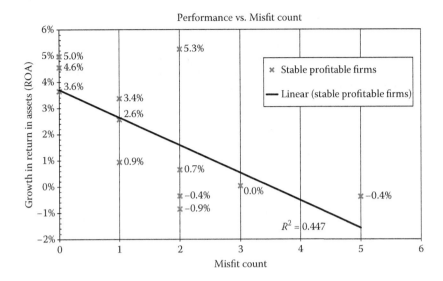

FIGURE 5.4   Correlation between adherence to IVA recommendation and ROA growth.

the organizational structure to add or subtract formalization, centralization, bureaucracy, and micro-management, and thereby to nudge the learning mode from distributed to centralized, or vice versa.

REFERENCES

Adler, P.S., and Clark, K.B., 1991. Behind the learning curve: A sketch of the learning process. *Management Science* 37(3): 267–281.

Ancker, C.J., and Gafarian, A.V., 1962. Queuing with impatient customers who leave at random. *Journal of Industrial Engineering* 13: 84–90.

Barabási, A.L., 2003. *Linked: How everything is connected to everything else and what it means.* Plume: Penguin Group.

Coase, R.H., 1988. *The firm, the market and the law.* University of Chicago Press: Chicago.

Davidow, W.H., 1992. *The virtual corporation: Structuring and revitalizing the corporation for the 21st century.* New York: Harper Business.

Denison, D.R., 1996. What is the difference between organizational culture and organizational climate? A native's point of view on a decade of paradigm wars. *Academy of Management Review* 21(3): 1–36.

Huberman, B.A., and Hogg, T., 1995. Communities of practice. *Computational and Mathematical Organization Theory* 1(1): 73–92.

Mintzberg, H., 1989. *Mintzberg on management.* New York: The Free Press.

Nasrallah, W., 2006. When does management matter in a dog-eat-dog world: An "interaction value analysis" model of organizational climate. *Computational and Mathematical Organization Theory* 12(4): 339–359.

Nasrallah, W.F., and Al-Qawasmeh, S.J., 2009. Comparing *n*-dimensional contingency fit to financial performance of organizations. *European Journal of Operations Research,* 194(3): 911–921.

Nasrallah, W.F., and Levitt, R.E., 2001. An interaction value perspective on firms of differing size. *Computational and Mathematical Organization Theory* 7(2): 113–144.

Nasrallah, W.F., Levitt, R.E., and Glynn, P., 1998. Diversity and popularity in organizations and communities. *Computational and Mathematical Organization Theory* 4(4): 347–372.

Nasrallah, W.F., Levitt, R.E., and Glynn, P., 2003. Interaction value analysis: When structured communication benefits organizations. *Organization Science* 14(5): 541–557.

Williamson, O.E., 1979. Transaction-cost economics: The governance of contractual relations. *Journal of Law and Economics* 22(2): 233–261.

6 Log-Linear and Non-Log-Linear Learning Curve Models for Production Research and Cost Estimation*

Timothy L. Smunt

CONTENTS

INTRODUCTION

Learning curve models attempt to explain the phenomenon of increasing productivity with experience. The first reported use of the learning curve phenomenon was by Wright (1936) and since then an extensive number of papers have reported its use in industrial applications and research settings (e.g., see Adler and Clark 1991; Gruber 1992; Bohn 1995; Rachamadugu and Schriber 1995; Epple et al. 1996; Mazzola and McCardle 1996). Wright's model, which assumed that costs decrease by a certain percentage as the number of produced units doubled, is still widely used and forms the basis for most other adaptations of the learning curve concept. Extensions of Wright's model to account for work in progress (Globerson and Levin 1995) and for use in project control (Globerson and Shtub 1995) have also been proposed to consider typical data-gathering problems and scenarios in industry settings.

    Wright's learning curve model, $y = ax^{-b}$, a log-linear model is often referred to as the "cumulative average" model since y represents the average cost of all units produced up

---

* Reprinted from *International Journal of Production Research* 1999, vol. 37, no. 17, 3901–3911.

to the xth unit. Crawford (see Yelle 1979) developed a similar log-linear model, using the same function as shown for the cumulative average model. However, in this case, y represents the unit cost for the particular x unit. For this reason, the Crawford approach is often referred to as the "unit cost" model. Both of these simple log-linear models are discrete in unit time for at least one type (unit cost or cumulative average cost) of cost calculation. In either the cumulative average or unit cost approach, an approximation is required to convert one type of cost to the other. This approximation can create difficulties both in empirical studies of production costs and in formulating analytical models for production planning, including learning. However, the use of the continuous form of the log-linear model, explained in detail in the next section, overcomes this discrete formulation problem. By making the assumption that learning can occur continuously, learning curve projections can be made from mid-units, thus eliminating any approximation error. This model, sometimes called the "mid-unit" approach, also provides for the reduction of computation time, which can become important in empirical analysis of manufacturing cost data due to the numerous items typically found in such datasets.

There has been a fair amount of confusion in the literature and operations management textbooks concerning the appropriate use of either a unit or a cumulative average learning curve for cost projections. The main objectives of this chapter are to illustrate that no conflicts exist in this regard under the assumption of continuous learning, and to provide guidelines for this model's application in research and cost estimation. Further to this, we also present variations of the log-linear models that are designed to analyze production cost data when the log-linear assumption may not hold, especially during production start-up situations.

This chapter is organized as follows. First, the continuous log-linear (mid-unit) model is presented. Second, examples of the use of the mid-unit model are provided to illustrate the relative ease of this learning curve approach for estimating both unit and cumulative average costs on labor effort. Finally, we discuss the relative advantages and disadvantages of using these models, including more complicated non-log-linear models, in the empirical research of productivity trends and in proposing analytical models for normative research in production.

MID-UNIT LEARNING CURVE MODEL

If learning can occur continuousl—that is, if learning can occur within unit production as well as across unit production—a continuous form of the log-linear learning curve can be modeled. It is not necessary in this case, then, to force the unit cost and the cumulative average cost of the first unit to be identical. We note that within unit learning is a reasonable assumption, especially when the product is complex and requires a number of tasks for an operator to complete. In fact, this assumption was previously made by McDonnell Aircraft Company for the purposes of their cost estimates in fighter aircraft production. Furthermore, as will be illustrated later, most cost data is gathered by production batch and, therefore, projecting from a mid-unit of the batch to another batch mid-unit does not require the assumption of continuous learning within a single unit of production.

A continuous form of the unit cost function permits the use of learning curves to be more exact and convenient. According to the cumulative average model:

$$Y_{tc}(x) = (a)(x^{1-b}),$$ (6.1)

where

$$b = \left[\frac{\log(\text{learning rate})}{\log(2.0)} \right].$$

The rate of increase of a total cost function can be described by the first derivative of that total cost function. Because the rate of increase of a total cost function can be explained as the addition of unit costs over time, the unit cost function becomes:

$$Y_u(x) = \frac{d(tc)}{d(x)} = (1-b)(a)(x^{-b}).$$ (6.2)

Thus, the cumulative average cost is the same as in the cumulative average model:

$$Y_{ca}(x) = a(x^{-b}).$$ (6.3)

The mid-unit model is illustrated in Figure 6.1.

A question that frequently arises concerning the mid-unit learning model is: "How can the first unit cost have different values for the cumulative average cost than that for the unit cost?" This disparity can easily be resolved by taking into consideration the relationship between the rates of total cost increase and the learning curve concept. The first derivative provides a continuous unit cost function. The continuous cost function assumes that learning exists, either within a unit or within a batch of units (or both).

For example, consider a "production batch of units" experiencing a 90% learning curve, and then consider a case of no learning effect where all units require 10 h each to produce. If no learning takes place within the first five units, the midpoint of effort would occur at the middle of the third unit. This midpoint, also known as the

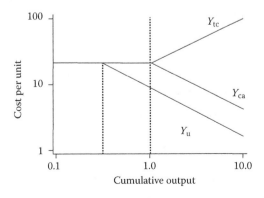

FIGURE 6.1 Mid-unit model.

"mid-unit," is the point of a unit or batch of units where the average cost theoretically occurs. However, when learning takes place, the initial units will cost more than the later units. If the first unit requires 10 h to produce the expected costs for the next four units, assuming a 90% learning curve, then we have the costs shown in Table 6.1.

In Table 6.1, the total cost for the batch of five units is 38.3 h. The mid-unit in production is reached in 19.1 (38.3 h/2) total hours, which occurs somewhere between unit 2 and unit 3. Interestingly, the unit cost for the first unit plotted at the midpoint of unit 1 is identical to the average cost to produce the whole unit (Figure 6.1).

DERIVATION OF MID-UNIT FORMULA

The use of the mid-unit model requires the determination of the mid-unit for most calculations and for regression analysis. Normally, the mid-unit is calculated for a production batch so that average costs can be projected from one batch to another. In essence, the average cost for a batch is the unit cost for the mid-unit of the batch. Therefore, the projection of a previous batch average cost simply requires that unit costs be projected from one mid-unit to another mid-unit.

To derive the mid-unit formula, it is important to consider two ways of calculating batch costs. First, within any given batch where x_2 is the quantity at the end of the batch, and x_1 is the quantity prior to the beginning of the batch, the total cost of the batch (utilizing the cumulative cost equation) would be:

$$\text{Total batch cost} = (a)(x_2^{1-b}) - (a)(x_1^{1-b})$$

$$= a[x_2^{1-b} - x_1^{1-b}]. \tag{6.4}$$

Alternatively, the unit cost is defined as:

$$Y_u = (1-b)(a)(x^{-b}). \tag{6.5}$$

The unit cost at the midpoint of effort, multiplied by the quantity in the batch results in the total cost for the batch. Defining x_M as the midpoint or mid-unit of the batch, then:

TABLE 6.1
90% Learning Curve Calculations

Unit	Calculation	Unit Cost (h)	Total Cost (h)
1st unit	10.0 h	10.0	
2nd unit	$(0.848)10\,h \times 2^{-0.152}$	7.6	17.6
3rd unit	$(0.848)10\,h \times 2^{-0.152}$	7.2	24.8
4th unit	$(0.848)10\,h \times 2^{-0.152}$	6.9	31.7
5th unit	$(0.848)10\,h \times 2^{-0.152}$	6.6	38.3

$$\text{Batch cost} = (1-b)(a)(x_M^{-b})(x_2 - x_1). \tag{6.6}$$

Setting Equations 6.4 and 6.6 for the batch cost equal to each other and solving for x_M:

$$(1-b)(a)(x_M^{-b})(x_2 - x_1) = (a)[x_2^{1-b} - x_1^{1-b}], \tag{6.7}$$

or

$$x_M = \left[\frac{(1-b)(x_2 - x_1)}{x_2^{1-b} - x_1^{1-b}} \right]^{1/b}. \tag{6.8}$$

PROJECTIONS FROM BATCH TO BATCH

Predicting batch costs at some future point from the most current cost data is accomplished in a similar manner to projecting from unit to unit. The mid-unit formula is used to determine "from" and "to" units for the cost projection. Figure 6.2 shows a unit cost curve based on an 80% learning curve ratio, with unit 50 having a cost of 10 labor hours. To project a cost for unit 100 we use the formula:

$$Y_u(x_2) = \left(\frac{x_2^{-b}}{x_1^{-b}} \right) (Y_u(x_1)). \tag{6.9}$$

When the learning curve rate is 80%,

$$b = -\frac{\log(0.80)}{\log(2.0)} = 0.3219.$$

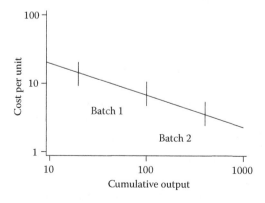

FIGURE 6.2   Mid-unit model example.

Therefore,

$$Y_u(100) = \left(\frac{100^{-0.3219}}{50^{-0.3219}} \right)(10) = 8 \text{ labor hours.}$$

Similarly, batch costs are projected by:

$$\text{Batch average}_2 = \left(\frac{x_{M_2}^{-b}}{x_{M_1}^{-b}} \right)(\text{batch average}_1). \tag{6.10}$$

Assume that the most current production batch shows a total cost of 200 labor hours for units 11–20 for a batch average of 20 labor hours per unit. The task then becomes one of projecting the cost for next month's production from unit 101 to unit 500. First, find the corresponding mid-units of each batch. Assuming an 80% learning curve:

$$x_{M_1} = \left[\frac{(0.6781)(20-10)}{20^{0.6781} - 10^{0.6781}} \right]^{1/0.3219} = 14.6,$$

$$x_{M_2} = \left[\frac{(0.6781)(500-100)}{500^{0.6781} - 100^{0.6781}} \right]^{1/0.3219} = 266.5.$$

Then,

$$\text{Batch average (units } 101-500) = \left(\frac{266.5^{-0.3219}}{14.6^{-0.3219}} \right)(20) = 7.85 \text{ h.}$$

The total cost for the batch is 7.85 h/unit × 400 units = 3140 h.

Dog-leg Learning Curves

Frequently, a production process will experience a change in the learning rate. When a break in the learning curve is expected, a method to project beyond this breaking point to a different learning curve is needed. Again, this cost projection takes place on the unit cost line. To derive a formula to project on a dog-leg learning curve, it is possible to make two separate projections for the same purpose. First, projecting from Batch 1 to the breaking point, BP:

$$Y_u(\text{BP}) = \left(\frac{\text{BP}^{-b}}{x_{M-1}} \right)(\text{average batch cost}_1). \tag{6.11}$$

Then, projecting from BP to Batch 2:

$$Y_u(x_{M_2}) = \left(\frac{x_{M_2}^{-b}}{BP^{-b}} \right) (Y_u(BP)). \tag{6.12}$$

Expanding Equation 6.12 to include Equation 6.11 we have,

$$Y_u(x_{M_2}) = \left(\frac{x_{M_2}^{-b}}{BP^{-b}} \right) \left(\frac{BP^{-b}}{x_{M_1}^{-b}} \right) (\text{average batch cost}_1). \tag{6.13}$$

Substituting the mid-unit formula (Equation 6.8) into Equation 6.13, and defining QP_1 and QT_1 as the previous and total units for the first batch and QP_2 and QT_2 as the previous and total units for the second batch, we have:

$$Y_u(x_{M_2}) = \left(\frac{\left[\frac{(1-b_2)(QT_2 - QP_2)^{1/b_2 - b_2}}{QT_2^{1-b_2} - QP_2^{1-b_2}} \right]}{BP^{-b_2}} \right) \left(\frac{BP^{-b_1}}{\left[\frac{(1-b_1)(QT_1 - QP_1)^{1/b_1 - b_1}}{QY_1^{1-b_1} - QP_1^{1-b_1}} \right]} \right)$$

(average batch cost$_1$). $\tag{6.14}$

Simplifying:

$$Y_u(x_{M_2}) = \left(\frac{(1-b_1)(QT_1 - QP_1)(BP^{b_2})(QT_2^{1-b_2} - QP_2^{1-b_2})}{(1-b_2)(QT_2 - QP_2)(BP^{b_1})(QT_1^{1-b_1} - QP_1^{1-b_1})} \right)$$

(average batch cost$_1$) $\tag{6.15}$

To illustrate the use of this formula, assume that we wish to estimate the cost of a batch from unit 201 to unit 500. Furthermore, assume that the average cost for an earlier batch (from unit 11 to unit 20) was 20 h/unit, the learning curve for the first leg was 80%, and the second leg learning curve is projected to be 90%. The break point is 100 units. The average cost for units 201–500 can then be calculated as:

$$\left(\frac{(0.6781)(20 - 10)(100^{0.152})(500^{0.848} - 200^{0.848})}{(0.848)(500 - 200)(100^{0.3219})(20^{0.6781} - 10^{0.6781})} \right) (20 \text{ h/unit}) = 8.95 \text{ h/unit}.$$

While the derivation of the above formula may have appeared complicated, note that the application is quite straightforward as shown in the above example. It is also possible to decompose this formula into smaller parts in order to reduce its apparent complexity. For example, projections can be first made from the mid-unit of the production batch to the dog-leg point. Projections can be then made from the dog-leg point to the projection batch using the new learning curve.

ADVANTAGES OF THE MID-UNIT MODEL
IN REGRESSION ANALYSIS

The advantages of a continuous form of a log-linear learning curve model, the mid-unit model, fall into two categories: (1) computation speed and (2) computation accuracy. The computational speed advantage is the result of the ability to project all types of cost; unit, batch average, cumulative average, and total costs, from the continuous unit cost function. For example, assume that the learning rate and unit cost for x_1 is known, and projections are required for the average cost of the production batch from units x_2 to x_3, for the cumulative average cost of x_3 units, and for the total cost for x_2 to x_3 units. Using the continuous form of the learning curve, the average cost for x_2 to x_3 is simply the unit cost projection from x_1 to the midpoint of $x_2 - x_3$. This could be accomplished using the unit cost model, but the unit cost model would necessitate the summation of x_3 units to determine the cumulative average cost of x_3.*

An additional computational problem arises when the underlying learning behavior is log-linear with the cumulative average, but regression analysis is performed on actual unit cost data without adjusting for the mid-unit of effort. For example, using an 80% learning curve and the cumulative average model, cumulative average costs are calculated from units 2 to 200. The difference between total costs from $x_n - x_n - 1$ becomes the unit costs for this model. The unit cost function is non-log-linear due to its discrete nature. An estimation of the learning curve using the unit costs from units 2 to 50 (call this our historical database) would indicate approximately a 79% learning curve. The upper half of Table 6.2 illustrates the projection error from performing regression analysis on unit cost data in this fashion. We also calculated the true unit costs when the cumulative average costs follow 70% and 90% learning curves. In all three cases, regression analyses on the actual unit costs provide slope estimates that are steeper than those actually occurring. Note that the cost projection error using these incorrect slope and intercept estimates increases as the cost estimate is made for production units further down the learning curve. The maximum error (7.24%) occurs for the largest unit number for the estimate (500) and for the steepest learning curve (70%).

Clearly, this estimate error is most pronounced when the actual unit costs are taken from the earliest production units. The lower half of Table 6.2 illustrates the estimation error when actual unit costs for units 26–50 are used (the latter half of the historical data in this example). The maximum percent error is reduced considerably using this approach (1.24%), but is still sufficiently large to induce substantial profit margin reductions if pricing has been based on these projections. Therefore, we see

---

* We note, however, that if the true underlying learning behavior follows a discrete pattern, as assumed in either the cumulative average or unit cost approach, using the mid-unit approach can induce approximation error (see Camm et al. 1987). In prior research projects, however, the author has found that when the mid-unit formulation with a continuous learning assumption is used to initially estimate learning curve parameters vis-à-vis least squares regression analysis, the subsequent application of the mid-unit approach eliminates the potential approximation error as discussed by Camm et al. (1987).

TABLE 6.2
Projection Error Caused by the Use of the Discrete Unit Learning Curve Model

Historical Data	Learning Curve	Unit Number for Estimate	Error %
2–50	70%	200	4.84
		300	5.96
		400	6.99
		500	7.24
	80%	200	3.09
		300	3.81
		400	4.32
		500	4.71
	90%	200	1.43
		300	1.77
		400	2.03
		500	2.22
26–50	70%	200	0.67
		300	0.89
		400	1.11
		500	1.24
	80%	200	0.42
		300	0.60
		400	0.70
		500	0.73
	90%	200	0.20
		300	0.28
		400	0.36
		500	0.40

that the use of the mid-unit approach is most important in reducing projection errors in low- to medium-volume production environments. Projection errors due to the use of discrete versions of learning curve models become negligible for high-volume production (e.g., appliance and automobile manufacturing).

Perhaps most important is that the ability to project from mid-unit to mid-unit also provides for an efficient regression analysis of typical historical cost data found in most firms, regardless of the production volume. Cost accounting systems usually capture the total cost needed to produce a batch of units, and therefore, the average cost for the batch. Also typical of "real" data is the omission of early production cost data, or of incorrect production cost reporting later in the production cycle. When these problems cause missing data points (i.e., missing batch averages), cumulative averages cannot be calculated without approximations, thus forcing the use of the unit cost function.

Of course, the use of tables or careful adjustments to reduce approximation errors is possible, but the use of the mid-unit model eliminates any such effort. The ability to formulate continuous cost functions for both unit and cumulative average costs provides fast and accurate learning curve analysis of most production cost databases.

GUIDELINES FOR FUTURE RESEARCH

This chapter illustrated the mid-unit approach for log-linear productivity trend analysis. The mid-unit model is a continuous form of the log-linear learning curve that allows production cost projections from both the cumulative average cost function and the unit cost functions. Cost projection errors caused by the discrete unit formulations for either the cumulative average or unit cost functions are eliminated. The formulation of the model requires negligible computational capabilities to accomplish even the most difficult learning curve projections.

Generally, this chapter has shown that a log-linear learning curve model provides good "fits" of empirical data for many products and processes. In other cases, typically start-up phases of production, it may be found that non-log-linear learning curve models better estimate productivity trends. Here, additional terms are typically used to change the log-linear shape. Note, however, that the more complex the model, the more difficult it is to compute the terms' coefficients. There are five basic patterns of non-log-linear learning curves, including plateau, hybrid or "dog-leg," S-shaped, convex-asymptotic, and concave-asymptotic.

The first two patterns are simple adaptations to the log-linear model in order to consider specific effects. The plateau effect, discussed by Baloff (1970, 1971), assumes that the learning effect is associated with the "start-up phase" of a new process, and that a steady-state phase, which exhibits negligible productivity improvements, occurs as a distinct interruption of the start-up phase. The hybrid model, or "dog-leg," is similar to the plateau model since it is a combination of two or more log-linear segments. Normally, the slopes of the succeeding segments are flatter. The change in slopes can be explained by the implementation of more automated processes for labor-intensive processes over time. The change in slopes indicates different learning rates associated with various production processes. The use of the mid-unit approach for these special cases remains valid. Other models explicitly use non-log-linear functions to estimate the change in learning rates over time.

For example, an S-shaped curve was proposed by Carlson (1973) and provides a robust approach to fitting data with both log-concave and log-convex characteristics. Levy (1965) also proposed an S-shaped learning curve model, later discussed by Muth (1986). Other researchers, including Knecht (1974) and De Jong (1957), have investigated convex-asymptotic learning curves that project a convergence to a steady-state as more units are produced. Still others (e.g., Garg and Milliman 1961) proposed learning curve models that are concave-asymptotic. Finally, it should be pointed out that non-linearities might be caused by inherent forgetting in the process resulting from interruptions in the performance of tasks. Dar-El et al. (1995a, 1995b) provide modifications to the log-linear model that specifically address the potential for forgetting in both short and long cycle time tasks.

Researchers utilizing empirical data from company databases should initially consider the use of the mid-unit log-linear model for two main reasons. First, the mid-unit approach is appropriate if the database of production costs holds either batch total cost or batch average cost information and has at least one missing data point—which is common in most industrial databases. Because the mid-unit approach does not require all unit cost data to be available in order to determine both the cumulative average and unit costs, regression analysis utilizing the mid-unit approach provides the most accurate method to estimate the intercepts and slopes of the component production costs. Second, the mid-unit approach is appropriate when the productivity trend is assumed to be log-linear, but the learning rate has the potential to change at identified time points (i.e., events), resulting in a dog-leg learning curve scenario.

Normative research can also benefit from the use of the mid-unit approach. For example, many of the studies using analytical approaches to the optimal lot-sizing problem have used the continuous learning model, but assume that the unit cost occurs at the unit number minus 0.5. This approximation will always result in some level of error, no matter how learning is assumed to occur (i.e., either continuously or discretely). (See Camm et al. [1987] for a discussion of the errors caused under the discrete learning assumption.) The use of the exact mid-unit calculation eliminates any error in such normative studies. In addition, the previously published normative studies of optimal lot-sizing under learning and forgetting can be extended to dog-leg learning conditions using the mid-unit log-linear approach in order to examine the impact of either start-up or new technology implementations.

It should be understood, however, that non-log-linear learning curve approaches might be most appropriate for special conditions, especially during the start-up phase of production. Prior research indicates that these non-log-linear trends have been found to exist for highly aggregated data—such as the total production costs for complex products (e.g., aircraft). Ultimately, the choice of using either the mid-unit log-linear approach or one of the non-log-linear models should be made with full recognition that it is sometimes impossible to determine the true, underlying learning behavior of a production process. In the end, simplicity may prove to be the best guide since production cost estimation can become complicated due to a number of other confounding factors.

REFERENCES

Adler, P.S., and Clark, K.B., 1991. Behind the learning curve: A sketch of the learning process. *Management Science* 37(3): 267–281.

Baloff, N., 1970. Start-up management. *IEEE Transactions on Engineering Management, EM* 17(4): 132–141.

Baloff, N., 1971. Extension of the learning curve – some empirical results. *Operations Research Quarterly* 22(4): 329–340.

Bohn, R.E., 1995. Noise and learning in semiconductor manufacturing. *Management Science* 41(1): 31–42.

Camm, J.D., Evans, J.R., and Womer, N.K., 1987. The unit learning curve approximation of total cost. *Computers and Industrial Engineering* 12(3): 205–213.

Carlson, J.G., 1973. Cubic learning curves: Precision tool for labor estimating. *Manufacturing Engineering Management* 71(5): 22–25.

Dar-El, E.M., Ayas, K., and Giland, I., 1995a. A dual-phase model for the individual learning process in industrial tasks. *IIE Transactions* 27(3): 265–271.

Dar-El, E.M., Ayas, K., and Giland, I., 1995b. Predicting performance times for long cycle time tasks. *IIE Transactions* 27(3): 272–281.

De Jong, J.R., 1957. The Effects of increasing skill on cycle time and its consequences for time standards. *Ergonomics* 1(1): 51–60.

Epple, D., Argote, L., and Murphy, K., 1996. An empirical investigation of the microstructure of knowledge acquisition and transfer through learning by doing. *Operations Research* 44(1): 77–86.

Garg, A., and Milliman, P., 1961. The aircraft progress curve modified for design changes. *Journal of Industrial Engineering* 12(1): 23–27.

Globerson, S., and Levin, N.L., 1995. A learning curve model for an equivalent number of units. *IIE Transactions* 27(3): 716–721.

Globerson, S., and Shtub, A., 1995. Estimating the progress of projects. *Engineering Management Journal* 7(3): 39–44.

Gruber, H., 1992. The learning curve in the production of semiconductor memory chips. *Applied Economics* 24(8): 885–894.

Knecht, G.R., 1974. Costing technological growth and generalized learning curves. *Operations Research Quarterly* 25(3): 487–491.

Levy, F., 1965. Adaptation in the production process. *Management Science* 11(6): 136–154.

Mazzola, J.B., and McCardle, K.F. 1996. A Bayesian approach to managing learning-curve uncertainty. *Management Science* 42(5): 680–692.

Muth, J.F., 1986. Search theory and the manufacturing progress function. *Management Science* 32(8): 948–962.

Rachamadugu, R., and Schriber, T.J., 1995. Optimal and heuristic policies for lot sizing with learning in setups. *Journal of Operations Management* 13(3): 229–245.

Yelle, L.E., 1979. The learning curve: Historical review and comprehensive survey. *Decision Sciences* 10(2): 302–328.

Wright, T.P., 1936. Factors affecting the cost of airplanes. *Journal of Aeronautical Sciences* 3(2): 122–128.

7 Using Parameter Prediction Models to Forecast Post-Interruption Learning[*]

Charles D. Bailey and Edward V. McIntyre

CONTENTS

---

[*] Reprinted from *IIE Transactions*, 2003, vol. 35, 1077–1090.

INTRODUCTION

Industrial learning curves (LCs) have found widespread use in aerospace and military cost estimation since World War II, and use in other sectors is increasing (e.g., Bailey 2000). Given a stable production process, such curves are excellent forecasting tools.

However, a number of factors can disturb the stable environment and therefore raise issues for research. One such issue is how to deal with interruptions of production and subsequent forgetting and relearning. The earliest advice in the area involved "backing up" the LC to a point that would represent forgetting—assuming forgetting to be the reverse of learning—and then resuming progress down the same curve (e.g., Anderlohr 1969; Adler and Nanda 1974; Carlson and Rowe 1976; Cherrington et al. 1987; Globerson et al. 1989). Subsequent research has tried to refine the modeling of both forgetting and relearning and has constructed relearning curves (RLCs) to describe the relearning of skills that have diminished during breaks in production. In addition, a limited body of research exists on predicting the parameters of RLCs using parameter prediction models (PPMs). The current chapter develops and tests PPMs for RLCs as a procedure for providing early estimates of post-interruption production times. If managers wish to use these PPMs to predict RLC parameters, they must somehow estimate the parameter values of the PPMs, which they may be able to do based on past experience. This estimation problem is similar to the one that managers face when they first employ an LC; they must somehow estimate the LC parameters. The analysis presented here should provide information on the relevant variables and forms of equations that are useful in predicting the parameters of RLCs. Candidate predictive variables include all information available when production resumes, such as the original LC parameters, the amount of learning that originally occurred, the length of the production interruption, and the nature of the task.

The potential usefulness of PPMs is that they can provide a means of estimating post-break production times well *before* post-break data become available. Other procedures, such as fitting an RLC to post-break data, or backing up a pre-break LC to a point that equals the time for the first post-break iteration, can only be employed *after* post-break data are available.

PREVIOUS RESEARCH

This section briefly discusses the nature and variety of LCs/RLCs, and then reviews the LC literature relevant to forgetting, relearning, and parameter prediction.

LEARNING CURVES

LCs are well-established cost-estimation tools, and a variety of curves have been developed to model the empirically observed improvement that comes from practice at a task (Yelle 1979). Almost all of the models reflect a pattern of large early improvements followed by slower returns to practice. In the popular "Wright" log-log LC (Wright 1936; Teplitz 1991; Smith 1989), the cumulative average time (and related cost) per unit produced declines by a constant factor each time cumulative production doubles, so that it is also called the "cumulative average" model. It is a power curve, in the form:

$$y = ax^b, \tag{7.1}$$

where y = the cumulative average time (or related cost) after producing x units

a = the hours (or related cost) to produce the first unit
x = the cumulative unit number
$b = \log R / \log 2$ = the learning index (<0 under conditions of learning)
R = the learning rate, often expressed as a percentage
$1 - R$ = the progress rate

Taking the natural logs of both sides of Equation 7.1 yields the following linear relationship (a useful feature because ordinary-least-squares regression can estimate the parameters):

$$\ln y = \ln a + b \ln x, \tag{7.1a}$$

Another approach is the "individual unit," or "Crawford" model (after the author, J. R. Crawford, in a Lockheed in-house training manual circa 1944 [Teplitz 1991; Smith 1989]). It applies this same form of power curve to forecast *individual unit times*.[*]

This curve, as well as a family of curves depicting rapid early improvement followed by diminishing returns to practice, can describe either organizational learning, as originally applied in the airframe industry, or individual learning. For a clear description of the fundamental shapes of power, hyperbolic, logistic, and exponential curves, see Swezey and Llaneras (1997). Zangwill and Kantor (1998), in an attempt to develop a more scientific theory of continuous improvement and the LC, accept the prevalence of the power law, but also recommend the exponential curve as a basis for management to establish improvement goals. They introduce a new "finite" model that allows time or cost to go to zero, as when certain steps in a work process totally disappear through learning. Regardless, power curves appear to dominate because of their robust descriptive ability (Newell and Rosenbloom 1981).

An observed difference in the nature (shape) of the LCs and RLCs (Bailey 1989; discussed below) prompted the study by Bailey and McIntyre (1992), in which they introduced the following power curve form, which fits better as both RLCs and LCs for the mechanical-assembly tasks used in the research reported in this chapter:

$$\ln y = a(\ln(x+1))^b. \tag{7.2}$$

Like curve (1), curve (2) is linearized by taking logs:

$$\ln(\ln y) = \ln a + b \ln(\ln(x+1)). \tag{7.2a}$$

---

[*] Confusion between these two forms of the model is common (see Chen and Manes 1985). Dar-El et al. Ayas and Gilad (1995) incorrectly identify the Wright model as using individual unit times. Arzi and Shtub (1997) commit this same error and then attribute the average form of the model to Hancock (1967). According to Newell and Rosenbloom (1981, 2), "the *log-linear learning law* or the *power law of practice* [which they find to be ubiquitous]… apparently showed up first in Snoddy's (1926) study of mirror tracing of visual mazes."

In Equations 7.2 and 7.2a, *y* represents the estimated *marginal* time of the *x*th unit. The *a* parameter no longer represents the estimated time to complete the first unit. Instead, the estimated time for the first unit is $\exp(a(\ln 2)^b)$. Also, the progress rate, *R*, now represents one minus the percentage reduction in $\ln y$ expected when $\ln(x + 1)$ doubles; but *b* is still interpreted as an improvement parameter.

In a subsequent study, using a less time-consuming experimental task for which they could obtain more data, Bailey and McIntyre (1997) compared the predictive ability of Equation 7.2 to that of the Wright and Crawford forms of Equation 7.1 in both learning and relearning settings. The curves that fit better also tended to predict better, defying the paradoxical result seen in some other settings, where the best-fitting curve is not necessarily the best predictor (Lorek et al. 1983). In particular, the form of the curve in Equation 7.2 performed well for both learning and relearning.

Globerson and Levin (1987) address the theoretical and technical aspects of introducing forgetting into LCs, emphasizing organizational forgetting. They model post-break relearning by introducing an additional parameter called the "cycle coefficient," *c*, into a "generalized power function" with slope dependent on *c*. However, we note that the resulting set of curves can be described well by an ordinary log-linear power function (R^2 ranging from .983 to .997 for the curve examples shown in their Figure 1). Thus, the benefit derived at the cost of an additional degree of freedom seems questionable, especially during the early phase of production with only a few observations available. These authors also discuss potential curves to model forgetting. They include an exponential function, reflecting rapid initial decline followed by diminishing losses, but also introduce an S-shaped function under the assumption that losses might be negligible for a short break but then increase sharply after, say, two weeks, and later level off as with the exponential function. While the S-shaped function may have applications, an exponential or power function displaying upward convexity seems more representative of the psychological literature on forgetting (Farr 1987; Wickelgren 1972), and thus more relevant to individual forgetting or to organizational forgetting viewed as an aggregate of individual forgetting. Globerson and Levin proceed to introduce models combining learning and forgetting under two sets of assumptions limited to constant time intervals (such as preventive maintenance performed every three months, or work shifts alternating with time off).

The current study addresses learning and relearning at the individual level, while other recent studies have addressed similar factors at the organizational level (Benkard 2000; Argote et al. 1990). Although the earliest studies of industrial LCs used aggregated data (Wright 1939), the industrial LC literature has long recognized that the same phenomenon applies at the individual-worker level as well (e.g., Hancock 1967; Newell and Rosenbloom 1981). Uzumeri and Nembhard (1998) describe organizational learning as a composite of individual LCs, and advocate a 3-parameter hyperbolic model (Mazur and Hastie 1978) as the best descriptor. However, the requirement of estimating an additional parameter militates against its use when only a few observations are available.

RELEARNING CURVES AND FORGETTING

RLCs are similar to LCs, but are used to estimate the improvement that occurs while relearning skills that have been forgotten during periods of interruption. Compared to the literature on LCs, little has been written on them. In terms of the curves discussed above, two aspects of RLCs are important: (1) the relearning rate, reflected by the b parameter; and (2) the initial production time after the break, reflected by the a parameter. We will discuss the related "forecasting" literature under these headings. Subscript L refers to parameters from the pre-break LC, and subscript R refers to the post-break RLC.

Factors Affecting Loss of Skill and thus Relearning Curve Parameter a_R

Bailey (1989) investigated the factors affecting forgetting (loss of performance capability) in a laboratory setting, using a mechanical assembly task. His experimental data included both learning and relearning series, so that separate curves could be computed for both. He found that forgetting is a multiplicative function of the amount learned before the break and the length of the break (regression $R^2 = .71$ with 31 observations, $F = 72.7$, $p < .0001$), but not a function of the learning rate. This result contradicts earlier speculations that the learning rate and forgetting rate would be associated.

Globerson, Levin, and Shtub (1989) conducted an experiment in which subjects repeated a data-entry task 16 times, and then returned after a period varying from 1 to 82 days to repeat the exercise. They tried 7 curve forms to predict the performance time for repetition 17, all using two variables: F, the calculated performance time of repetition 17 using the power model derived from the first session; and D, the break time between repetitions 16 and 17. For the relearning model, they use the Stanford B model (e.g., see Teplitz 1991), modeling its "previous experience" parameter as a function of F, D, and the learning rate.

Shtub et al. (1993), studying a data-entry task, found the a parameter of the RLC to be a function of the skill level achieved during learning and the length of the break ($R^2 = .70$). The skill level achieved is measured by the estimated time for the first post-break unit, assuming no break occurred, ET_{m+1}. It would seem that the *amount learned*, rather than the absolute skill level achieved would be the relevant variable, since only such *learning* can be lost. In a regression substituting the amount learned in place of absolute skill, Shtub et al. (1993) obtained a lower R^2 (.43). Perhaps, in fact, *both* variables are relevant: the amount learned, because such learning can be lost; and the skill level achieved, because it represents the reference point along the performance scale.

In a different approach, which is appropriate to their field data, Nembhard (2000), and Nembhard and Uzumeri (2000) model learning and forgetting as concurrent processes by adding a fourth parameter to their parabolic model of learning. They do not assume the distinct learning and relearning series found in the laboratory data of Bailey (1989), Bailey and McIntyre (1997), and Shtub, Levin, and Globerson (1993). Thus, although they find a correlation between the learning rate and forgetting rate parameters within the same series, this result is difficult to compare with the laboratory findings of no such correlation between a learning rate from continuous production and a forgetting rate during a subsequent idle period.[*] However,

---

[*] Farr (1987, S-3) notes that "most relevant scientific studies [in this area] were methodologically flawed, lacking a common metric for measuring the degree of learning and the rate of forgetting."

Bailey (1989) notes that forgetting should be operationalized in terms of a difference between predicted performance times in the absence of an interruption and actual performance times following the interruption. He then operationalizes the forgetting *rate* in terms of the proportion of learning lost. The learning–forgetting correlation that Nembhard and Uzumeri report may result from their parameters reflecting absolute rather than proportional changes: a worker who has learned more, inherently can forget more.

Arzi and Shtub (1997) report a correlation between the learning rate and skill decrement during a break, using a measure of "intensity of forgetting... [defined as] the ratio between performance time after the break and the corresponding predicted time assuming no break". This is not the same measure as Bailey's (1989), and it does not control for the amount learned before the interruption. For example, a measure of 110/100 = 1.1 could relate to a subject who had learned much (improving from, e.g., 200 sec/unit to 100 sec/unit) and regressed slightly, or from one who initially performed at 110 sec/unit, learned only a little, and regressed completely. Furthermore, a correlation between this measure of forgetting and the learning rate is hard to interpret, because the learning rate is confounded with the amount learned, a problem common to most research on individual differences in forgetting (Arthur et al. 1998). Because of this confounding, any apparent relationship between forgetting and the learning rate could be attributed to a correlation with the cognitive skills of the worker (as Arzi and Shtub imply), *or* to the better consolidation of knowledge that comes through advanced practice. This latter explanation is more consistent with the psychological literature on forgetting (Arthur et al. 1998).

Dar-El et al. (1995) undertake to predict LC parameters, particularly the initial starting time for a new task, *before* the start of production, in the absence of empirical data. Accordingly, they focus on between-task differences in "cognitive" and "motor" skill content and the sequencing of sub-tasks. They address the forgetting phenomenon over periods of one, two, and seven days to "replicate the conditions of a long task, where a sub-task is done just once over the entire cycle" (p. 279). They show that the learning rate was steeper for cognitive elements than for motor elements, and that the learning rates for both kinds of tasks were reduced as a power function of the time between sub-tasks.

Factors Affecting the Relearning Rate, Parameter b_R

Bailey (1989) found the relearning rate to be uncorrelated with the learning rate. This result is not intuitive, because one would expect the "ability to acquire skill" to be stable across time. Farr (1987, 34) concludes, however, that "from the existing evidence about relearning, it appears that, for any particular skilled performance, we cannot predict its rate of relearning as a function of its [original learning]." Bailey (1989, 349) noted that "a review of all of the relearning curves, plotted on log-log scales... indicated that they may not exhibit the same log-linear relationship as the LCs. Efforts to improve upon the RLC fit did not, however, change the finding of virtually no learning-relearning rate correlation." In a regression to explain the relearning rate, nonetheless, he found the two main explanatory variables to be the amount of original learning and the amount of lost skill. This is consistent with the fact that the recovery of lost skill requires fewer repetitions than during the original acquisition period (Farr 1987), and the amount of variance due to this effect may have swamped the underlying correlation between the learning and relearning rates.

Shtub et al. (1993), in a different environment (a data-entry task), confirmed the lack of a direct, simple correlation between original learning and relearning rates. However, in a stepwise regression with the relearning slope as the dependent variable, the original learning rate did enter the equation after two other variables had entered. These other variables were the *a* parameter of the RLC, and the ratio of that parameter to the same parameter of the LC.

Globerson et al. (1998) examine forgetting of both predominantly motor and cognitive tasks, with pilots using flight simulators and being brought back to perform the same tasks after intervals ranging from 14 to 154 days. They found forgetting intensity to be greater for the cognitive task than for the motor task, and that the power model describes forgetting well. Similarly, Arzi and Shtub (1997) (discussed above for their metrics of forgetting) found a greater decrement for their "mental" task than for their "mechanical" task.

CHOICE OF LEARNING CURVE FORMS AND DEVELOPMENT OF PARAMETER PREDICTION MODELS

This study required two sets of models: LCs/RLCs to model improvement in the experimental task, and PPMs to predict the parameters of the RLCs. In this section, we give the logic for the choice of LCs/RLCs and discuss in more detail the development of the PPMs.

CHOICE OF CURVE FORMS

To reduce the potentially large number of LCs, RLCs, and PPMs, we predict parameters for only two RLCs. Our criterion in selecting RLCs on which to test PPMs was to use RLCs that were good fits to the empirical data used in the analysis. Prior research on this data set shows that two forms fit the data better than seven other models tested, both as LCs and RLCs (R^2s of .82 to .85 as LCs and .57 to .72 as RLCs [Bailey and McIntyre 1997]).[*] These two forms are Equations 7.1 and 7.2. While it is possible that some other RLC model not tested might have fit the data slightly better,[†] the two RLC models that we used provide good fits to the data and are appropriate for tests of PPMs on this data set.

---

[*] Equation 7.2a eliminates serial correlation with relearning data that exists with Equation 7.1a, which has difficulty reflecting the rapid initial improvement. The Durbin–Watson statistics for our data averaged 2.1 for Equation 7.2, very close to the optimal 2.0 that reflects no serial correlation.

[†] We fitted exponential curves to all 116 sets of learning data and to all 116 sets of relearning data. For only 8 learning curves and 16 relearning curves did the exponential curve fit better than a power curve. Likewise, we fitted the 3-parameter hyperbolic curves advocated by Mazur and Hastie (1978) to 6 each of randomly chosen learning and relearning series. In only 2 did the hyperbola fit better than a power curve, based on adjusted R^2; average adjusted R^2s were .488 for the hyperbola versus .648 for the power curve.

Our tests are based on marginal time estimates from each model. More specifically, in the average-time power model:

$$\text{Marginal time}_x = \text{total time}_x - \text{total time}_{x-1} = ax^{b+1} - a(x-1)^{b+1}, \qquad (7.3)$$

and in the Bailey-McIntyre model

$$\text{Marginal time}_x = \exp\{a(\ln(x+1))^b\}. \qquad (7.4)$$

MODELS TO PREDICT THE REAL LEARNING CURVE PARAMETERS

As discussed above, little empirical or theoretical work exists to guide our choice of PPMs. We have chosen to model PPMs as log or power functions in order to capture the expected diminishing marginal effects of the independent variables. Furthermore, as mentioned above, previous studies have shown a similarity in patterns of learning and forgetting. Since log and power functions have been used successfully to model both learning and forgetting, we believe it is also proper to use these patterns to predict the RLC parameters, which are functions of both previous learning and forgetting. Finally, we wish to use equations that can be estimated with linear regression, since we think that this property would make the models more accessible to potential users. These criteria apply to our choice of PPMs for both RLC parameters, a_R and b_R. We develop and describe the PPMs below. For each model, we specify the values of the PPM parameters required to impart the expected marginal effects.

In our PPMs for both a_R and b_R, we incorporate the effects of prior experience acquired during the pre-break performance of this task by including the amount learned during this time, L, as an independent variable. No other information is available regarding related prior experience; however, the recruitment of subjects from the student community gives some assurance that such past experience would be negligible and randomly distributed.

Variables in Parameter Prediction Model to Predict Parameter a_R

For either of the LCs/RLCs described above, we define a_R as the a parameter of the RLC. Drawing on Bailey (1989), we use parameter a from the pre-break LC, the length of the break, and the amount learned as the independent variables. Thus,

$$a_L = a \text{ in the learning curve (pre-break)}$$
$$B = \text{length of production break in days}$$
$$L = \text{amount learned} = ET_1 - ET_{m+1}$$

where ET_1 is the estimated time for first pre-break unit determined from the pre-break LC, and ET_{m+1} is estimated time for the first post-break unit, $m + 1$, assuming no break, based on the m-unit learning curve.

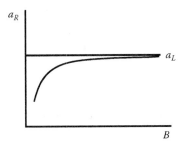

FIGURE 7.1   Relationship between a_R and break, B.

Graphs of expected relationships

In the traditional model, a is the estimated time for the first post-break unit; in the Bailey-McIntyre model, a is an increasing function of initial post-break time.* Thus, as shown in Figure 7.1, we would expect the corresponding parameter in the RLC, a_R, to increase as the length of the break, B (and hence the degree of forgetting) increases. However, one would not expect a_R to exceed a_L. Although not simply the reverse of learning, forgetting does follow a similar exponential pattern of larger early losses and declining later losses (Carlson and Rowe 1976; Globerson et al. 1989; Jaber and Bonney 1997).

Further, for a given length of break, we would expect that the time for the first post-break unit, and therefore the value of a_R, would decrease as the amount that has been learned increases (see Figure 7.2). When nothing has been learned ($L = 0$), we would expect $a_R = a_L$. Finally, we expect a positive association between a_R and a_L.

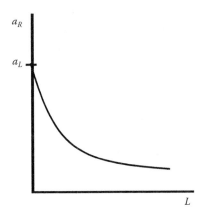

FIGURE 7.2   Relationship between a_R and amount learned, L.

---

* In the Bailey-McIntyre model, the value of a for $x = 1$ is $a = \ln y/(\ln 2)^b$; hence, the larger the value of y (the marginal time of the first unit), the larger the value of a.

Equations to Predict Relearning Curve Parameter a_R

The first PPM that we test, models an interaction between L and B in the same way that Bailey (1989) found to be useful in explaining loss of skill.

$$a_R = q_0 + q_1 L \ln B. \tag{7.5}$$

Figures 7.1 and 7.2 imply different signs for the separate effects of L (which reduces task time) and B (which increases task time). The product, however, is directly related to the amount of learning lost (Bailey 1989), so that we expect a positive effect on a_R with $q_1 > 0$. The greater the loss of skill, the higher should be the value of a_R.

Equation 7.6 combines a_L, L, and B in a log form, while Equation 7.7 combines them in a power form. Signs required for the expected marginal effects are shown below each equation.

$$a_R = q_0 + q_1 \ln(a_L) + q_2 \ln(L) + q_3 \ln(B), \tag{7.6}$$

$$q_1, q_3 > 0,$$

$$q_2 < 0.$$

$$a_R = q_0 (a_L)^{q_1} (L)^{q_2} (B)^{q_3}, \tag{7.7}$$

$$q_0 > 0,$$

$$q_2 < 0,$$

$$0 < q_1, \; q_3 < 1.$$

Equation 7.7 is made linear by a log transformation; that is,

$$\ln a_R = \ln q_0 + q_1 \ln(a_L) + q_2 \ln(L) + q_3 \ln(B). \tag{7.7a}$$

Equation 7.8 is the "best of seven" curves tested by Globerson et al. (1989):

$$a_R = q_0 (ET_{m+1})^{q_1} (B)^{q_2}, \tag{7.8}$$

which, in linear form, is

$$\ln a_R = \ln q_0 + q_1 \ln(ET_{m+1}) + q_2 \ln(B), \tag{7.8a}$$

where, as noted above, ET_{m+1} is the estimated time for the first post-break unit assuming no break, based on the LC. Globerson et al. (1989) found parameter values in the following ranges:

$$q_0 > 0,$$

$$0 < q_1, \ q_2 < 1.$$

Equation 7.8 uses as an independent variable the predicted next time, ET_{m+1}; however, the predicted initial time from the LC, ET_1, may also have information content, because the difference between the two reflects the improvement, as $L = ET_1 - ET_{m+1}$. This observation is in keeping with our comment, above, that perhaps both the amount learned and the skill level achieved should be relevant. This leads to the following simple extension of Equations 7.8 and 7.8a:

$$a_R = q_0 (ET_{m+1})^{q_1} (B)^{q_2} (ET_1)^{q_3}, \qquad (7.9)$$

$$q_0 > 0,$$

$$0 < q_2 < 1,$$

and

$$\ln a_R = \ln q_0 + q_1 \ln(ET_{m+1}) + q_2 \ln(B) + q_3 \ln(ET_1). \qquad (7.9a)$$

Equations 7.9 and 7.9a use as separate variables the two components of the amount learned ($L = ET_1 - ET_{m+1}$), allowing for the independent variability of these components across subjects. In these equations, the expected values of q_1 and q_3 are less clear. In other PPMs we expect a negative coefficient on L, but various combinations of values for q_1 and q_2 are consistent with this result. For example, a positive coefficient on both q_1 and q_2 would be consistent with a negative coefficient on L if the coefficient on ET_{m+1} is greater than the coefficient on ET_1 (which is what we find). The positive coefficient on ET_{m+1} is consistent with the findings of Globerson et al. (1989).

Variables in Parameter Prediction Models to Predict Parameter b_R

Drawing again on the findings of Bailey (1989), we use the amount learned (L) and the length of break (B) as independent variables to estimate the relearning rate. Bailey (1989) did not use L and B directly to estimate the relearning rate, but used "lost time," which was a function of L and B. Finally, although that earlier work found no relationship between the learning rate (b_L) and the relearning rate (b_R), we believe that this relationship is worth testing again in some other functional form and with the benefit of the Bailey-McIntyre LC, (Equation 7.2), which models this type of task better.

Amount Learned Before Break (L)

Since we use a start-anew RLC, the relearning rate will depend on the number of repetitions performed before the break. For example, if the change between units 1 and 2 reflects an 80% LC, an RLC after no interruption (or a short interruption) will not reflect this same 80% for the tenth and eleventh units (renumbered 1 and 2 in an RLC), but will show a much slower rate. This effect suggests an inverse relationship between the amount learned (L) and the steepness of the RLC. *Ceteris paribus*, the more learned before the break, the less the opportunity for learning after the break, which implies higher values for the relearning rate and for b_R. However, greater learning during the pre-break period implies a faster learning rate, which may also be reflected in a faster relearning rate (a lower value of b_R). This effect would produce a positive relationship between b_R and L. Since we do not know which of these effects is stronger, we do not specify expected signs on the coefficient of L.

Length of Break (B)

We expect a positive relationship between the length of the break and the "steepness" of the RLC. A substantial amount of learning places the worker on a relatively flat part of the curve. As forgetting occurs, regression "back up the curve" places the worker on a steeper part of the curve. After a short break, the individual should resume learning in the relatively flat part of the curve, and an RLC fit to post-break observations should reflect slower learning. A longer break would move the worker back into the steeper portion of the curve and an RLC fitted to these observations should show a steeper relearning rate—that is, more rapid relearning, implying lower values for the relearning rate and b_R. Moreover, because memory traces of original skills are retained, *re*learning can be faster than initial learning of the same material by a person starting at that same performance level (Farr 1987). Thus, when all learning appears to be forgotten, as measured by time on the first post-break performance, the improvement between units 1, 2, 3, and so on, will appear much steeper than for the corresponding original units. Thus, because we use a start-anew RLC beginning with units designated 1, 2, 3, and so forth, b_R can easily be more negative than b_L. Figure 7.3 shows the expected relationship between b_R and B. Additionally, we include terms for interaction between B and L, since each may alter the effect of the other; e.g., a worker who has learned a great deal may still be "down the curve" despite a long break.

Pre-Break Value of b (b_L)

The nature of the relationship between b_R and b_L is ambiguous. One might argue for a positive association reflecting the innate ability of the individual. On the other hand,

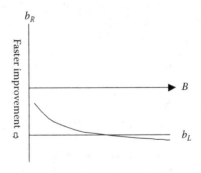

FIGURE 7.3   Expected relationship between b_R and B.

the faster one learns, the further down the curve he or she will be, and by the arguments above, and other things equal, the slower will be the rate of further learning. Because either or both of these effects are possible, we do not specify an expected sign on the coefficients for b_L.

Equations to Predict the Parameter b_R

Using the same criteria specified earlier, we test the following PPMs for b_R. The first candidate uses logs to show the expected diminishing marginal effects and includes all three independent variables plus a term that allows for interaction between B and L. Because signs of the coefficient on L are not specified, we do not specify expected signs for the coefficients of interaction terms.

$$b_R = q_0 + q_1 \ln B + q_2 \ln L + q_3(\ln B \times \ln L) + q_4 b_L, \tag{7.10}$$

$$q_1 < 0.$$

A slight modification, using the interaction term with L in its arithmetic form, which worked well in Bailey (1989), is:

$$b_R = q_0 + q_1 \ln B + q_2 \ln L + q_3(\ln B \times L) + q_4 b_L, \tag{7.11}$$

$$q_1 < 0.$$

We also test several power forms that model the expected diminishing marginal effects. The first is:

$$b_R / b_L = q_0 B^{q_1} L^{q_2}, \tag{7.12}$$

$$q_0 > 0,$$

$$0 < q_1 < 1,$$

which in log form becomes linear:

$$\ln (b_R / b_L) = \ln q_0 + q_1 \ln B + q_2 \ln L. \tag{7.12a}$$

An alternative construction is to treat b_L as an independent variable, yielding:

$$b_R + 1 = q_0 B^{q_1} L^{q_2} (b_L + 1)^{q_3}, \tag{7.13}$$

$$q_0 > 0,$$

$$q_1 < 0,$$

where the addition of the constant, +1, is necessary in order to take logs. In log form, this becomes:

$$\ln(b_R + 1) = \ln q_0 + q_1 \ln B + q_2 \ln L + q_3 \ln(b_L + 1). \tag{7.13a}$$

Finally, we include Shtub, Levin, and Globerson's (1993) equation (20):

$$b_R = q_0 + q_1 b_L + q_2 a_R' + q_3(a_R'/a_L) + q_4 \ln B, \tag{7.14}$$

where a_R' is estimated using their equation, which is our Equation 7.8. Shtub et al. found that:

$$q_0 > 0,$$

$$q_1, q_2, q_3, q_4 < 0.$$

METHOD

The PPMs described above were tested on 116 pairs of LCs and RLCs constructed from experimental data as reported in Bailey and McIntyre (1992) (61 subjects), and Bailey and McIntyre (1997) (55 subjects). The experimental task involved repeatedly assembling a product (a crane, a dump truck, or a radar station) from an erector set for a period of approximately four hours. After breaks of approximately two to five weeks, the subjects repeated the exercise. Assembly times to the nearest second were recorded by one of the authors or a research assistant. The average number of assembly repetitions was 9.9 before the break and 10.0 after the break. Subjects were undergraduate and graduate students of normal working age. They were paid a realistic wage using either a flat rate, a piece rate, or a fixed fee plus bonus. The average pay under all methods was approximately $5 per hour, and supplementary analysis indicated that the method of payment did not affect the results reported here. The two articles referred to above provide further descriptions of the experiments and report statistical data on the LCs and RLCs that were fit to the experimental data.

The data available to test PPMs were parameters from the 116 LCs fit to the pre-break data and from the 116 RLCs fit to the post-break data, the length of the break (B) for each subject, and the amount learned (L) during the pre-break period. We compute both goodness-of-fit statistics and measures of predictive ability for the PPMs. The results are reported below.

RESULTS

REGRESSIONS FOR a_R AND b_R

Because the dependent variables differ across our PPMs for a_R and b_R, we make the comparisons consistent by computing an adjusted R^2 value for the correlation between the RLC values of a_R (b_R) and the PPM estimates of a_R (b_R) for each PPM (see Kvålseth 1985). Table 7.1 summarizes the results of the PPMs used to estimate a_R. All the models are highly statistically significant (F-statistics $\leq 1.0E-16$). All of the PPMs, except Equation 7.5, fit the data very well, especially when used with the Bailey-McIntyre model. Equation 7.9a, however, has the highest adjusted R^2 for both equation forms.

TABLE 7.1

Fit and Significance Levels of Coefficients of PPMs for a_R

Equation	RLC Form	Significance Levels of Independent Variables						Adjusted R^2
		ln B	ln L	$\ln(ET_{m+1})$	$\ln(ET_1)$	$L \times \ln B$	ln a_L	
7.5	LL					0.0000		0.539
	BM					0.0000		0.463
7.6	LL	0.0000	0.0000				0.0000	0.743
	BM	0.0000	0.0000				0.0000	0.887
7.7a	LL	0.0000	0.0000				0.0000	0.828
	BM	0.0000	0.0000				0.0000	0.902
7.8a	LL	0.0000		0.0000				0.831
	BM	0.0000		0.0000				0.913
7.9a	LL	0.0000		0.0000	0.0000			0.866
	BM	0.0000		0.0000	0.0002			0.921

Note: Adjusted R^2 is computed consistently across equations based on the arithmetic terms a_R and estimated a_R, using Kvålseth's (1985) recommended "Equation #1."

The details of the two regressions for Equation 7.9a appear in Table 7.2. The coefficients in Panel B reflect that, for the Bailey-McIntyre curve, $a_R = \exp\{-0.1243 + 0.2827 \ln(ET_{m+1}) + 0.0437 \ln B + 0.0698 \ln(ET_1)\}$. The signs of the coefficients of the intercept and ln B are as predicted. As discussed in the "Equations to predict RLC parameter a_R" section, the expected signs of q_1 and q_3 are somewhat ambiguous. The difference, $q_3 - q_1$, is negative for PPM Equation 7.9a applied to both forms of RLCs, which is consistent with a negative coefficient on $L = ET_1 - ET_{m+1}$ and supports our predicted inverse relationship between L and a_R.

TABLE 7.2

Regression Results for PPM Equation 7.9a, for both RLC forms

Panel A: Log-log RLC

	Coeff.	S.E.	t Stat.	p-value
Intercept	−0.3188	0.1374	−2.3196	0.0222
$\ln(ET_{m+1})$	0.6359	0.0552	11.5223	0.0000
ln B	0.1638	0.0242	6.7720	0.0000
$\ln(ET_1)$	0.2967	0.0576	5.1471	0.0000

Panel B: Bailey-McIntyre RLC

	Coeff.	S.E.	t Stat.	p-value
Intercept	−0.1243	0.0384	−3.2345	0.0016
$\ln(ET_{m+1})$	0.2827	0.0189	14.9590	0.0000
ln B	0.0437	0.0067	6.4963	0.0000
$\ln(ET_1)$	0.0698	0.0184	3.8011	0.0002

The coefficients of Equation 7.9a for the log-log form of RLC (Table 7.2, Panel A) are directionally similar. They imply that the *a* parameter for the RLC is strongly related to the expected position on the original LC, ET_{m+1}, but is higher as a function of elapsed break time and also higher as a function of the original starting time.

The results of the regressions to predict the relearning slope parameter, b_R, are summarized in Table 7.3. Although the R^2 values are lower than for a_R, all of the models are highly statistically significant (*F*-statistics $\leq 1.0\text{E-}7$). Equation 7.12a, which uses the ratio of b_R to b_L, performs poorly; it assumes a fixed relationship between b_R and b_L that does not appear to be appropriate. The remaining equations, Equations 7.10, 7.11, 7.13a, and 7.14 are competitive alternatives. Equation 7.10, which expresses most simply and directly the relationships between relearning rate and length of break, amount learned, and their interaction, produces the highest adjusted R^2, so that the variations introduced in Equations 7.11 and 7.13a are not helpful for our data. Most of the coefficients of Equation 7.14 are not significant because the regressions suffer from multicollinearity; a'_R/a_L is correlated highly with a'_R, $\ln(B)$, and b_L (R^2 ranging from .60 to .80).

The regression results for Equation 7.10, for both curve forms, appear in Table 7.4. The negative coefficients for ln *B* are consistent with faster relearning after "regressing" up the curve. The negative coefficient on ln *L* suggests that, of the two possible effects of *L* on b_R discussed previously, the second is the stronger. That is, a larger amount learned in the pre-break period indicates faster learning abilities, and that these abilities are also reflected in a faster relearning rate (i.e., a lower value of b_R). Regression results for Equation 7.10 display some multicollinearity, especially between the interaction term and each of its separate components.*

Despite the effects of this multicollinearity, which biases against finding significance, each of the independent variables in Equation 7.10 is significant at traditional levels when Equation 7.10 is used with the Bailey-McIntyre form of RLC. Furthermore, although multicollinearity limits the usefulness of the coefficients separately, it does not adversely affect predictions of the dependent variable made using the whole equation. Thus, despite the multicollinearity, we believe each of the independent variables should remain in Equation 7.10.

An Example

As an example of how these PPMs may be used, suppose that a Bailey–McIntyre-form LC was fit to a worker's (or team's) initial production times for seven units, with the following results:

$$\ln y = 3.6166\left[\ln(x+1)\right]^{-0.2144}.$$

* Correlation Coefficients Amog Independent Variables:

	b_L	$\ln(B)$	$\ln(L)$
$\ln(B)$	−0.036		
$\ln(L)$	−0.716	0.210	
$\ln(B) \times \ln(L)$	−0.454	0.792	0.750

TABLE 7.3

Fit and Significance Levels of Coefficients of PPMs for b_R

Equation	RLC Form	Significance Levels of Independent Variables									
		$\ln B$	$\ln L$	$\ln B \times \ln L$	$L \times \ln B$	b_L	$\ln(b_L + 1)$	$b_L \times \ln B$	a'_R	a'_R/a_L	Adjusted R^2
7.10	LL	0.0071	0.0643	0.1237		0.6738					0.3166
	BM	0.0003	0.0514	0.0120		0.0000					0.3168
7.11	LL	0.0000	0.2591		0.6274	0.6634					0.3032
	BM	0.0000	0.7690		0.1366	0.0001					0.2911
7.12a	LL	0.0000	0.0030								0.0824
	BM	0.0004	0.0000					0.0000			0.1999
7.13a	LL	0.0000	0.1398				0.7839				0.3071
	BM	0.0000	0.0607				0.0001				0.2760
7.14	LL	0.0090				0.4147			0.4018	0.8129	0.2760
	BM	0.0004				0.0207			0.1426	0.6475	0.2763

Note: Adjusted R^2 is computed consistently based on the arithmetic terms b_R and estimated b_R using Kvålseth's (1985) recommended "Equation 1."

TABLE 7.4

Regression Results for PPM Equation 7.10, for Both RLC Forms

Panel A: Log-log curve

	Coefficients	S.E.	t Stat	p-value
Intercept	0.2518	0.1268	1.9861	0.0495
$\ln(B)$	−0.0995	0.0363	−2.7430	0.0071
$\ln(L)$	−0.0870	0.0465	−1.8687	0.0643
$\ln(B) \times \ln(L)$	0.0196	0.0126	1.5510	0.1237
b_L	0.0351	0.0832	0.4221	0.6738

Panel B: Bailey–McIntyre curve

	Coefficients	S.E.	t Stat	p-value
Intercept	0.2926	0.1156	2.5307	0.0128
$\ln B$	−0.1242	0.0332	−3.7447	0.0003
$\ln L$	−0.0821	0.0417	−1.9690	0.0514
$\ln B \times \ln L$	0.0295	0.0115	2.5527	0.0120
b_L	0.3215	0.0757	4.2497	0.0000

Management wishes to estimate production time following a break of 56 days. To use Equations 7.9a and 7.10, the following data is required:

$$a_L = 3.6166;$$

$$b_L = -0.2144;$$

$$B = \text{planned break} = 56 \text{ days};$$

$$ET_1 = \exp\left(3.6166(\ln 2)^{-0.2144}\right) = 50.0 \text{ hours};$$

$$ET_{m+1} = ET_8 = \exp\left(3.6166(\ln 9)^{-0.2144}\right) = 21.2 \text{ hours};$$

$$L = ET_1 - ET_8 = 28.8 \text{ hours}.$$

Using regression results for Equations 7.9a and 7.10 applied to the Bailey–McIntyre curve, as reported in Panel B of Tables 7.2 and 7.4, respectively, we obtain:

$$\ln a_R = -0.1243 + 0.2827 \ln 21.2 + 0.0437 \ln 56 + 0.0698 \ln 50$$

$$= 1.880,$$

or,

$$\hat{a}_R = 3.2806,$$

and

$$b_R = 0.2926 - 0.1242 \ln 56 - 0.0821 \ln 28.8$$

$$+ 0.0295(\ln 56)(\ln 28.8) + 0.3215\,(-0.2144)$$

$$= -0.1531.$$

The estimated RLC after a 56-day break is therefore:

$$\ln y = 3.2806\left[\ln(x+1)\right]^{-0.1531}.$$

Substituting $x = 1, 2, 3, 4$, and so on, into the above equation yields estimated marginal post-break production hours of 32.1, 25.4, 22.7, 21.7, and so forth. In addition to providing early estimates of post-break production times, these equations can be used to estimate the costs (in terms of increased production times) of breaks of various lengths.

PREDICTIVE ABILITY

Tests of predictive ability are an important evaluation method for models that will be used for prediction. We test the predictive ability of our PPMs in two ways. First, we take the best combination of our PPMs and use them to predict RLC parameters, and then use these RLCs to predict post-break marginal times. We compare these predictions to the actual post-break times and compute the mean absolute percentage error (MAPE).[*] While this comparison provides a test of the ability of the PPMs to predict useful RLC parameters, it is not a test of predictive ability in the strictest sense because the tests use the same post-break data that were used to determine the PPM parameter values. Therefore, we conduct a second test of predictive ability using a hold-out sample. The results of these two tests are reported below.

First Test of Predictive Ability

Based on adjusted R^2 values, the best combination of PPMs is Equation 7.9a and 7.10. Using these two PPMs, we constructed the two forms of RLCs for each subject and used each RLC to predict post-break marginal times. The results were an average overall MAPE of 12.5% for the log-log RLC and 11.59% for the Bailey-McIntyre RLC.

Because the best-fitting model does not always predict the most accurate (e.g., Lorek et al. 1983), we also compared the MAPEs using the next-best combination of PPMs (Equations 7.8a and 7.11). Table 7.5 reports the MAPEs for all combinations

---

TABLE 7.5
Comparisons of MAPEs for Combinations of PPMs

Log-Log Curve

PPM for a_R	PPM for b_R	
	7.10	7.11
7.8a	14.12	14.10
7.9a	12.51	12.70

Bailey–McIntyre Curve

PPM for a_R	PPM for b_R	
	7.10	7.11
7.8a	11.86	11.84
7.9a	11.59	11.61

---

[*] MAPE = $100 \times$ averaged(|projected time − actual time|/actual time).

of these four PPMs. For both types of RLCs, the MAPEs are lower for the 7.9a–7.10 combination than for any other combination of PPMs. For the log-log RLC, the MAPEs of all three alternative combinations are significantly higher that the 7.9a–7.10 combination at $p < .001$ in paired t-tests. The Bailey-McIntyre RLC appears more robust to the choice of PPMs, with only the 7.8a–7.10 and 7.8a–7.11 combinations having significantly higher MAPEs than the 7.9a–7.10 combination (with both p values $\approx .01$). Overall, this finding of the best-predicting model being the same as the best-fitting model is consistent with Bailey and McIntyre (1997).

Next, we compare the MAPEs computed above for the combination 7.9a and 7.10 to MAPEs from two alternative approaches: (1) a "start-anew" RLC using the available relearning data to forecast future relearning times, and (2) a backing-up procedure in which we use the *forecasted* first relearning time from our "best" Equation 7.9 to establish the nearest corresponding iteration on the *original* log-linear LC, then use that original curve to forecast future relearning times, in accordance with Anderlohr (1969). Figure 7.4 provides this comparison of MAPEs. The percentage on the y axis is the MAPE for predicting all future performance times after having completed the number of relearning iterations shown on the x axis. Because an RLC requires a minimum of two data points, the start-anew RLC begins with the second iteration. The MAPE for a two-unit RLC is based on its forecast errors for units 3 to n; the MAPE for a three-unit RLC is based on its forecast errors for units 4 to n, and so on. The RLCs constructed from PPM-based forecasts, however, as well as the backing-up approach using forecasted time on resumption, can begin with zero relearning observations, at which point they can predict units 1, 2, 3, and so on. The PPM-based forecasts exhibit fairly stable MAPEs across iterations. The start-anew RLC forecasts are based on the RLC that performed best as reported by Bailey and McIntyre (1997); that is, a curve in the form of Equation 7.2 fitted to the relearning data.

Based on MAPEs for predicting times 1 through n (i.e., after zero relearning iterations), the PPMs performed better when using the Bailey-McIntyre model

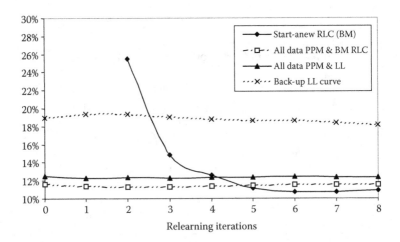

FIGURE 7.4   MAPEs for PPMs vs. back-up RLC and start-anew RLC.

(MAPE = 11.59%) than when using the traditional log-log model (MAPE = 12.51%), paired $t = 7.11$, $p < .0001$. The advantage continues at a similar level for predicting 2 through n, 3 through n, and so forth, as more data become available. The performance of the start-anew RLC depends on the number of new data observations available to fit it. After two observations, both PPM-based curves are markedly better than the new RLC. After three units, both PPM-based curves retain a clear advantage ($p < .0001$ in a two-tailed paired t-test); after four units, the difference is still significant only for the Bailey–McIntyre curve ($p = .025$). After five observations, the start-anew curve is significantly better than the log-log-based PPM curve ($p = .005$) but not the Bailey-McIntyre-based curve. After six units, the advantage has shifted marginally in favor of the newly-fitted curve (two-tailed $p = .065$, or one-tailed .033 assuming the direction should be in favor of the new RLC). Finally, the backing-up approach produces poor results, because the original LC slope does not reflect the more rapid improvement during relearning.

Given that the relearning slope will be steeper, we also tried a naïve model in which all subjects' b_L parameters were multiplied by a constant factor $k > 1$. Optimizing k for this data set ($k \approx 1.15$), to minimize overall MAPE, reduced the MAPE for this forecasting approach from about 18% to about 15%, still substantially worse than the PPM-based results. Although we label this model a naïve model, by using the value of k that minimizes the overall MAPE, the model uses data that are computed *after* the predictions have been made, giving it an advantage over any actual forecasting model. Even with this advantage, it does not predict as well as our PPM-based RLCs.

Second Test of Predictive Ability

Our second test of predictive ability involves a hold-out sample so that our predictions are for post-break times that were not used in the computation of PPM parameters. We randomly selected 20 subjects from our 116, fit curves 7.9a and 7.10 to their data, used the resulting PPMs to specify RLCs, and forecast post-break times for the remaining 96 subjects. The use of a much smaller estimation sample to test predictive ability for the hold-out sample represents a rigorous test of predictive ability despite the fact that the forms of the PPMs were derived from the full sample. The results of these tests appear in Figure 7.5. This figure also shows, for comparison, the MAPEs from the two PPM-based RLCs and the start-anew RLC from Figure 7.4. The average MAPEs of the 96 Bailey-McIntyre RLCs, using parameter values derived from the 20-observation PPMs, are very close to the average MAPEs of both the start-anew RLCs and the RLCs that used parameters from PPMs developed using all 116 subjects. Much higher average MAPEs are evident for the 96 log-log RLCs developed from the 20-observations PPMs. Note that in Table 7.3 the coefficients for Equation 7.10, developed using the log-log form of LC, are not significant, except for ln B, at the .05 level, whereas the coefficients developed using the Bailey-McIntyre LC are all significant at levels of .00 to .05. Thus, it is not surprising that the predictive ability of the log-log curve suffers much more from the reduction in sample size.

The robustness of forecasts using the Bailey-McIntyre model when using a smaller sample indicates that users could derive useful PPM-based forecasts from a relatively small number of experiences. Further, for manual assembly tasks similar

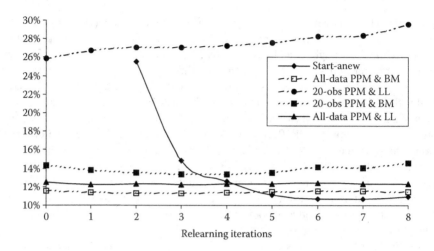

FIGURE 7.5 MAPEs for 20-observation PPMs vs. new RLCs.

to the ones used in our experiments, the parameter values in Tables 7.2 and 7.4 represent potential starting points for users who wish to employ the PPMs of Equations 7.9a and 7.10 with either the Wright or Bailey-McIntyre form of RLC.

DISCUSSION, CONCLUSIONS, AND IMPLICATIONS

PPMs emerge from this study as good predictors of RLC parameters and, subsequently, as good predictors of future performance during the relearning phase. The PPM for the a_R parameter is surprisingly good. It is better for the Bailey-McIntyre curve than for the log-log curve, probably because the former was developed using this type of production assembly data, but also because it is better able to model relearning. Past research indicates that the progress of learning is quite predictable, given that the model captures the nature of the underlying learning process. Unexplained variations will certainly occur from one production time to the next, but since the a parameter captures the "average" level of performance for the learning (or relearning) phase, it is plausible that a_R and a_L would be as closely related as the PPMs indicate.

Although the relearning rate (b_R) is less predictable, the best PPM shows that it is related to the original learning rate and the length of the break. The relatively low explanatory power of this PPM indicates potential for further research; however, Equation 7.10 provides a better fit to the data than four other equations that we tested, including the equation suggested by Shtub et al. (1993). Thus, our PPM provides an incremental benefit relative to these other models, and could serve as a benchmark for future research. Furthermore, the RLCs derived using our best PPMs for both a_R and b_R show good predictive ability, indicating that they have the potential to improve predictive accuracy of post-break production time. Since the purpose of PPMs is ultimately to provide useful estimates of post-break times, our best PPMs meet this objective despite the differences in the R^2 values of the two models. In contrast to other approaches, PPMs provide a basis for estimating post-break production times well before post-break data become available. At some point after production resumes and observations of relearning data become available, a curve fitted to the new data will prevail. Our research suggests that even when several data points are available (about five for our data), the PPM-based estimates are more accurate, and they continue to be competitive after that point. Additionally, the predicted parameters would provide a benchmark against which users could evaluate the early results of the relearning process. Specifically, users may be able to apply judgment, supported by the PPM estimates, to avoid projecting unrealistic trends (Bailey and Gupta 1999).

Since our PPM parameters were estimated from a common mechanical-assembly task setting, the parameters of our PPMs might be directly applicable to similar settings, where users do not have sufficient data to develop their own PPMs. This can only be determined empirically as the research progresses, but experience in applying the method could lead to tabulated PPM coefficients similar to the tabulated learning rates currently available for over 40 industrial tasks (Konz and Johnson 2000).

Our study tested PPMs after only one interruption. The success of our PPMs for repeated asynchronous interruptions is an empirical question for which we have no direct evidence; however, to the extent that the forgetting-relearning process remains relatively stable, we would expect the models to retain their usefulness. An interesting question for future research is whether, in repeated applications, estimates from the LC that are used as independent variables in the PPMs should come from the original LC or from the immediately preceding RLC.

Future research should address the consistency of the PPMs across tasks. That is, can the parameters estimated for our PPMs be relied on for a variety of tasks? Bohlen and Barany (1976) identified seven characteristics of operators (workers) and five characteristics of operations (tasks) that they used to predict the parameters of LCs for new bench assembly operations. These variables might also moderate the prediction of RLC parameters. The meta-analytical review by Arthur et al. (1998) is an important resource related to future research on skill decay, which relates to the a_R parameter.

Undoubtedly, further progress is possible in refining the PPMs, particularly for the learning rate. One additional possibility is to incorporate actual relearning observations into the PPMs as such observations become available (Towill 1973).

REFERENCES

Adler, G. L., and Nanda, R. 1974. The effects of learning on optimal lot size determination – Single product case. *AIIE Transactions* 6(1): 14–20.

Anderlohr, G., 1969. What production breaks cost. *Industrial Engineering* 20(9): 34–36.

Argote, L., Beckman, S.L., and Epple, D., 1990. The persistence and transfer of learning in industrial settings. *Management Science* 36(2): 140–154.

Arthur, W., Jr., Bennett, W., Jr., Stanush, P.L., and McNelly, T.L., 1998. Factors that influence skill decay and retention: A quantitative review and analysis. *Human Performance* 11(1): 57–101.

Arzi, Y., and Shtub, A., 1997. Learning and forgetting in mental and mechanical tasks: A comparative study. *IIE Transactions* 29(9): 759–768.

Bailey, C.D., 1989. Forgetting and the learning curve: A laboratory study. *Management Science* 35(3): 340–352.

Bailey, C.D., 2000. Learning-curve estimation of production costs and labor hours using a free Excel plug-in. *Management Accounting Quarterly* 1(4): 25–31.

Bailey, C.D., and Gupta, S., 1999. Judgment in learning-curve forecasting: A laboratory study. *Journal of Forecasting* 18(1): 39–57.

Bailey, C.D., and McIntyre, E.V., 1992. Some evidence on the nature of relearning curves. *The Accounting Review* 67(2): 368–378.

Bailey, C.D., and McIntyre, E.V., 1997. The relation between fit and prediction for alternative forms of learning curves and relearning curves. *IIE Transactions* 29(6): 487–495.

Benkard, C.L., 2000. Learning and forgetting: The dynamics of aircraft production. *The American Economic Review* 90(4): 1034–1054.

Bohlen, G.A., and Barany, J.W., 1976. A learning curve prediction model for operators performing industrial bench assembly operations. *International Journal of Production Research* 14(2): 295–303.

Carlson, J.G., and Rowe, A.J., 1976. How much does forgetting cost? *Industrial Engineering* 8(9): 40–47.

Chen, J.T., and Manes, R.P., 1985. Distinguishing the two forms of the constant percentage learning curve model. *Contemporary Accounting Research* 1(2): 242–252.

Cherrington, J.E., Lippert, S., and Towill, D.R., 1987. The effect of prior experience on learning curve parameters. *International Journal of Production Research* 25(3): 399–411.

Conway, R., and Schultz, A., 1959. The manufacturing progress function. *Journal of Industrial Engineering* 10(1): 39–53.

Dar-El, E.M., Ayas, K., and Gilad, I., 1995. Predicting performance times for long cycle time tasks. *IIE Transactions* 27(3): 272–281.

Farr, M.J., 1987. *The long-term retention of knowledge and skills.* New York: Springer-Verlag.

Globerson, S., and Levin, N., 1987. Incorporating forgetting into learning curves. *International Journal of Operations & Production Management* 7(4): 80–94.

Globerson, S., Levin, N., and Shtub, A., 1989. The impact of breaks on forgetting when performing a repetitive task. *IIE Transactions* 10(3): 376–381.

Globerson, S., Nahumi, A., and Ellis, S., 1998. Rate of forgetting for motor and cognitive tasks. *International Journal of Cognitive Ergonomics* 2(3): 181–191.

Hancock, W.M., 1967. The prediction of learning rates for manual operations. *The Journal of Industrial Engineering* 18(1): 42–47.

Jaber, M.Y., and Bonney, M., 1997. A comparative study of learning curves with forgetting. *Applied Mathematical Modeling* 21(8): 523–531.

Konz, S., and Johnson, S., 2000. *Work design: Industrial ergonomics,* 5th ed., Scottsdale: Holcomb Hathaway.

Kvålseth, T.O., 1985. Cautionary note about R^2. *The American Statistician* 39(4): 279–285.

Lorek, K.S., Icerman, J.D., and Abdulkader, A.A., 1983. Further descriptive and predictive evidence on alternative time-series models for quarterly earnings. *Journal of Accounting Research* 21(1): 317–328.

Mazur, J.E., and Hastie, R., 1978. Learning as accumulation: A reexamination of the learning curve. *Psychological Bulletin* 85(6): 1256–1274.

Nembhard, D.A., 2000. The effects of task complexity and experience on learning and forgetting: A field study. *Human Factors* 42(2): 272–286.

Nembhard, D.A., and Uzumeri, M.V., 2000. Experiential learning and forgetting for manual and cognitive tasks. *International Journal of Industrial Ergonomics* 25(4): 315–326.

Newell, A., and Rosenbloom, P.S., 1981. Mechanisms of skill acquisition and the law of practice. In *Cognitive skills and their acquisition,* ed. J. R. Anderson, pp. 1–55. Hillsdale: Lawrence Erlbaum.

Shtub, A., Levin, N., and Globerson, S., 1993. Learning and forgetting industrial tasks: An experimental model. *International Journal of Human Factors in Manufacturing* 3(3): 293–305.

Smith, J., 1989. *Learning curve for cost control.* Norcross: Industrial Engineering and Management Press.

Snoddy, G.S., 1926. Learning and stability. *Journal of Applied Psychology* 10(1): 1–36.

Swezey, R.W., and Llaneras, R.E., 1997. Models in training and instruction. In *Handbook of human factors and ergonomics,* 2nd ed., ed. G. Salvendy, pp. 514–577, New York: Wiley.

Teplitz, C.J., 1991. *The learning curve deskbook: A reference guide to theory, calculations, and applications.* New York: Quorum Books.

Towill, D.R., 1973. An industrial dynamics model for start-up management. *IEEE Transactions on Engineering Management EM*-20(2): 44–51.

Uzumeri, M., and Nembhard, D., 1998. A population of learners: A new way to measure organizational learning. *Journal of Operations Management* 16: 515–528.

Wickelgren, W.A., 1972. Trace resistance and the decay of long-term memory. *Journal of Mathematical Psychology* 9(4): 418–455.

Wright, T.P., 1936. Factors affecting the cost of airplanes. *Journal of the Aeronautical Sciences* 3(2): 122–128.

Yelle, L.E., 1979. The learning curve: Historical review and comprehensive survey. *Decision Sciences* 10(2): 302–328.

Zangwill, W.I., and Kantor, P.B., 1998. Toward a theory of continuous improvement and the learning curve. *Management Science* 44(7): 910–920.

8 Introduction to Half-Life Theory of Learning Curves

Adedeji B. Badiru

CONTENTS

INTRODUCTION

The military is very much interested in training troops fast, thoroughly, and effectively. Team training is particularly important as a systems approach to enhancing military readiness. Thus, the prediction of team performance is of great importance in any military system. In military training systems that are subject to the variability and complexity of interfaces, the advance prediction of performance is useful for designing training programs for efficient knowledge acquisition and the sustainable retention of skills. Organizations invest in people, work processes, and technology for the purpose of achieving increased and enhanced production capability. The systems nature of such an investment strategy requires that the investment is a carefully planned activity, stretching over multiple years. Learning curve analysis is one method through which system enhancement can be achieved in terms of cost, time, and performance vis-à-vis the strategic investment of funds and other assets. The predictive capability of learning curves is helpful in planning for system performance enhancement and resilience.

Formal analysis of learning curves first emerged in the mid-1930s in connection with the analysis of the production of airplanes (Wright 1936). Learning refers to the improved operational efficiency and cost reduction obtained from the repetition of a task. This has a big impact on training and design of work. Workers learn and improve by repeating operations. Thus, a system's performance and resilience are dependent on the learning characteristics of its components; with workers being a major component of the system. Learning is time dependent and externally controllable. The antithesis of learning is *forgetting*. Thus, as a learning curve leads to increasing performance through cost reduction, forgetting tends to diminish performance. Considering the diminishing impact of forgetting, the half-life measure will be of interest for assessing the resilience and sustainability of a system. Derivation of the half-life equations of learning curves can reveal more about the properties of the various curves that have been reported in the literature. This chapter, which is based on Badiru and Ijaduola (2009), presents the half-life derivations for some of the classical learning curve models available in the literature.

Several research and application studies have confirmed that human performance improves with reinforcement or with frequent and consistent repetitions. Badiru (1992, 1994) provides a computational survey of learning curves as well as their industrial application to productivity and performance analysis. Reductions in operation processing times achieved through learning curves can directly translate to cost savings. The wealth of literature on learning curves shows that they are referred to by several names, including progress function, cost-quantity relationship, cost curve, production acceleration curve, performance curve, efficiency curve, improvement curve, and learning function. In all of these different perspectives, a primary interest is whether or not a level of learning, once achieved, can be sustained. The sustainability of a learning curve is influenced by several factors such as natural degradation, forgetting, and reduction due to work interruption. Thus, it is of interest to predict the future state and behavior of learning curves. In systems planning and control, the prediction of performance is useful for determining the line of corrective action that should be taken. Learning curves are used extensively in business, science, technology, engineering, and industry to predict performance over time. Thus, there has been a big interest in the behavior of learning curves over the past several decades.

This chapter introduces the concept of the half-life analysis of learning curves as a predictive measure of system performance. Half-life is the amount of time it takes for a quantity to diminish to half its original size through natural processes. The quantity of interest may be cost, time, performance, skill, throughput, or productivity. Duality is of natural interest in many real-world processes. We often speak of "twice as much" and "half as much" as benchmarks for process analysis. In economic and financial principles, the "rule of 72" refers to the length of time required for an investment to double in value. These common "double" or "half" concepts provide the motivation for the proposed half-life analysis.

The usual application of half-life analysis is in natural sciences. For example, in physics, the half-life is a measure of the stability of a radioactive substance. In practical terms, the half-life attribute of a substance is the time it takes for one-half of the atoms in an initial magnitude to disintegrate. The longer the half-life of a substance, the more stable it is. This provides a good analogy for modeling learning curves with the aim of increasing performance or decreasing cost with respect to the passage of time. This approach provides another perspective to the body of literature on learning curves. It has application not only in the traditional production environment, but also in functions such as system maintenance, safety, security skills, marketing effectiveness, sports skills, cognitive skills, and resilience engineering. The positive impacts of learning curves can be assessed in terms of cost improvement, the reduction in production time, or the increase in throughput time. The adverse impacts of forgetting can be assessed in terms of declining performance. We propose the following formal definitions:

For learning curves: Half-life is the production level required to reduce the cumulative average cost per unit to half its original size.

For forgetting curves: Half-life is the amount of time it takes for a performance to decline to half its original magnitude.

LITERATURE ON LEARNING CURVES

Although there is an extensive collection of classical studies of the *improvement* due to learning curves, only very limited attention has been paid to performance *degradation* due to the impact of forgetting. Some of the classical works on process improvement due to learning include Smith (1989), Belkaoui (1976, 1986), Nanda (1979), Pegels (1969), Richardson (1978), Towill and Kaloo (1978), Womer (1979, 1981, 1984), Womer and Gulledge (1983), Camm et al. (1987), Liao (1979), McIntyre (1977), Smunt (1986), Sule (1978), and Yelle (1979, 1980, 1983). It is only in recent years that the recognition of "forgetting" curves has begun to emerge, as can be seen in more recent literature (Badiru, 1995), Jaber and Sikström (2004), Jaber et al. (2003), Jaber and Bonney (2003, 2007), and Jaber and Guiffrida (2008). The new and emerging research on the forgetting components of learning curves provides the motivation for studying the half-life properties of learning curves. Performance decay can occur due to several factors, including a lack of training, a reduced retention of skills, lapses in performance, extended breaks in practice, and natural forgetting.

ANALYSIS OF SYSTEM PERFORMANCE AND RESILIENCE

Resilience engineering is an emerging area of systems analysis that relates to the collection of activities designed to develop the ability of a system (or organization) to continue operating under extremely adverse conditions such as a "shock" or an attack. Thus, a system's resilience is indicative of the system's level of performance under shock. If the learning characteristic of the system is stable and retainable, then the system is said to be very resilient. The ability to predict a system's performance using learning curve analysis provides an additional avenue to develop corrective strategies for managing a system. For example, suppose that we are interested in how fast an IT system responds to service requests from clients. We can model the system's performance in terms of its response time with respect to the passage of time. In this case, it is reasonable to expect the system to improve over time because of the positive impact of the learning curve of the IT workers. Figure 8.1 shows a graphical representation of the response time as a function of time. The response time decreases as time progresses, thus indicating increasing levels of performance. The shorter the response time, the more resilient we can expect the system to be in the event of an attack or shock to the system. Typical learning curves measure cost or time reduction, but the reduction can be translated to, and represented as, performance improvement. Consequently, computing the half-life of the system can be used to measure how long it will take the system's response time to reduce to half its starting value. Figure 8.2 shows a generic profile for the case where the performance metric (e.g., number of requests completed per unit time) increases with respect to the passage of time.

HALF-LIFE PROPERTY OF NATURAL SUBSTANCES

The half-life concept of learning curves measures the amount of time that it takes for performance to degrade by half. Degradation of performance occurs both through natural and imposed processes. The idea of using the half-life approach comes from physics, where half-life is a measure of the stability of a radioactive substance. The longer the half-life of a substance, the more stable it is. By analogy, the longer the half-life of a learning curve model, the more sustainable the fidelity of the learning

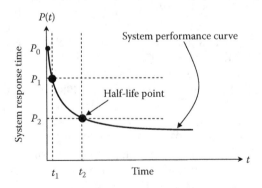

FIGURE 8.1 Representation of system response time with respect to passage of time.

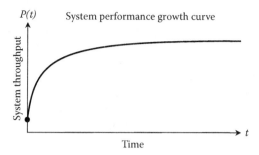

FIGURE 8.2   System performance growth curve.

curve effect. If learning is not very sustainable, then the system will be more vulnerable to the impact of learning curve decline brought on by such random events as system interruptions. To appreciate the impact of half-life computations, consider an engineering reactor that converts the relatively stable uranium 238 into the isotope plutonium 239. After 15 years, it is determined that 0.043% of the initial amount A_0 of the plutonium has disintegrated. We are interested in determining the half-life of the isotope. In physics, the initial value problem is stated as:

$$\frac{dA}{dt} = kA,$$

with $A(0) = A_0$. This has a general solution of the form:

$$A(t) = A_0 e^{kt}.$$

If 0.043% of the atoms in A_0 have disintegrated, then 99.957% of the substance remains. To find k, we will solve:

$$\alpha A_0 = A_0 e^{15k},$$

where α is the remaining fraction of the substance. With $\alpha = 0.99957$, we obtain $k = -0.00002867$. Thus, for any time t, the amount of plutonium isotope remaining is represented as:

$$A(t) = A_0 e^{-0.00002867t}.$$

This has a general decay profile similar to the plot of $P(t)$ in Figure 8.1. Now we can compute the half-life as a corresponding value at time t for which $A(t) = A_0/2$. That is:

$$\frac{A_0}{2} = A_0 e^{-0.00002867t},$$

which yields a t (half-life) value of 24,180 years. With this general knowledge of the half-life, several computational analyses can be done to predict the behavior and magnitude of the substance over time. The following examples further illustrate the utility of half-life computations. Let us consider a radioactive nuclide that has a half-life of 30 years. Suppose that we are interested in computing the fraction of an initially pure sample of this nuclide that will remain undecayed at the end of a time period of, for example, 90 years. From the equation of half-life, we can solve for k:

$$\frac{A_0}{2} = A_0\, e^{-kt_{\text{half-life}}},$$

$$k = \frac{\ln 2}{t_{\text{half-life}}},$$

which yields $k = 0.0231049$. Now, we can use this value of k to obtain the fraction we are interested in computing. That is,

$$\frac{A_0}{A} = e^{-(0.0231049)(90)} = 0.125.$$

As another example, let us consider a radioactive isotope with a half-life of 140 days. We can compute the number of days it would take for the sample to decay to one-seventh of its initial magnitude. Thus:

$$\frac{A_0}{2} = A_0\, e^{-kt_{\text{half-life}}},$$

$$k = \frac{\ln 2}{t_{\text{half-life}}},$$

which yields $k = 0.004951$. Now, using the value of k obtained above, we need to find the time for

$$A = 1/7\, A_0,$$

That is:

$$\frac{1}{7} A_0 = A_0\, e^{-kt},$$

$$t = \frac{\ln 7}{k} = 393\,\text{days}.$$

For learning curves, analogous computations can be used to predict the future system performance level and to conduct a diagnostic assessment of previous performance levels given a present observed performance level. Since there are many alternate models of learning curves, each one can be analyzed to determine its half-life. Thus, a comparative analysis of the different models can be conducted. This general mathematical approach can become the de-facto approach for the computational testing of learning curve models.

HALF-LIFE APPLICATION TO LEARNING CURVES

Learning curves present the relationship between cost (or time) and the level of activity on the basis of the effect of learning. An early study by Wright (1936) disclosed the "80% learning" effect, which indicates that a given operation is subject to a 20% productivity improvement each time the activity level or production volume *doubles*. The proposed half-life approach is the antithesis of the double-level milestone. A learning curve can serve as a predictive tool for obtaining time estimates for tasks that are repeated within a project's life cycle. A new learning curve does not necessarily commence each time a new operation is started, since workers can sometimes transfer previous skills to new operations. The point at which the learning curve begins to flatten depends on the degree of similarity of the new operation to previously performed operations. Typical learning rates that have been encountered in practice range from 70% to 95%. Several alternate models of learning curves have been presented in the literature. Some of the classical models are:

- Log-linear model
- S-curve model
- Stanford-B model
- DeJong's learning formula
- Levy's adaptation function
- Glover's learning formula
- Pegels' exponential function
- Knecht's upturn model
- Yelle's product model

The basic log-linear model is the most popular learning curve model. It expresses a dependent variable (e.g., production cost) in terms of some independent variable (e.g., cumulative production). The model states that the improvement in productivity is constant (i.e., it has a constant slope) as output increases. That is:

$$C(x) = C_1 x^{-b},$$

or

$$\log C(x) = -b(\log x) + \log C_1,$$

where:

$C(x)$ = cumulative average cost of producing x units
C_1 = cost of the first unit

x = cumulative production unit

b = learning curve exponent.

Notice that the expression for $C(x)$ is practical only for $x > 0$. This makes sense because the learning effect cannot realistically kick in until at least one unit ($x \geq 1$) has been produced. For the standard log-linear model, the expression for the learning rate, p, is derived by considering two production levels where one level is double the other. For example, given the two levels x_1 and x_2 (where $x_2 = 2x_1$), we have the following expressions:

$$C(x_1) = C_1(x_1)^{-b}.$$

$$C(x_2) = C_1(2x_1)^{-b}.$$

The percent productivity gain, p, is then computed as:

$$p = \frac{C(x_2)}{C(x_1)} = \frac{C_1(2x_1)^{-b}}{C_1(x_1)^{-b}} = 2^{-b}.$$

The performance curve, $P(x)$, shown earlier in Figure 8.1 can now be defined as the reciprocal of the average cost curve, $C(x)$. Thus, we have:

$$P(x) = \frac{1}{C(x)},$$

which will have an increasing profile compared to the asymptotically declining cost curve. In terms of practical application, learning to drive is one example where a maximum level of performance can be achieved in a relatively short time compared with the half-life of performance. That is, learning is steep, but the performance curve is relatively flat after steady state is achieved. The application of half-life analysis to learning curves can help address questions such as the ones below:

- How fast and how far can system performance be improved?
- What are the limitations to system performance improvement?
- How resilient is a system to shocks and interruptions to its operation?
- Are the performance goals that are set for the system achievable?

DERIVATION OF HALF-LIFE OF THE LOG-LINEAR MODEL

Figure 8.3 shows a pictorial representation of the basic log-linear model, with the half-life point indicated as $x_{1/2}$. The half-life of the log-linear model is computed as follows: Let:

C_0 = initial performance level

$C_{1/2}$ = performance level at half-life

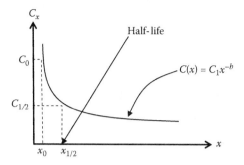

FIGURE 8.3   General profile of the basic learning curve model.

$$C_0 = C_1 x_0^{-b} \quad \text{and} \quad C_{1/2} = C_1 x_{1/2}^{-b}.$$

However, $C_{1/2} = 1/2\, C_0$. Therefore, $C_1 x_{1/2}^{-b} = 1/2\, C_1 x_0^{-b}$, which leads to $x_{1/2}^{-b} = 1/2 x_0^{-b}$,

which, by taking the $(-1/b)$th exponent of both sides, simplifies to yield the following expression as the general expression for the standard log-linear learning curve model:

$$x_{1/2} = \left(\frac{1}{2}\right)^{-(1/b)} x_0, \quad x_0 \geq 1,$$

where $x_{1/2}$ is the half-life and x_0 is the initial point of operation. We refer to $x_{1/2}$ (Figure 8.3) as the *first-order half-life*.

The *second-order half-life* is computed as the time corresponding to half the preceding half. That is:

$$C_1 x_{1/2(2)}^{-b} = \frac{1}{4} C_1 x_0^{-b},$$

which simplifies to yield:

$$x_{1/2(2)} = \left(\frac{1}{2}\right)^{-(2/b)} x_0.$$

Similarly, the *third-order half-life* is derived to obtain:

$$x_{1/2(3)} = \left(\frac{1}{2}\right)^{-(3/b)} x_0.$$

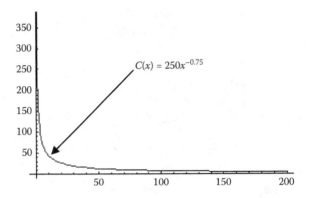

FIGURE 8.4 Learning curve with $b = -0.75$.

In general, the kth-order half-life for the log-linear model is represented as:

$$x_{1/2(k)} = \left(\frac{1}{2}\right)^{-(k/b)} x_0.$$

The characteristics of half-life computations are illustrated in Figures 8.4 and 8.5.

COMPUTATIONAL EXAMPLES

Figures 8.2 and 8.3 show examples of log-linear learning curve profiles with $b = 0.75$ and $b = 0.3032$, respectively. The graphical profiles reveal the characteristics of learning, which can dictate the half-life behavior of the overall learning process. Knowing the point where the half-life of each curve occurs can be very useful in assessing learning retention for the purpose of designing training programs or designing work.

For Figure 8.4 ($C(x) = 250x^{-0.75}$), the first-order half-life is computed as:

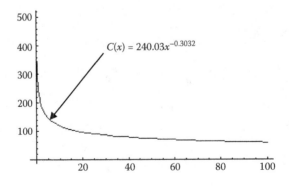

FIGURE 8.5 Learning curve with $b = -0.3032$.

$$x_{1/2} = \left(\frac{1}{2}\right)^{-(1/0.75)} x_0, \quad x_0 \geq 1.$$

If the above expression is evaluated for $x_0 = 2$, the first-order half-life yields x1/2 = 5.0397, which indicates a fast drop in the value of $C(x)$. Table 8.1 summarizes values of $C(x)$ as a function of the starting point, x_0. The specific case of $x_0 = 2$ is highlighted in Table 8.1. It shows $C(2) = 148.6509$ corresponding to a half-life of 5.0397. Note that $C(5.0397) = 74.7674$, which is about half of 148.6509. The arrows in the table show how the various values are linked. The conclusion from this analysis is that if we are operating at the point $x = 2$, we can expect this particular curve to reach its half-life decline point at $x = 5$.

For Figure 8.5 ($C(x) = 250x^{-0.3032}$), the first-order half-life is computed as:

$$x_{1/2} = \left(\frac{1}{2}\right)^{-(1/0.3032)} x_0, \quad x_0 \geq 1$$

TABLE 8.1

Numeric Calculation of Half-Lives for $C(x) = 250x^{-0.75}$

x_0	$C(x_0)$	$x_{1/2}$
1	250	2.519842
2	148.6508894	5.039684
3	109.6728344	7.559526
4	88.38834765	10.07937
5	74.76743906	12.59921
10	44.45698525	25.19842
15	32.79982785	37.79763
20	26.43428159	50.39684
25	22.36067977	62.99605
30	19.50289433	75.59526
35	17.37356628	88.19447
40	15.71791787	100.7937
45	14.38900036	113.3929
50	13.29573974	125.9921
55	12.37849916	138.5913
60	11.59649035	151.1905
65	10.92081352	163.7897
70	10.33038432	176.3889
100	7.90569415	251.9842
120	6.895314416	302.3811
150	5.832725853	377.9763

If we evaluate the above function for $x_0 = 2$, the first-order half-life yields $x1/2 = 19.6731$. This does not represent as precipitous a drop as in Figure 8.4. These numeric examples agree with the projected profiles of the curves in Figures 8.2 and 8.3, respectively.

COST EXPRESSIONS FOR THE LOG-LINEAR MODEL

For the log-linear model, using the basic expression for cumulative average cost, the *total cost* of producing units is computed as:

$$TC(x) = (x)C_x = xC_1 x^{-b} = C_1 x^{(-b+1)}.$$

The *unit cost* of producing the xth unit is given by:

$$UC(x) = C_1 x^{(-b+1)} - C_1 (x-1)^{(-b+1)}$$
$$= C_1 \left[x^{-b+1} - (x-1)^{-b+1} \right].$$

The *marginal cost* of producing the xth unit is given by:

$$MC(x) = \frac{dTC_x}{dx} = (-b+1)C_1 x^{-b}.$$

If desired, one can derive half-life expressions for the cost expressions above. For now, we will defer those derivations for interested readers. An important application of learning curve analysis is the calculation of expected production time as illustrated by the following examples. Suppose, in a production run of a complex technology component, it was observed that the cumulative hours required to produce 100 units is 100,000 hours with a learning curve effect of 85%. For future project planning purposes, an analyst needs to calculate the number of hours required to produce the fiftieth unit. Following the standard computations, we have the following: $p = 0.85$ and $x = 100$ units. Thus, $0.85 = 2^{-b}$, which yields $b = 0.2345$. Consequently, we have $1,000 = C_1(100)^{-b}$, which yields $C_1 = 2,944.42$ hours. Since b and C_1 are now known, we can compute the cumulative *average* hours required to produce 49 and 50 units, respectively, to obtain $C(49) = 1,182.09$ hours and $C(50) = 1,176.50$ hours. Consequently, the *total* hours required to produce the fiftieth unit is $50[C(50)] - 49[C(49)] = 902.59$ (approximately 113 work days). If we are interested in knowing when these performance metrics would reach half of their original levels in terms of production quantity, we would use half-life calculations.

ALTERNATE FORMULATION OF THE LOG-LINEAR MODEL

An alternate formulation for the log-linear model is called the *unit cost model*, which is expressed in terms of the specific cost of producing the xth unit, instead of the conventional cumulative average cost expression. The unit cost formula specifies that the individual cost per unit will decrease by a constant percentage as cumulative production doubles. The functional form of the unit cost model is the same as for the average cost model except that the interpretations of the terms are different. It is expressed as:

$$UC(x) = C_1 x^{-b},$$

where:
$UC(x)$ = cost of producing the xth unit
C_1 = cost of the first unit
x = cumulative production count
b = the learning curve exponent, as discussed previously.

From the unit cost formula, we can derive expressions for the other cost elements. For the discrete case, the total cost of producing units is given by:

$$TC(x) = \sum_{q=1}^{x} UC_q = C_1 \sum_{q=1}^{x} q^{-b},$$

and the cumulative average cost per unit is given by:

$$C(x) = \frac{C_1}{x} \sum_{q=1}^{x} q^{-b}.$$

The marginal cost is found as follows:

$$MC(x) = \frac{d[TC(x)]}{dx} = \frac{d\left[C_1 \sum_{i=1}^{x} (i)^{-b}\right]}{dx} = C_1 \frac{d\left[1 + 2^{-b} + 3^{-b} + \cdots + x^{-b}\right]}{dx} = C_1 b x^{-b-1}.$$

For the continuous case, the corresponding cost expressions are:

$$TC(x) = \int_0^x UC(z)\,dz = C_1 \int_0^x z^{-b}\,dz = \frac{C_1 x^{(-b+1)}}{-b+1},$$

$$C(x) = \left(\frac{1}{x}\right) \frac{C_1 x^{(-b+1)}}{-b+1},$$

$$MC(x) = \frac{d[TC(x)]}{dx} = \frac{d\left[\frac{C_1 x^{(-b+1)}}{-b+1}\right]}{dx} = C_1 x^{-b}.$$

As in the previous illustrations, the half-life analysis can be applied to the forego-ing expressions to determine when each cost element of interest will decrease to half its starting value. This information can be useful for product pricing purposes, par-ticularly for technology products which are subject to rapid price reductions due to declining product cost. Several models and variations of learning curves have been reported in the literature (see Badiru, 1992; Jaber and Guiffrida, 2008). Models are developed through one of the following approaches:

1. Conceptual models
2. Theoretical models
3. Observational models
4. Experimental models
5. Empirical models

HALF-LIFE ANALYSIS OF SELECTED CLASSICAL MODELS

S-CURVE MODEL

The S-curve (Towill and Cherrington, 1994) is based on an assumption of a gradual start-up. The function has the shape of the cumulative normal distribution function for the start-up curve and the shape of an operating characteristics function for the learn-ing curve. The gradual start-up is based on the fact that the early stages of production are typically in a transient state, with changes in tooling, methods, materials, design, and even changes in the work force. The basic form of the S-curve function is:

$$C(x) = C_1 + M(x+B)^{-b},$$

$$MC(x) = C_1 \left[M + (1-M)(x+B)^{-b} \right],$$

where:
$C(x)$ = learning curve expression
b = learning curve exponent
$M(x)$ = marginal cost expression
C_1 = cost of first unit
M = incompressibility factor (a constant)
B = equivalent experience units (a constant).

Assumptions about at least three out of the four parameters (M, B, C_1, and b) are needed in order to solve for the fourth one. Using the $C(x)$ expression and derivation procedure outlined earlier for the log-linear model, the half-life equation for the S-curve learning model is derived to be:

$$x_{1/2} = (1/2)^{-1/b} \left[\frac{M(x_0+B)^{-b} - C_1}{M} \right]^{-1/b} - B,$$

where:
   $x_{1/2}$ = half-life expression for the S-curve learning model
   x_0 = initial point of evaluation of performance on the learning curve.

In terms of practical applications of the S-curve model, consider when a worker begins learning a new task. The individual is slow, initially at the tail end of the S-curve. However, the rate of learning increases as time goes on, with additional repetitions. This helps the worker to climb the steep-slope segment of the S-curve very rapidly. At the top of the slope, the worker is classified as being proficient with the learned task. From then on, even if the worker puts much effort into improving on the task, the resultant learning will not be proportional to the effort expended. The top end of the S-curve is often called the slope of *diminishing returns*. At the top of the S-curve, workers succumb to the effects of *forgetting* and other performance-impeding factors. As the work environment continues to change, a worker's level of skill and expertise can become obsolete. This is an excellent reason for the application of half-life computations.

STANFORD-B MODEL

An early form of a learning curve is the Stanford-B model, which is represented as:

$$UC(x) = C_1 (x + B)^{-b},$$

where:
   $UC(x)$ = direct cost of producing the xth unit
   b = learning curve exponent
   C_1 = cost of the first unit when $B = 0$
   B = slope of the asymptote for the curve ($1 < B < 10$).

This is equivalent to the units of previous experience at the start of the process, which represents the number of units produced prior to first unit acceptance. It is noted that when $B = 0$, the Stanford-B model reduces to the conventional log-linear model. Figure 8.6 shows the profile of the Stanford-B model with $B = 4.2$ and $b = -0.75$. The general expression for the half-life of the Stanford-B model is derived to be:

$$x_{1/2} = (1/2)^{-1/b}(x_0 + B) - B,$$

where:
   $x_{1/2}$ = half-life expression for the Stanford-B learning model
   x_0 = initial point of the evaluation of performance on the learning curve.

DERIVATION OF HALF-LIFE FOR BADIRU'S MULTI-FACTOR MODEL

Badiru (1994) presents applications of learning and forgetting curves to productivity and performance analysis. One empirical example presented used production data to

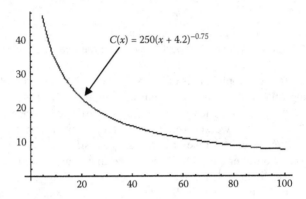

FIGURE 8.6 Stanford-B model with parameters $B = 4.2$ and $b = -0.75$.

develop a predictive model of production throughput. Two data replicates are used for each of 10 selected combinations of cost and time values. Observations were recorded for the number of units representing double production levels. The resulting model has the functional form below and the graphical profile shown in Figure 8.7.

$$C(x) = 298.88 x_1^{-0.31} x_2^{-0.13},$$

where:
  $C(x) =$ cumulative production volume
  $x_1 =$ cumulative units of Factor 1
  $x_2 =$ cumulative units of Factor 2
  $b_1 =$ first learning curve exponent $= -0.31$
  $b_2 =$ second learning curve exponent $= -0.13$.

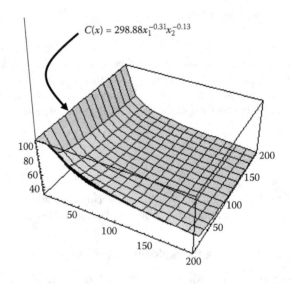

FIGURE 8.7 Bivariate model of learning curve.

A general form of the modeled multi-factor learning curve model is:

$$C(x) = C_1 x_1^{-b_1} x_2^{-b_2},$$

and the half-life expression for the multi-factor learning curve was derived to be:

$$x_{1(1/2)} = (1/2)^{-1/b_1} \left[\frac{x_{1(0)} x_{2(0)}^{b_2/b_1}}{x_{2(1/2)}^{b_2/b_1}} \right]^{-1/b_1},$$

$$x_{2(1/2)} = (1/2)^{-1/b_2} \left[\frac{x_{2(0)} x_{1(0)}^{b_1/b_2}}{x_{1(1/2)}^{b_2/b_1}} \right]^{-1/b_2},$$

where:

$x_{i\,(1/2)}$ = half-life component due to factor i ($i = 1, 2$)

$x_{i\,(0)}$ = initial point of factor i ($i = 1, 2$) along the multi-factor learning curve.

Knowledge of the value of one factor is needed in order to evaluate the other factor. Just as in the case of single-factor models, the half-life analysis of the multi-factor model can be used to predict when the performance metric will reach half a starting value.

DeJong's Learning Formula

DeJong's learning formula is a power function that incorporates parameters for the proportion of manual activity in a task. When operations are controlled by manual tasks, the time will be compressible as successive units are completed. If, by contrast, machine cycle times control operations, then the time will be less compressible as the number of units increases. DeJong's formula introduces an incompressible factor, M, into the log-linear model to account for the man-machine ratio. The model is expressed as:

$$C(x) = C_1 + Mx^{-b},$$

$$MC(x) = C_1 \left[M + (1 - M) x^{-b} \right],$$

where:

$C(x)$ = learning curve expression

$M(x)$ = marginal cost expression

b = learning curve exponent

C_1 = cost of first unit

M = incompressibility factor (a constant).

When $M = 0$, the model reduces to the log-linear model, which implies a completely manual operation. In completely machine-dominated operations, $M = 1$. In that case, the unit cost reduces to a constant equal of C_1, which suggests that no learning-based cost improvement is possible in machine-controlled operations. This represents a condition of high incompressibility. Figure 8.8 shows the profile of DeJong's learning formula for hypothetical parameters of $M = 0.55$ and $b = -0.75$. This profile suggests impracticality at higher values of production. Learning is very steep and the average cumulative production cost drops rapidly. The horizontal asymptote for the profile is below the lower bound on the average cost axis, suggesting an infeasible operating region as the production volume increases.

The analysis above agrees with the fact that no significant published data is available on whether or not DeJong's learning formula has been successfully used to account for the degree of automation in any given operation. Using the expression, $MC(x)$, the marginal cost half-life of DeJong's learning model is derived to be:

$$x_{1/2} = (1/2)^{-1/b} \left[\frac{(1-M)x_0^{-b} - M}{2(1-M)} \right]^{-1/b},$$

where:
$x_{1/2}$ = half-life expression for DeJong's learning curve marginal cost model
x_0 = initial point of the evaluation of performance on the marginal cost curve.

If the $C(x)$ model is used to derive the half-life, then we obtain the following derivation:

$$x_{1/2} = (1/2)^{-1/b} \left[\frac{Mx_0^{-b} - C_1}{M} \right]^{-1/b},$$

where:
$x_{1/2}$ = half-life expression for DeJong's learning curve model
x_0 = initial point of the evaluation of performance on DeJong's learning curve.

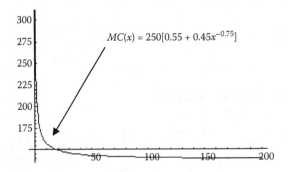

$$MC(x) = 250[0.55 + 0.45x^{-0.75}]$$

FIGURE 8.8   DeJong's learning formula with $M = 0.55$ and $b = -0.75$.

LEVY'S ADAPTATION FUNCTION

Recognizing that the log-linear model does not account for the leveling off of production rate and the factors that may influence learning, Levy (1965) presented the following learning cost function:

$$MC(x) = \left[\frac{1}{\beta} - \left(\frac{1}{\beta} - \frac{x^{-b}}{C_1} \right) k^{-kx} \right]^{-1},$$

where:
β = production index for the first unit
k = constant used to flatten the learning curve for large values of x.

The flattening constant, k, forces the curve to reach a plateau instead of continuing to decrease or turning in the upward direction. Figure 8.9 shows alternate profiles

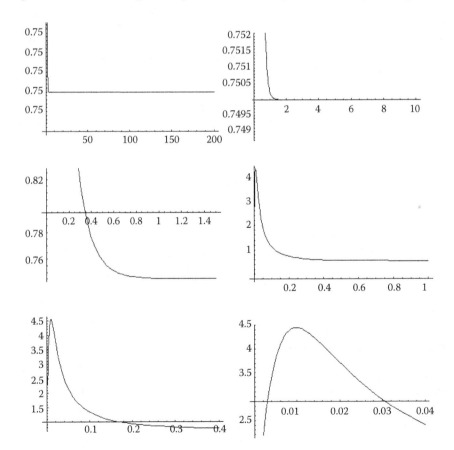

FIGURE 8.9 Profiles of Levy's adaptation over different production ranges.

of Levy's adaptation function over different ranges of production for $\beta = 0.75$, $k = 5$, $C_1 = 250$, and $b = 0.75$. The profiles are arranged in an increasing order of ranges of operating intervals. The half-life expression for Levy's learning model is a complex non-linear expression derived as shown below:

$$(1/\beta - x_{1/2}^{-b}/C_1)k^{-kx_{1/2}} = 1/\beta - 2\left[1/\beta - (1/\beta - x_0^{-b}/C_1)k^{-kx_0}\right],$$

where:
   $x_{1/2}$ = half-life expression for Levy's learning curve model
   x_0 = initial point of the evaluation of performance on Levy's learning curve.

Knowledge of some of the parameters of the model is needed in order to solve for the half-life as a closed-form expression.

GLOVER'S LEARNING MODEL

Glover's (1966) learning formula is a learning curve model that incorporates a work commencement factor. The model is based on a bottom-up approach that uses individual worker learning results as the basis for plant-wide learning curve standards. The functional form of the model is expressed as:

$$\sum_{i=1}^{n} y_i + a = C_1 \left(\sum_{i=1}^{n} x_i\right)^m,$$

where:
   y_i = elapsed time or cumulative quantity
   x_i = cumulative quantity or elapsed time
   a = commencement factor
   n = index of the curve (usually $1 + b$)
   m = model parameter.

This is a complex expression for which a half-life expression is not easily computable. We defer the half-life analysis of Levy's learning curve model for further research by interested readers.

PEGELS' EXPONENTIAL FUNCTION

Pegels (1976) presented an alternate algebraic function for the learning curve. His model, a form of an exponential function of marginal cost, is represented as:

$$MC(x) = \alpha a^{x-1} + \beta,$$

where α, β, and a are parameters based on empirical data analysis. The total cost of producing x units is derived from the marginal cost as follows:

$$TC(x) = \int \left(\alpha\, a^{x-1} + \beta\right) dx = \frac{\alpha\, a^{x-1}}{\ln(a)} + \beta x + c,$$

where c is a constant to be derived after the other parameters have been found. The constant can be found by letting the marginal cost, total cost, and average cost of the first unit all be equal. That is, $MC_1 = TC_1 = AC_1$, which yields:

$$c = \alpha - \frac{\alpha}{\ln(a)}.$$

The model assumes that the marginal cost of the first unit is known. Thus,

$$MC_1 = \alpha + \beta = y_0.$$

Pegels' also presented another mathematical expression for the total labor cost in start-up curves, which is expressed as:

$$TC(x) = \frac{a}{1-b}\, x^{1-b},$$

where:
  $x =$ cumulative number of units produced
  $a, b =$ empirically determined parameters.

The expressions for marginal cost, average cost, and unit cost can be derived, as shown earlier, for other models. Figure 8.10 shows alternate profiles of Pegels' exponential function for $\alpha = 0.5$, $\beta = 125$, and $a = 1.2$. The functions seem to suggest an unstable process, probably because the hypothetical parameters are incongruent with the empirical range for which the model was developed.

Using the total cost expression, $TC(x)$, we derive the expression for the half-life of Pegels' learning curve model to be as shown below:

$$x_{1/2} = (1/2)^{-1/(1-b)}\, x_0.$$

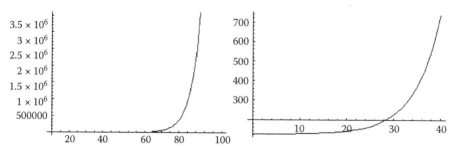

FIGURE 8.10   Alternate forms of Pegels' exponential function for $\alpha = 0.5$, $\beta = 125$, and $a = 1.2$.

KNECHT'S UPTURN MODEL

Knecht (1974) presents a modification to the functional form of the learning curve in order to analytically express the observed divergence of actual costs from those predicted by learning curve theory when the number of units produced exceeds 200. This permits the consideration of non-constant slopes for the learning curve model. If UC_x is defined as the unit cost of the xth unit, then it approaches 0 asymptotically as x increases. To avoid a zero limit unit cost, the basic functional form is modified. In the continuous case, the formula for cumulative average costs is derived as:

$$C(x) = \int_0^x C_1 z^b dz = \frac{C_1 x^{b+1}}{(1+b)}.$$

This cumulative cost also approaches zero as x goes to infinity. Knecht alters the expression for the cumulative curve to allow for an upturn in the learning curve at large cumulative production levels. He suggested the functional form below:

$$C(x) = C_1 x^{-b} e^{cx},$$

where c is a second constant. Differentiating the modified cumulative average cost expression gives the unit cost of the xth unit as shown below. Figure 8.11 shows the cumulative average cost plot of Knecht's function for $C_1 = 250$, $b = 0.25$, and $c = 0.25$.

$$UC(x) = \frac{d}{dx}\left[C_1 x^{-b} e^{cx}\right] = C_1 x^{-b} e^{cx}\left(c + \frac{-b}{x}\right).$$

The half-life expression for Knecht's learning model turns out to be a non-linear complex function as shown below:

$$x_{1/2} e^{-cx_{1/2}/b} = (1/2)^{-1/b} e^{-cx_0/b} x_0,$$

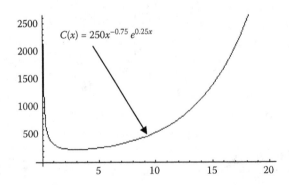

FIGURE 8.11 Knecht's cumulative average cost function for $c_1 = 250$, $b = 0.25$, and $c = 0.25$.

where:

$x_{1/2}$ = half-life expression for Knecht's learning curve model

x_0 = initial point of the evaluation of performance on Knecht's learning curve.

Given that x_0 is known, iterative, interpolation, or numerical methods may be needed to solve for the half-life value.

YELLE'S COMBINED PRODUCT LEARNING CURVE

Yelle (1976) proposed a learning curve model for products by aggregating and extrapolating the individual learning curve of the operations making up a product on a log-linear plot. The model is expressed as shown below:

$$C(x) = k_1 x_1^{-b_1} + k_2 x_2^{-b_2} + \cdots + k_n x_n^{-b_n},$$

where:

$C(x)$ = cost of producing the xth unit of the product

n = number of operations making up the product

$k_i x_i^{-bi}$ = learning curve for ith operation.

The deficiency of Knecht's model above is that a product-specific learning curve seems to be a more reasonable model than an integrated product curve. For example, an aggregated learning curve with a 96.6% learning rate obtained from individual learning curves with the respective learning rates of 80%, 70%, 85%, 80%, and 85% does not appear to represent reality. If this type of composite improvement is possible, then one can always improve the learning rate for any operation by decomposing it into smaller integrated operations. The additive and multiplicative approaches of reliability functions support the conclusion of impracticality of Knecht's integrated model.

CONTEMPORARY LEARNING–FORGETTING CURVES

Several factors can, in practice, influence the learning rate. A better understanding of the profiles of learning curves can help in developing forgetting intervention programs and for assessing the sustainability of learning. For example, shifting from learning one operational process to another can influence the half-life profile of the original learning curve. Important questions that half-life analysis can address include the following:

1. What factors influence learning retention and for how long?
2. What factors foster forgetting and at what rate?
3. What joint effects exist to determine the overall learning profile for worker performance and productivity?
4. What is the profile and rate of decline of the forgetting curve?

The issues related to the impact of forgetting in performance and productivity analysis are brought to the forefront by Badiru (1994, 1995) and all the references therein. Figure 8.12 shows some of the possible profiles of the forgetting curve. The impact of forgetting can occur continuously over time or discretely over bounded intervals of time. Also, forgetting can occur as random interruptions in the system performance or as scheduled breaks (Anderlohr 1969). The profile of the forgetting curve and its mode of occurrence can influence the half-life measure. This is further evidence that the computation of half-life can help distinguish between learning curves, particularly if a forgetting component is involved.

Recent literature has further highlighted the need to account for the impact of forgetting. Because of the recognition of the diminishing impacts of forgetting curves, these curves are very amenable to the application of the half-life concept. Jaber and Sikström (2004) present the computational comparisons of three learning and forgetting curves based on previous models available in the literature:

1. LFCM (learn–forget curve model) provided by Jaber and Bonney (1996)
2. RC (recency model) provided by Nembhard and Uzumeri (2000)
3. PID (power integration diffusion) provided by Sikström and Jaber (2002).

All three models assume that learning conforms to the original log-linear model presented by Wright (1936) and denoted here as Wright's learning curve (WLC):

$$T(x) = T_1 x^{-b},$$

where $T(x)$ is the time to produce the xth unit, T_1 is the time to produce the first unit, x is the cumulative production unit, and b is the learning curve constant ($0 < b < 1$).

JABER–BONNEY LEARN–FORGET CURVE MODEL (LFCM)

Jaber and Bonney (1996) present the learn–forget curve model (LFCM), which suggests that the forgetting curve exponent could be computed as:

$$f_i = \frac{b(1-b)\log(u_i + n_i)}{\log(1 + D/t(u_i + n_i))},$$

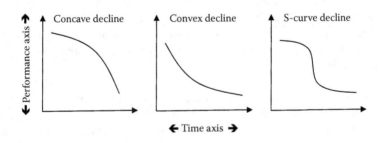

FIGURE 8.12   Alternate profiles declining impact of forgetting.

where $0 \le f_i \le 1$, n_i is the number of units produced in cycle i up to the point of interruption, D is the break time for which total forgetting occurs, and u_i is the number of units producible due to retained learning at the beginning of cycle i from producing x_{i-1} in previous $i-1$ cycles. Note that in i production cycles, there are $i-1$ production breaks, where $x_{i-1} = \sum_{j=1}^{i-1} n_j$ and $0 < u_i < x_{i-1}$. That is, if the learning process is interrupted at the time of length D, then the performance reverts to a threshold value, usually equivalent to T_1. Denote $t(u_i + n_i)$ as the time to produce $u_i + n_i$ units (equivalent units of cumulative production accumulated by the end of cycle i), and b is the learning curve constant. Then, $t(u_i + n_i)$ is computed as presented by Jaber and Sikström (2004):

$$t\left(u_i + n_i\right) = \sum_{x=1}^{n_i} T_1\left(u_i + x\right)^{-b} \cong \int_0^{u_i + n_i} T_1 x^{-b}dx = \frac{T_1}{1-b}\left(u_i + n_i\right)^{1-b}.$$

The above function is plotted in Figure 8.13 for $t(u_i + n_i)$, for $T_1 = 25$, and for $b = 0.65$.

The number of units produced at the beginning of cycle $i+1$ is given from Jaber and Bonney (1996) as:

$$u_{i+1} = \left(u_i + n_i\right)^{(1+f_i/b)y_i^{-f_i/b}},$$

where $u_1 = 0$, and y_i is the number of units that would have been accumulated, if the production was not ceased for d_i units of time, y_i is computed as:

$$y_i = \left\{\frac{1-b}{T_1}\left[t\left(u_i + n_i\right) + d_i\right]\right\}^{1/(1-b)}.$$

When total forgetting occurs, we have $u_{i+1} = 0$. However, $u_{i+1} \to 0$ as $y_i \to +\infty$; or alternatively, as $d_i \to +\infty$, where all the other parameters are of non-zero positive values. Thus, we deduce that total forgetting occurs only when d_i holds a very large

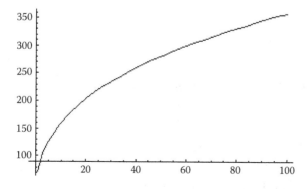

FIGURE 8.13   Plot of Jaber–Sikström's learn–forget model.

value. This does not necessarily contradict the assumption of finite value of D to which total forgetting occurs. By doing so, $u_{i+1} < 1$, when $d_i = D$, and it flattens out at zero for increasing values of $d_i > D$. Anderlohr (1969), McKenna and Glendon (1985), and Globerson et al. (1998) reported findings of the impact of production breaks through empirical studies. As reported by Jaber and Sikström (2004), the intercept of the forgetting curve could be determined as:

$$\hat{T}_{1i} = T_1 \left(u_i + n_i \right)^{-(b+f_i)}.$$

The time to produce the first unit in cycle i could then be predicted as:

$$\tilde{T}_{1i}^{\text{LFCM}} = T_1 \left(u_i + n_i \right)^{-b}.$$

Nembhard–Uzumeri Recency (RC) Model

The recency (RC) model presented by Nembhard and Uzumeri (2000) has the capability of capturing multiple breaks. Nembhard and Uzumeri (2000) modified the three hyperbolic learning functions of Mazur and Hastie (1978) by introducing the measure—"recency" of experiential learning, R. For each unit of cumulative production, x, Nembhard and Uzumeri (2000) determined the corresponding recency measure, R_x, by computing the ratio of the average elapsed time to the elapsed time of the most recent unit produced. Nembhard and Osothsilp (2001) suggested that R_x could be computed as:

$$R_x = 2 \frac{\sum_{i=1}^{x} \left(t_i - t_0 \right)}{x \left(t_x - t_0 \right)},$$

where x is the accumulated number of produced units, t_x is the time when units x are produced, t_0 is the time when the first unit is produced, t_i is the time when unit i is produced, and $R_x \in (1,2)$. The performance of the first unit after a break could be computed as:

$$\tilde{T}_{1i}^{\text{RC}} = T_1 \left(x R_x^{\alpha} \right)^{-b},$$

where α is a fitted parameter that represents the degree to which the individual forgets the task.

Sikström–Jaber Power Integration Diffusion Model

The power integration diffusion (PID) model presented by Sikström and Jaber (2002) advocates that each time a task is performed, a memory trace is formed. The strength of this trace decays as a power function over time. For identical repetitions of a task, an aggregated memory trace could be found by integrating the strength of the memory trace over the time interval of the repeated task. The integral of the power function memory trace is a power function. Therefore, the memory strength of an uninterrupted set of repetitions can be described as the difference between a power function of the retention interval at the start of the repetitions and a power function of the retention interval at the end of repetitions. The time it takes to perform a task is determined by "a diffusion" process where the strength of the memory constitutes the signal. To simplify the calculation, the noise in the diffusion process is disregarded and the time to perform a task is the inverse of the aggregated memory strength plus a constant reflecting the start time of the diffusion process. The strength of the memory trace follows a power function of the retention interval since training is given. That is, the strength of a memory trace (at which t time units have passed between learning and forgetting) encoded during a short time interval (dt) is:

$$S'(t) = S_0 t^{-a} dt,$$

where a is the forgetting parameter, $a \in (0,1)$, S_0 is a scaling parameter >0 (to be compared with the parameter in other models that represents the time to produce the first unit). The strength of a memory trace encoded for an extended time period is $S(t_{e,1}, t_{e,2})$, where $t_{e,1}$ time units passed since the start of encoding of unit e and $t_{e,2}$ time units passed since the end of encoding of unit e and $t_{e,1} > t_{e,2}$. This memory strength can be calculated by the integral over the time of encoding.

$$S = (t_{e,1}, t_{e,2}) = \int_{t_{e,1}}^{t_{e,2}} S'(t) dt = \frac{S_0}{1-a} \left[t_{e,2}^{1-a} - t_{e,1}^{1-a} \right].$$

The profile of the above function is plotted in Figure 8.14 for hypothetical values of $S_0 = 20$ and $a = 0.35$.

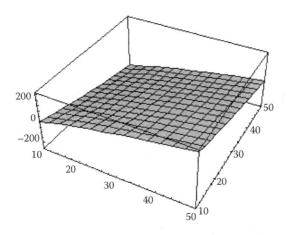

FIGURE 8.14   Plot of Jaber-Sikström's power integration diffusion model.

The strength of the memory trace following encoding during N time intervals is the sum over these intervals, and it is determined as:

$$S = \left(t_{e,1}, t_{e,2}\right) = \frac{S_0}{1-a} \sum_{e=1}^{N} \left[t_{e,2}^{1-a} - t_{e,1}^{1-a} \right].$$

The time to produce a unit is calculated with a diffusion model where the strength of the memory trace is conceived of as a signal. For simplification, the noise in the diffusion process is set to zero. The time to produce a unit is the inverse of the memory strength. The start time of the diffusion process constitutes a constant (t_0) that is added to the total time to produce a unit:

$$T\left(t_r\right) = S\left(t_{e,1}, t_{e,2}\right)^{-1} + t_0$$

$$= \frac{1-a}{S_0} \left\{ \sum_{e=1}^{N} \left[t_{e,2}^{1-a} - t_{e,1}^{1-a} \right] \right\}^{-1} + t_0$$

$$= S_0' \left\{ \sum_{e=1}^{N} \left[t_{e,1}^{a'} - t_{e,1}^{a'} \right] \right\}^{-1} + t_0,$$

where $S_0' = [(1 - a)/S_0]$, which is a rescaling of S_0, and $a' = 1 - a$, $a' \in (0,1)$ is a rescaling of a. The rescaling of the parameters is introduced for convenience to simplify the final expression. Sikström and Jaber (2002) showed that without production breaks, the predictions of PID are a good approximation of Wright's learning curve model. That is, the predictions are identical, given that the accumulated time to produce a unit can be approximated as:

$$t\left(x\right) = T_1 \sum_{n=1}^{x} n^{-b} \approx T_1 \int_0^x n^{-b} = T_1 x^{1-b} / \left(1-b\right).$$

Thus, in this approximation, Wright's original learning curve model is a special case of PID where:

$$T_x = dt\left(x\right)/dx = \left\{ \left[\left(1+a'\right) S_0' \right]^{1/(1+a')} \right\} \left(x^{-a'/(1+a')} \right) / \left(1+a'\right)$$

$$= T_1 x^{-b} \quad \text{and} \quad t_0 = 0,$$

from which Jaber and Sikström (2004) deduce the following relationships, between T_1, a and S_0, and a and b, respectively, as:

$$T_1 = \frac{\left[\left(1+a'\right)S_0'\right]^{1/(1+a')}}{1+a'}, \quad \text{where } x = 1,$$

and

$$b = \frac{a'}{1+a'}, \text{ for every } x > 1,$$

where $0 < b < 1/2$ for $0 < a' < 1$, with $a' = 1 - a$ and $S_0' = (1 - a)/S_0$.

The early studies of learning curves did not address the forgetting function. In this case, the contemporary functions that address the impact of forgetting tend to be more robust and more representative of actual production scenarios. These models can be further enhanced by carrying out a half-life analysis on them.

POTENTIAL HALF-LIFE APPLICATION TO HYPERBOLIC DECLINE CURVES

Over the years, the decline curve technique has been extensively used by the oil industry to evaluate future oil and gas predictions. These predictions are used as the basis for economic analysis to support development, property sale or purchase, industrial loan provisions, and also to determine if a secondary recovery project should be carried out. It is expected that the profile of hyperbolic decline curves can be adapted for an application to learning curve analysis. The graphical solution of the hyperbolic equation is through the use of a log-log paper, which sometimes provides a straight line that can be extrapolated for a useful length of time to predict future production levels. This technique, however, sometimes fails to produce the straight line needed for the extrapolation required for some production scenarios. Furthermore, the graphical method usually involves some manipulation of data, such as shifting, correcting and/or adjusting scales, which eventually introduce bias into the actual data.

In order to avoid the noted graphical problems of hyperbolic decline curves and to accurately predict future performance of a producing well, a non-linear least-squares technique is often considered. This method does not require any straight line extrapolation for future predictions. The mathematical analysis proceeds as follows. The general hyperbolic decline equation for oil production rate (q) as a function of time (t) can be represented as:

$$q(t) = q_0 \left(1 + m D_0 t\right)^{-1/m}$$

$$0 < m < 1,$$

where:
  $q(t) =$ oil production at time t
  $q_0 =$ initial oil production

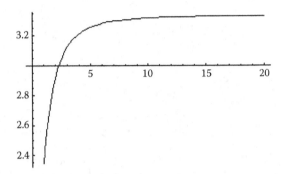

FIGURE 8.15   Plot of hyperbolic decline curve for cumulative production over time.

D_0 = initial decline
m = decline exponent.

Also, the cumulative oil production at time t, $Q(t)$ can be written as:

$$Q(t) = \frac{q_0}{(m-1)D_0}\left[(1+mD_0t)^{(m-1)/m} - 1\right],$$

where $Q(t)$ is the cumulative production as of time t.

By combining the above equations and performing some algebraic manipulations, it can be shown that:

$$q(t)^{1-m} = q_0^{1-m} + (m-1)D_0q_0^{-m}Q(t),$$

which shows that the production at time t is a non-linear function of its cumulative production level. By rewriting the equations in terms of cumulative production, we have:

$$Q(t) = \frac{q_0}{(1-m)D_0} + q(t)^{1-m}\frac{q_0^m}{(m-1)D_0}.$$

The above function is plotted in Figure 8.15. It is evident that the model can be investigated in terms of conventional learning curve techniques, forgetting decline curve, and half-life analysis in a procedure similar to the techniques presented earlier in this chapter.

CONCLUSIONS

Degradation of performance occurs naturally, either due to internal processes or due to externally imposed events, such as extended production breaks. For productivity assessment purposes, it may be of interest to determine the length of time it takes for a production metric to decay to half of its original magnitude. For example, for a career planning strategy, one may be interested in how long it takes for skill sets to degrade by half in relation to the current technological needs of the workplace. The half-life phenomenon may be due to intrinsic factors, such as forgetting, or due to external factors, such as a shift in labor requirements. Half-life analysis can have application in intervention programs designed to achieve the reinforcement of learning. It can also have application for assessing the sustainability of skills acquired through training programs. Further research on the theory of half-life of learning curves should be directed to topics such as the following:

- Half-life interpretations
- Learning reinforcement program
- Forgetting intervention and sustainability programs.

In addition to the predictive benefits of half-life expressions, they also reveal the ad-hoc nature of some of the classical learning curve models that have been presented in the literature. We recommend that future efforts to develop learning curve models should also attempt to develop the corresponding half-life expressions to provide full operating characteristics of the models. Readers are encouraged to explore the half-life analyses of other learning curve models not covered in this chapter.

REFERENCES

Anderlohr, G., 1969. What production breaks cost. *Industrial Engineering* 20(9): 34–36.

Badiru, A.B., 1992. Computational Survey of univariate and multivariate learning curve models. *IEEE Transactions on Engineering Management* 39(2): 176–188.

Badiru, A.B., 1994. Multifactor learning and forgetting models for productivity and performance analysis. *International Journal of Human Factors in Manufacturing* 4(1): 37–54.

Badiru, A.B., 1995. Multivariate analysis of the effect of learning and forgetting on product quality. *International Journal of Production Research* 33(3): 777–794.

Badiru, A.B., and Ijaduola, A., 2009. Half-life theory of learning curves. In *Handbook of military industrial engineering*, eds. A.B. Badiru and M.U. Thomas, 33-1–33-28. Boca Raton: CRC Press/Taylor and Francis.

Belkaoui, A., 1976. Costing through learning. *Cost and Management* 50(3): 36–40.

Belkaoui, A., 1986. *The learning curve*. Westport: Quorum Books.

Camm, J.D., Evans, J.R., and Womer, N.K., 1987. The unit learning curve approximation of total costs. *Computers and Industrial Engineering* 12(3): 205–213.

Globerson, S., Nahumi, A., and Ellis, S., 1998. Rate of forgetting for motor and cognitive tasks. *International Journal of Cognitive Ergonomics* 2(3): 181–191.

Glover, J.H., 1966. Manufacturing progress functions: An alternative model and its comparison with existing functions. *International Journal of Production Research* 4(4): 279–300.

Jaber, M.Y., and Bonney, M., 1996. Production breaks and the learning curve: The forgetting phenomenon. *Applied Mathematics Modeling* 20(2): 162–169.

Jaber, M.Y., and Bonney, M., 2003. Lot sizing with learning and forgetting in set-ups and in product quality. *International Journal of Production Economics* 83(1): 95–111.

Jaber, M.Y., and Bonney, M., 2007. Economic manufacture quantity (EMQ) model with lot size dependent learning and forgetting rates. *International Journal of Production Economics* 108(1–2): 359–367.

Jaber, M.Y., and Guiffrida, A., 2008. Learning curves for imperfect production processes with reworks and process restoration interruptions. *European Journal of Operational Research* 189(1): 93–104.

Jaber, M.Y., Kher, H.V., and Davis, D., 2003. Countering forgetting through training and deployment. *International Journal of Production Economics* 85(1): 33–46.

Jaber, M.Y., and Sikström, S., 2004. A numerical comparison of three potential learning and forgetting models. *International Journal of Production Economics* 92(3): 281–294.

Knecht, G., 1974. Costing, technological growth and generalized learning curves. *Operations Research Quarterly* 25(3): 487–491.

Levy, F., 1965. Adaptation in the production process. *Management Science* 11(6): 136–154.

Liao, W.M., 1979. Effects of learning on resource allocation decisions. *Decision Sciences* 10(1):116–125.

Mazur, J.E., and Hastie, R., 1978. Learning as accumulation: A re-examination of the learning curve. *Psychological Bulletin* 85(6): 1256–1274.

McIntyre, E., 1977. Cost-volume-profit analysis adjusted for learning. *Management Science* 24(2): 149–160.

McKenna, S.P., and Glendon, A.I., 1985. Occupational first aid training: Decay in cardio-pulmonary resuscitation (CPR) skills. *Journal of Occupational Psychology* 58(2): 109–117.

Nanda, R., 1979. Using learning curves in integration of production resources. *Proceedings of 1979 IIE Fall Conference*, 376–380.

Nembhard, D.A., and Osothsilp, N., 2001. An empirical comparison of forgetting models. *IEEE Transactions on Engineering Management* 48(3): 283–291.

Nembhard, D.A., and Uzumeri, M.V., 2000. Experiential learning and forgetting for manual and cognitive tasks. *International Journal of Industrial Ergonomics* 25(3): 315–326.

Pegels, C., 1969. On start-up or learning curves: An expanded view. *AIIE Transactions* 1(3): 216–222.

Richardson, W.J., 1978. Use of learning curves to set goals and monitor progress in cost reduction programs. *Proceedings of 1978 IIE Spring Conference*, 235–239. Norcross, GA: Institute of Industrial Engineers.

Sikström, S., and Jaber, M.Y., 2002. The power integration diffusion (PID) model for production breaks. *Journal of Experimental Psychology: Applied* 8(2): 118–126.

Smith, J., 1989. *Learning curve for cost control*. Norcross, GA: Industrial Engineering and Management Press.

Smunt, T.L., 1986. A comparison of learning curve analysis and moving average ratio analysis for detailed operational planning. *Decision Sciences* 17(4): 475–495.

Sule, D.R., 1978. The effect of alternate periods of learning and forgetting on economic manu-facturing quantity. *AIIE Transactions* 10(3): 338–343.

Towill, D.R., and Cherrington, J.E., 1994. Learning curve models for predicting the per-formance of advanced manufacturing technology. *International Journal of Advanced Manufacturing Technology* 9(3):195–203.

Towill, D.R., and Kaloo, U., 1978. Productivity drift in extended learning curves. *Omega* 6(4): 295–304.

Womer, N.K.,1979. Learning curves, production rate, and program costs. *Management Science* 25(4): 312–219.

Womer, N.K., 1981. Some propositions on cost functions. *Southern Economic Journal* 47(4): 1111–1119.

Womer, N.K., 1984. Estimating learning curves from aggregate monthly data. *Management Science* 30(8): 982–992.

Womer, N.K., and Gulledge, T.R., Jr., 1983. A dynamic cost function for an airframe produc-tion program. *Engineering Costs and Production Economics* 7(3): 213–227.

Wright, T.P., 1936. Factors affecting the cost of airplanes. *Journal of the Aeronautical Sciences* 3(2): 122–128.

Yelle, L.E., 1979. The learning curve: Historical review and comprehensive survey. *Decision Sciences* 10(2): 302–328.

Yelle, L.E., 1980. Industrial life cycles and learning curves: Interaction of marketing and production. *Industrial Marketing Management* 9(2): 311–318.

Yelle, L.E., 1983. Adding life cycles to learning curves. *Long Range Planning* 16(6): 82–87.

9 Influence of Breaks in Learning on Forgetting Curves

Sverker Sikström, Mohamad Y. Jaber, and W. Patrick Neumann

CONTENTS

INTRODUCTION

Forgetting is one of the most fundamental aspects of cognitive performance. Typically, forgetting is studied across time, where performance on a memory test plotted as a function of time is called a "forgetting curve." This forgetting is essential to all forms of behavioral studies, because it greatly influences how much information can be accessed. On very short timescales, most, if not all, information is retained in the cognitive system, where this information is rapidly lost over time with a fraction of the original memory maintained over an extended period. These memories are maintained in different memory systems that span across various timescales, are related to different types of interference, and are possibly based on different neurophysiological substrates. On very short timescales, memory is maintained in a modality-specific perceptual system that holds the incoming sensory information so that it can be processed and summarized by the cognitive systems. The details of these sensory systems depend on the modality, where, for example, the visual system maintains the information more briefly than the auditory system (Martin and Jones 1979). In fact, the visual-sensory system holds information for so short a period of time that the content typically disappears before it can be efficiently reported by the subjects. Memories maintained for periods longer than the sensory memory are typically divided into short-term and long-term memory, where the retained information is increasingly based on semantic or episodic aspects of the studied material. Short-term memory and the similar concept of working memory are functionally seen as the memory systems required for processing and manipulating information that is directly related to the current task. The working memory can be further divided into systems maintaining visual/auditory information and a more central executive component. The visual and auditory systems may be seen as slave systems to the central executive system, where higher order and controlled processes occur, includ-ing elaboration of the information carried in the slave systems (Baddeley and Hitch 1974). Memories in the working memory are maintained for time periods measured in seconds, whereas the consolidation or the elaboration of memory promotes the storing of memories in a persistent long-term memory system, where information is maintained for hours, days, and years.

Perhaps the most important factor influencing forgetting is time. The most significant effect is that the more time that has elapsed since encoding, the less well the material learnt can be retrieved. However, a few exceptions to this rule exist, as will be elaborated on later. Another factor is the rate of forgetting, which depends largely on the age of the memory: an older memory is better maintained than a younger one. This empirical phenomenon has been described by Jost's law (1897), which suggests that two memories of equal strength have the same likelihood of successful retrieval. The law also indicates that younger memories are forgotten faster than older memo-ries. Other studies have also observed that older memories are less likely to be lost than newer memories (e.g., Ribot 1881; Squire et al. 2001). A common and impor-tant theoretical implication of these studies, as well as others in the literature, is that memories require time to consolidate in order to become persistent against loss.

Jost's law has important implications for how forgetting can be summarized. If the rate of forgetting is less for older than for younger memories, then this seems to rule out the possibility that a certain amount of the memory is forgotten at each time interval—or stated in mathematical terms, that forgetting does not follow an expo-nential function. However, before accepting this conclusion another possibility needs to be ruled out—namely, that the strengthening of memories decreases the rate of forgetting. That is, it could be the case that memories are forgotten at an exponential rate, but that older memories (possibly due to more efficient encoding) decay more slowly than younger (and weaker) memories. This hypothesis was investigated by Slamecka and McElree (1983). They studied forgetting following either a strong or a weak initial learning. It was found that initial learning neither influenced the shape of the forgetting curve, nor the rate of forgetting. This suggests that the forgetting curves were reasonably parallel across the retention interval, which is defined as the time from the last presentation of the material (the last trial) to the test. The rate of decay also decreased over the retention interval, so that 36% of the facts learned were lost on the first day, whereas the average daily rate of the subsequent five days decreased to 11.5%. This data support the suggestion that forgetting cannot be described by a simple exponential decay function.

Given that the rate of forgetting decreases with the age of the memory, a power function seems to be a good candidate for explaining forgetting data. Indeed, a number of empirical studies have found forgetting curves that are well described by power functions (Anderson and Schooler 1991; Wickelgren 1974, 1977; Wixted and Ebbesen 1991). In an extensive meta-review of over 210 datasets published over the span of a century, Rubin and Wenzel (1996) tested over 100 forgetting functions and concluded that these data were consistent with the power function. However, based on this data they were not able to distinguish this candidate from three other functions—namely, the logarithmic function, the exponential in the square root of time, and the hyperbola in the square root of time—thus indicating that more precise data are required for identifying a single forgetting function.

Another interpretation of forgetting curves is that memories are stored in different memory systems, where each system decays at an exponential rate with widely different half-time constants, and that forgetting that is found in behavioral data is based on an aggregation of these curves. This interpretation is appealing for at least three reasons. First, exponential curves are naturally occurring in most simple systems. For example, it is consistent with the idea that memories decay with a certain probability on each time step, while under different circumstances forgetting might occur due to interference following learning of other materials. In both cases, these theories are most easily understood as exponential functions. Second, it is consistent with the view of memories being contained in different memory systems—e.g., perceptual memory, short- and long-term memory, working memory, and so forth—where each memory system could be described by an exponential function with different time constants. Third, it is consistent with the theory that there are several biological mechanisms, operating at different timescales, which are relevant for memory and that these will affect forgetting in somewhat different ways. Each of these learning mechanisms may be forgotten at exponential rates with widely different time constants, giving rise to an aggregated forgetting curve on the behavioral level that is well summarized as a power function.

Given the theory that the forgetting curve is an aggregation of underlying memory traces with exponential decay, a significant question to consider is how these underlying exponential traces could be combined so that we can obtain a forgetting curve that describes the empirical data at the behavioral level. This question was studied in detail by Sikström (1999, 2002). He found that power-function forgetting curves can be obtained on the behavioral levels, provided that the probability distribution of the half-time constants (in the underlying exponential functions) was drawn from a power-function distribution. The parameters describing the distribution of half-time constants directly determine the rate of forgetting on the behavioral power-function forgetting curve. Furthermore, power functions with any rate of forgetting can be obtained with this model. Sikström (2002) implemented a forgetting model in a Hopfield neural network and found that it nicely reproduced empirical power-function forgetting curves. Therefore, it can be clearly deduced from the literature that forgetting curves are well described by power functions.

BREAKS AND FORGETTING

An interesting aspect of memory function is how breaks of a repetitive task can influence the learning of that task. Data show that two conditions that are studied over the same length of time may be remembered quite differently depending on whether the learning is spread out over time or massed into one session. This phenomenon suggests that repetitive learning of the same material cannot directly be described as a sum of learning of each independent learning episode.

Extensive research has shown that superior performance is found with spaced learning compared with massed learning. Continuous repetition of a to-be-remembered item yields lower performance on a delayed test compared to if a short break is introduced between the subsequent repetitions. Uninterrupted learning of material is typically referred to as "massed learning," whereas learning that is interrupted by short or longer breaks is called "spaced" or "distributed." The benefits of spaced learning have been observed for verbal tasks such as list recall, paragraph recall, and paired associates learning (Janiszewski et al. 2003), and during skill learning; for example, mirror tracing or video game acquisition (Donovan and Radosevich 1999). The effect is typically large and well reproduced (Cepeda et al. 2006). It is worth noting that massed and spaced learning have their parallel in industrial engineering as massed and intermittent production (or production with breaks), respectively, where studying the effects of production breaks on the productivity of production systems has gained considerable attention (Jaber 2006). However, boundary conditions of the beneficial effects of spaced learning are known, where massed repetition can sometimes be better than spaced repetition. The retention interval, or the time between the second presentation and retrieval, needs to be large in comparison with the break time. For example, a lower performance in the spaced condition can be obtained on immediate testing in combination with a very long break time, compared to a massed condition immediately followed by a test. In this case the massed condition will effectively function as an extended presentation time, where the first presentation in the spaced presentation is largely forgotten. In particular, when the performance is plotted against the retention interval, immediate testing tends to give superior performance for massed presentation; whereas the standard finding of better performance for spaced repetition occurs as the retention interval increases.

An important theoretical and practical question to consider is how to schedule learning so that people retain as much of the knowledge that they have learnt in previous learning sessions as is possible. Empirical data available in the literature suggest that there exists a length of break that maximizes performance (Sikstrom and Jaber, in revision). This finding was also supported by Glenberg and Lehmann (1980). For example, if the retention interval (storage of learning over a break) is less than one minute, then an inter-stimuli interval (ISI) of less than one minute maximizes performance; however, if the retention interval is six months or more, then an inter-stimuli interval of at least one month maximizes performance.

The scheduling of the ISI is also important for learning. This has been studied by comparing fixed and expanding ISIs. In a fixed ISI the same time passes between each relearning, so that subjects know more about the material for each repetition. Expanding ISIs are scheduled so that the time between each stimuli increases, and thus function to counteract the aggregation of memory strength over repeated learning. However, the empirical results for fixed and expanding ISI are mixed. Some researchers have shown that expanding ISIs are beneficial for long-term retention (Hollingworth 1913; Kitson 1921; Landauer and Bjork 1978; Pyle 1913), whereas other have found the opposite effect (Cull 1995, 2000).

THEORIES OF MASSED AND SPACED LEARNING

Massed and spaced effects have implications for theories of memory. The inattention theory (Hintzman 1974) suggests that when the ISIs are short, subjects will pay less attention, because the item to be remembered is more familiar. This theory thus accounts for inferior performance in massed conditions because less attention is paid at the second presentation. By contrast, Sikström and Jaber (in revision) have suggested that encoding requires resources, and that these resources are consumed, or depleted, during repetitive learning. In massed conditions, the second presentation has fewer resources available, leading to a poorer performance. This depletion model was implemented in a computer model where a memory trace is formed at every time unit of encoding. The depletion of encoding resources diminishes the overall strength of these traces. Furthermore, all traces that are formed at encoding are summed to an aggregated memory strength, which determines the performance at retrieval. Sikström and Jaber (in revision) fitted this computational model to several different datasets, including motoric and verbal learning, with short- and long-retention intervals. They found that the model fit the data well, providing support for the suggested theory. This suggests that the depletion of encoding resources could be an important mechanism for accounting for the differences observed between spaced and massed learning.

Another theory describing learning performance is the context variability theory (Glenberg 1979; Melton 1970), which suggests that performance is highly dependent on the contextual overlap at encoding and at retrieval. Furthermore, a context drift over time is assumed, so that performance drops when the differences between the encoding and retrieval contexts increase. A spaced superiority effect is predicted in this context variability model because, as the spacing increases between the two presentations of the item to be encoded, the likelihood that the encoding context matches at least one of them also increases. Cepeda et al. (2006) simulated a version of the context variability theory and were able to reproduce several of the basic findings regarding massed and spaced effects. However, the basic version of the context variability model makes specific predictions that are not directly supported by data. In particular, it predicts that the probability of retrieving two words in a list increase with the spacing between the words in the list. This prediction has not been supported, suggesting a problem with the context variability theory (Bellezza et al. 1975).

The consolidation theory (Wickelgren 1972) suggests that memories are first encoded in a fragile state and then, as time passes, they change to a relatively more stable state. This process is called "consolidation." This theory proposes that the memory generated on the second delayed presentation inherits the consolidated state, and therefore is less prone to forgetting. Finally, if the retention interval is too long (e.g., a year) then there will be no initial memory trace left and retention will be poor due to the lack of consolidation of the memories.

BREAKS AND THE BRAIN

The possibility of measuring, or manipulating, states of the brain has added understanding to how breaks influence forgetting. Muellbacher et al. (2002) directly investigated consolidation in a skilled motor task by applying either direct or delayed interference to the brain. Subjects learned a motoric task and were tested again after a 15-minute break. In a condition where subjects had rested during this break, performance was well maintained. However, if the break was filled with a repetitive transcranial magnetic stimulation (rTMS; a treatment tool for various neurological conditions) applied to the motor cortex, it was found that performance dropped to the level prior to the initial learning. By contrast, if the same rTMS treatment were given following a delay of 6 hours then no loss in performance was found. This experiment clearly shows that the memory moves from a fragile to a stable representation during an interval ranging from 15 minutes to 6 hours. This susceptibility to interference may also be introduced by performing a similar task during the retention interval. Brashers-Krug et al. (1996) showed that learning skills similar to the original task interfered with the original learning; however, this interference only occurred during the first 4 hours following learning, whereas later skill learning did not.

These studies indicate that what happens during the retention interval is critical to final learning. Beyond external influence during this interval, the brain may also restructure the representation of memories as time passes, indicating that "forgetting" may be a much more active process than what had previously been thought. For example, studies using functional imaging of the brain have shown that, although the performance in skill learning may be unaffected during a 6-hour period, the brain areas supporting this task may change significantly. Following this 6-hour delay in the waking state, activity was higher than it had been (during the initial learning) in the premotor, parietal, and cerebellar regions (Shadmehr and Brashers-Krug 1997), indicating that memory is reorganized in the brain, which may not be directly evident by only looking at the behavioral data.

SLEEPING BREAKS

Humans spend one-third of their lifetimes sleeping, so any break longer than a day will most likely include sleep. The question of how sleep influences performance therefore becomes very relevant. A fundamental theory of forgetting is that memories are lost because we learn new material that "overwrites," or interferes with, previously learned memories. One of the first tests of this theory was to compare sleeping with wakefulness. Results showed that memories are spared more during sleeping than during the waking state (Jenkins and Dallenbach 1924).

A perhaps more intriguing finding is that not only are memories preserved, but they can actually be enhanced during a retention interval. This phenomenon has been most clearly found in motoric tasks following sleeping. Walker et al. (2002, 2003) let subjects perform a finger tapping motoric task at 12-hour intervals, which either included sleeping or did not. Within the first session, subjects typically increased their performance level by 60%. The performance was largely unaffected by a 12-hour retention in the waking state, whereas an increase of 20% in speed, and 39% in accuracy, was found if the retention interval included sleep. Furthermore, these effects were not dependent on the time of day that the first training session occurred. If this session was in the morning, then the first 12 hours of wakefulness did not influence performance, whereas the following 12 hours of sleep did. If the first training occurred in the evening, then the first 12 hours of sleep improved performance, while the second 12 hours of wakefulness did not. These remarkable effects cannot be directly accounted for by the interference stemming from motoric activities that naturally occurs during the wakeful state. In a control condition, subjects wore mittens during the retention interval, which effectively eliminated hand movements and minimized the possibilities of interference. However, this manipulation did not improve the performance in the waking condition.

A number of additional findings provide further clues of what is happening during sleep: First, improvements in skill performance do not necessarily have to follow sleep during the night. Fischer et al. (2002) found similar benefits in performance during shorter sleeps in the day. Second, the largest amount of improvement in skill learning occurs after the first night; however, additional nights did show a small additional benefit in performance. Third, doubling the amount of training does not seem to influence the amount of improvement by sleep. Fourth, prior to sleeping, each additional session provides additional learning, whereas following sleeping this effect is diminished. Fifth, the amount of training-induced learning does not correlate with the amount of sleep-induced learning, indicating that they tap into different processes (Walker et al. 2003).

Given that sleep has been shown to improve performance, one may then ask what type of process in sleep provides this effect. Sleep can be divided into several levels, where perhaps the most salient difference occurs between the more superficial REM sleep and the deeper non-REM sleep. Gais et al. (2000) investigated a visual-perceptual task and found that both types of sleep are essential for consolidation as well as the order of the sleep types. The sleep pattern early in the night, which is dominated by non-REM and slow-wave sleep (SWS), are important for the initiation of consolidation, whereas the pattern later in the night, dominated by REM sleep, causes additional consolidation, which only occurs if the sleep patterns earlier in the night are preserved. This indicates that memory consolidation is dependent on at least two sleep-dependent processes that should occur in the right order.

Maquet et al. (2003) investigated sleep-deprived subjects in a procedural visual-motoric task and found that they did not show any improvements over time in this task, while the sleeping control group did. Furthermore, the control group showed increased activation in the superior temporal sulcus at the later test compared with the first test, while the sleep-deprived group did not show this increase.

Sleep is not necessary for increasing performance. Breaks of 5–15 minutes have been found to increase performance on a visuomotor task, a phenomena referred to as "reminiscence." This effect is short-lived, lasting for only a few minutes, and falls back to baseline as the break period becomes longer (Denny 1951). In comparison, a 24-hour delay including sleep also showed increased performance, which did not fall back to baseline during continuous testing (Holland 1963). This suggests that the short- and long-term enhancement of breaks depends on different processes, where the short-term increase in performance may be a result of the release of inhibition following repetitive testing, whereas the long-term sleep-dependent effects are dependent on consolidation.

CONCLUSION

How learning interacts with breaks is a complex but fascinating topic. In this chapter, we have shown that forgetting can be modeled as a power function. This function is influenced by the length of the study time, the length of the retention interval, the amount of interference, and that learning can be modified by sleep. These influencing phenomena are important because learning and breaks from tasks are continuously made in everyday life, and a deeper understanding of this field may provide opportunities to plan our lives so that we can retain more of the information that we encounter. A combination of methods based on behavioral data, neurophysiological measures, and computational modeling will shed further light on this field.

In particular, these findings have important real-life applications in an everyday work environment. All work schedules include breaks. These breaks may be long— e.g., vacations and changes between different work tasks or even jobs—or of intermediate length, such as weekends or subsequent work days, which typically involve sleep. Finally, short breaks such as coffee breaks, or interruptions such as telephone calls, may also influence the speed at which we gain or lose skills and memories. An understanding of these phenomena may thus have important real-life implications.

ACKNOWLEDGMENTS

The authors wish to thank the Social Sciences and Engineering Research Council (SSHRC) of Canada-Standard Grant, and the Swedish Research Council for supporting this research. They also wish to thank Dr. Frank Russo from the Department of Psychology at Ryerson University for his valuable comments and suggestions.

REFERENCES

Anderson, J.R., and Schooler, L.J., 1991. Reflections of the environment in memory. *Psychological Science* 2(6): 396–408.

Baddeley, A.D., and Hitch, G.J., 1974. Working memory. In *Recent advances in learning and motivation,* ed. G. Bower, Vol. VIII, 47–90. London: Academic Press.

Bellezza, F.S., Winkler, H.B., and Andrasik, F., Jr., 1975. Encoding processes and the spacing effect. *Memory and Cognition* 3(4): 451–457.

Brashers-Krug, T., Shadmehr, R., and Bizzi, E., 1996. Consolidation in human motor memory. *Nature* 382(6588): 252–255.

Cepeda, N.J., Pashler, H., Vul, E., Wixted, J.T., and Rohrer, D., 2006. Distributed practice in verbal recall tasks: A review and quantitative synthesis. *Psychological Bulletin* 132(3): 354–380.

Cull, W.L., 1995. *How and when should information be restudied?* Chicago: Loyola University.

Cull, W. L., 2000. Untangling the benefits of multiple study opportunities and repeated testing for cued recall. *Applied Cognitive Psychology* 14(3): 215–235.

Denny, L.M., 1951. The shape of the post-rest performance curve for the continuous rotary pursuit task. *Motor Skills Research Exchange* 3: 103–105.

Donovan, J.J., and Radosevich, D.J., 1999. A meta-analytic review of the distribution of practice effect: Now you see it, now you don't. *Journal of Applied Psychology* 84(5): 795–805.

Fischer, S., Hallschmid, M., Elsner, A.L., and Born, J., 2002. Sleep forms memory for finger skills. *Proceedings of the National Academy of Sciences USA* 99(18): 11987–11991.

Gais, S., Plihal, W., Wagner, U., and Born, J., 2000. Early sleep triggers memory for early visual discrimination skills. *Nature Neuroscience* 3(12): 1335–1339.

Glenberg, A.M., 1979. Component-levels theory of the effects of spacing of repetitions on recall and recognition. *Memory and Cognition* 7(2): 95–112.

Glenberg, A.M., and Lehmann, T.S., 1980. Spacing repetitions over one week. *Memory and Cognition* 8(6): 528–538.

Hintzman, D.L., 1974. Theoretical implications of the spacing effect. In *Theories in cognitive psychology: The Loyola symposium,* ed. R. L. Solso, 77–97. Erlbaum: Potomac.

Holland, H.C., 1963. Massed practice and reactivation inhibition, reminiscence and disinhibition in the spiral after-effect. *British Journal of Psychology* 54: 261–272.

Hollingworth, H.L., 1913. *Advertising and selling: Principles of appeal and response.* New York: D. Appleton.

Jaber, M.Y., 2006. Learning and forgetting models and their applications. In *Handbook of Industrial and Systems Engineering*, ed. A.B. Badiru, 1–27 (Chapter 32). Baco Raton, FL: CRC Press-Taylor and Francis Group.

Janiszewski, C., Noel, H., and Sawyer, A.G., 2003. A meta-analysis of the spacing effect in verbal learning: Implications for research on advertising repetition and consumer memory. *Journal of Consumer Research* 30(1): 138–149.

Jenkins, J.G., and Dallenbach, K.M., 1924. Obliviscence during sleep and waking. *American Journal of Psychology* 35(4): 605–612.

Jost, A., 1897. Die Assoziationsfestigkeit in ihrer Abhängigkeit von der Verteilung der Wiederholungen [The strength of associations in their dependence on the distribution of repetitions]. *Zeitschrift fur Psychologie und Physiologie der Sinnesorgane* 16: 436–472.

Kitson, H.D., 1921. *The mind of the buyer: A psychology of selling.* New York: MacMillan.

Landauer, T. K., and Bjork, R.A., 1978. Optimum rehearsal patterns and name learning. In *Practical aspects of memory,* eds. P.E.M.M. Gruneberg, P.N. Morris, and R.N. Sykes, 625–632. London: Academic Press.

Maquet, P., Schwartz, S., Passingham, R., and Frith, C., 2003. Sleep-related consolidation of a visuomotor skill: Brain mechanisms as assessed by functional magnetic resonance imaging. *Journal of Neuroscience* 23(4): 1432–1440.

Martin, M., and Jones, G.V., 1979. Modality dependency of loss of recency in free recall. *Psychological Research* 40(3): 273–289.

Melton, A.W., 1970. The situation with respect to the spacing of repetitions and memory. *Journal of Verbal Learning and Verbal Behavior* 9(5): 596–606.

Muellbacher, W., Ziemann, U., Wissel, J., Dang, N., Kofler, M., Facchini, S., Boroojerdi, B., Poewe, W., and Hallett, M., 2002. Early consolidation in human primary motor cortex. *Nature* 415(6872): 640–644.

Pyle, W.H., 1913. Economical learning. *Journal of Educational Psychology* 4(3): 148–158.

Ribot, T., 1881. *Les maladies de la memoire* [Diseases of memory]. Paris: Germer Bailliere.

Rubin, D.C., and Wenzel, A. E., 1996. One hundred years of forgetting: A quantitative description of retention. *Psychological Review* 103(4): 734–760.

Shadmehr, R., and Brashers-Krug, T., 1997. Functional stages in the formation of human long-term motor memory. *Journal of Neuroscience* 17(1): 409–419.

Sikström, S., 1999. A connectionist model for frequency effects in recall and recognition. In *Connectionist Models in Cognitive Neuroscience: The 5th Neural Computation and Psychology Workshop,* eds. D. Heinke, G.W. Humphreys, and A. Olson, 112–123. London: Springer Verlag.

Sikström, S., 2002. Forgetting curves: Implications for connectionist models. *Cognitive Psychology* 45(1): 95–152.

Sikström, S., and Jaber, M.Y. (in revision). The depletion, power, integration, diffusion model of spaced and massed repetition.

Slamecka, N.J., and McElree, B., 1983. Normal forgetting of verbal lists as a function of their degree of learning. *Journal of Experimental Psychology: Learning, Memory, and Cognition* 9(3): 384–397.

Squire, L.R., Clark, R.E., and Knowlton, B.J., 2001. Retrograde amnesia. *Hippocampus* 11(1): 50–55.

Walker, M.P., Brakefield, T., Morgan, A., Hobson, J.A., and Stickgold, R., 2002. Practice with sleep makes perfect: Sleep-dependent motor skill learning. *Neuron* 35(1): 205–211.

Walker, M. P., Brakefield, T., Seidman, J., Morgan, A., Hobson, J.A., and Stickgold, R., 2003. Sleep and the time course of motor skill learning. *Learning and Memory* 10(4): 275–284.

Wickelgren, W.A., 1972. Trace resistance and the decay of long-term memory. *Journal of Mathematical Psychology* 9(4): 418–455.

Wickelgren, W.A., 1974. Single-trace fragility theory of memory dynamics. *Memory and Cognition* 2(4): 775–780.

Wickelgren, W.A., 1977. *Learning and memory*: Englewood Cliffs, NJ: Prentice Hall.

Wixted, J.T., and Ebbesen, E.B., 1991. On the form of forgetting. *Psychological Science* 2(6): 409–415.

10 Learning and Forgetting: Implications for Workforce Flexibility in AMT Environments

Corinne M. Karuppan

CONTENTS

INTRODUCTION

For decades, workforce flexibility has been heralded as a strategic weapon to respond to and anticipate changing demand in a competitive marketplace. Despite the overall positive effects of workforce flexibility, there has been growing evidence that its deployment should be measured rather than unbounded. Not only has empirical research shown that "total" flexibility may not be a worthy pursuit, but it has also underscored its limitations when tasks are complex. These limitations have been mostly attributed to efficiency losses when individuals are slow to learn new tasks or when they have to relearn tasks that they had forgotten. In the backdrop of the deep recession, two realities cannot be ignored: (1) the demand for flexible, knowledge workers is more likely to increase rather than subside, and (2) the pressure for high returns on training investments will continue to intensify.

The dual purpose of this chapter is therefore to understand the nature of performance decrements related to workforce flexibility in environments with complex,
advanced manufacturing technologies (AMTs), and to highlight the reliable mechanisms that will expand an individual's flexibility potential with minimum drawbacks. The first part of the chapter describes the concept of workforce flexibility. The
nature of the relationships between flexibility and performance are then explained
in the context of activation theory. These relationships subsume learning and forgetting phenomena. A review of the factors affecting these phenomena is presented,
and mental models emerge as intervening factors in the relationship between train-
ing/learning and forgetting. Theories from various fields of inquiry and empirical
validations undergird the formulation of a conceptual framework. Implications for
research and practice conclude the chapter.

WORKFORCE FLEXIBILITY

Workforce flexibility is a strategic capability that organizations deploy to address the
realities of increased market diversity, shorter product life cycles, and fierce global
competition. In a nutshell, flexibility is the ability to adapt quickly to change with
minimum penalty (Gerwin 1987; Koste and Malhotra 1999). Based on a thorough
review of the literature, Koste and Malhotra (1999) found that the construct was
mapped along four elements: range-number (RN), range-heterogeneity (RH), mobil-
ity (MOB), and uniformity (UN). In the context of workforce flexibility, RN refers
to the number of tasks or operations performed by an individual, the number of
technologies that he/she operates, and so on. It is similar to the notion of horizontal
loading, job enlargement, or intradepartmental flexibility, according to which individuals master a larger sequence of upstream and/or downstream activities related
to their job. RH involves the degree of differentiation among the tasks, operations,
and technologies. It has been construed as interdepartmental flexibility and is often
used interchangeably with cross-training. MOB evokes effortlessness in terms of
cost and time expended to move from one option to another, that is, learning or
relearning tasks. UN is the consistency of performance outcomes (quality, time, cost,
etc.) under different options. The first two elements of workforce flexibility deal
with worker deployment, whereas the other two capture the notions of efficiency and
effectiveness.

The benefits of a flexible workforce have been praised extensively in the literature.
Employees with a greater variety of skills are better equipped to respond to demand
fluctuations and load unbalances on the shop floor. Let us assume there are two
departments, A and B. If the workload decreases in department A and increases in
department B, workers can be pulled from department A and re-assigned to department B (RH flexibility), while the remaining workers in department A can perform
a wider range of tasks (RN flexibility) previously performed by others. Empirical
evidence certainly supports the advantages of a flexible workforce. A diverse skill
set enables the production of a more varied mix (e.g., Parthasarthy and Sethi 1992;
Upton 1997) and is also credited with reduced inventories, lower lead times, and
better due date performance (e.g., Allen 1963; Fryer 1976; Hogg et al. 1977; Felan
et al. 1993).

Despite an overwhelmingly favorable picture of flexibility, Treleven (1989) questioned the early practice of simulating labor flexibility RH (interdepartmental) with no loss of efficiency or with a temporary drop followed by 100% efficiency, essentially implying that the results might be "too good to be true." This concern was confirmed by the results of a field study in a printed circuit-board plant (Karuppan and Ganster 2004) where greater levels of labor flexibility RN increased the probability of producing a varied mix, whereas labor flexibility RH had the opposite effect.

Further curbing the enthusiasm for large increases in both types of labor flexibility deployment are the diminishing returns observed for complex jobs. In the simulation of a dual resource constrained (DRC) job shop, Malhotra et al. (1993) demonstrated that high levels of workforce flexibility (captured as interdepartmental flexibility, or flexibility RH) could worsen shop performance when learning was slower and initial processing times were higher. In their simulation modeling of an assembly line, McCreery and Krajewski (1999) operationalized worker flexibility as the number of different adjacent tasks that each worker was trained to perform (flexibility RN). Task complexity was characterized by slower learning, more rapid forgetting, a higher proportion of learning possible in multiples of standard time—(time to produce first unit/standard time)—and product as opposed to process learning. Their results indicated that greater flexibility led to a mixed performance in an environment characterized by high variety and high complexity, as increases in throughput came at the expense of worker utilization. Extending the research to work teams, McCreery et al. (2004) concluded that high-complexity environments called for restrictive flexibility deployment approaches. The above studies suggest that high levels of both labor flexibility RN and RH are undesirable when jobs are complex. However, they offered little or no theoretical rationale for their findings.

Conceptually, very few researchers have considered the possibility that high levels of "good" job characteristics—such as flexibility or task variety—might have adverse rather than beneficial effects. Traditionally, these good job characteristics have been those identified in Hackman and Oldham's (1975; 1976) job characteristics model (JCM): task variety, task identity, skill utilization, feedback, and autonomy. The JCM postulated a strictly positive-linear relationship between these job characteristics and psychological responses to the job. As an extension of their work, Champoux (1978) suggested that very high levels of such "good" job characteristics might cause a dysfunctional, overstimulation effect. His rationale was based on activation theory.

ACTIVATION THEORY

Lindsley's (1951) neurophysiological approach to activation theory—subsequently refined by others—is based on the notion of activation level; namely, the degree of neural activity in the reticular activation system. As Malmo (1959, 368) explained, "[…] the greater the cortical bombardment, the higher the activation. Furthermore, the relation between activation and behavioral efficiency is described by an inverted U-curve." From low activation to an optimal level, performance increases monotonically. Beyond the optimal point, the relationship becomes non-monotonic with further increases in activation, resulting in performance decrements. In other words, low activation levels should produce low levels of performance; moderate activation levels should produce optimal performance; and high activation levels should result in low performance. Malmo recommended that the term "moderate" be construed in relative rather than absolute terms: low < moderate < high, with moderate/optimal levels varying across tasks and individuals. The optimal level of activation for an individual allows the cerebral cortex to operate most efficiently, resulting in improved behavioral (e.g., response time) or cerebral performance. Stimulation from internal or external sources may cause deviations from this optimal level and therefore diminish performance (Gardner and Cummings 1988).

When applying this theory to job design, both low and high levels of task variety/labor flexibility would be harmful—the former due to understimulation, the latter due to overstimulation (French and Caplan 1973; Martin and Wall 1989; Schaubroeck and Ganster 1993; Xie and Johns 1995). At low levels of variety, a lack of alertness or activation hampers performance, and variety increases are initially matched by increases in performance. A leveling off effect then occurs as variety increases no longer yield the proportionate increases in job outcomes. Eventually, overstimulation occurs, and further increases in variety become counterproductive (Singh 1998) as a result of an impaired information-processing capability (Easterbrook 1959; Humphreys and Revelle 1984) and perceptions of lack of focus and work/information overload, which are well known stressors (Kahn and Byosiere 1992). These arguments corroborate a critical review of the literature indicating that "the case for job enlargement has been drastically overstated and over generalized" (Hulin and Blood 1968, 50).

Several empirical findings support the curvilinear relationship between good job characteristics and: (a) psychological responses (e.g., Champoux 1980, 1992); (b) stress (e.g., French et al. 1982; Xie and Johns 1995); and (c) task performance (Gardner 1986). Karuppan (2008) tested this relationship for labor flexibility RN and RH. She found that labor flexibility RN and RH exhibited a significant quadratic relationship with the worker's production of a varied mix—that is, mix flexibility. Plots of curve estimations supported the inverted U pattern. The leveling effect occurred more rapidly for the production of a more highly varied mix (mix flexibility RH) for which more extensive product learning was required.

Increasing flexibility essentially amounts to increases in learning within and across knowledge domains. When learning new tasks, there is competition for cognitive and physiological resources (Sünram-Lea et al. 2002). Divided attention at this encoding stage leads to reductions in memory performance (e.g., Craik et al. 1996). The number of tasks learned concurrently further taxes resources by reducing the amount of *time* available for encoding (Naveh-Benjamin 2001). Supporting this argument in industrial settings, Nembhard (2000) and Nembhard and Uzumeri (2000) found that rapid learning was associated with rapid forgetting. Jaber and Kher (2002) studied this phenomenon further. They showed that the forgetting slope increased with the learning rate b in the range ($0 < b \leq 0.5$), meaning that as the learning rate goes from 99% to about 70%, faster forgetting occurs.

Complexity raises activation levels beyond those induced by flexibility (Gardner and Cummings 1988). This combination strains information-processing capacity, hampering both encoding and retrieval. It also leads to heightened perceptions of quantitative (too much work), qualitative (too complex), and role (too many responsibilities) overload, which are detrimental to workers' psychological and physiological well-being (e.g., Bolino and Turnley 2005; Schultz et al. 1998). As a result, the range of optimal performance is much broader for simple than for complex jobs (Gardner 1986). Energy should therefore be expended on training workers effectively and efficiently. As shown in the next section, individuals who build a coherent knowledge base are better able to integrate new, related knowledge and are less susceptible to forgetting.

LEARNING AND FORGETTING

In cognitive science, it has been widely acknowledged that human beings form mental representations of the world, of themselves, of their capabilities, and of the tasks that they need to perform (Johnson-Laird 1983; Norman 1983). Their interactions with the environment, with others, and with the artifacts of technology shape such mental conceptualizations or models. Essentially, mental models are knowledge representations that enable individuals to describe the purpose of a system, explain its functioning, and predict its future state (Rouse and Morris 1986). The formation of these cognitive structures (or knowledge-acquisition processes) has been described by several theories, one of which is assimilation theory. Assimilation theory postulates that a learner assimilates new knowledge by connecting it to knowledge that already exists in the learner's memory (Bransford 1979). This assimilation process involves three steps. At the first step, the reception step, the learner must pay attention to the new information so that it can reach short-term memory. At the second step, the availability step, prerequisite concepts must be stored in long-term memory for the new information to be assimilated. At the third step, the activation step, the learner must actively use the prerequisite knowledge to enable the connections between new and existing knowledge (Ausubel 1968; Mayer 1981). Assimilation theory has been deemed appropriate to describe the process of learning new information systems (IS) (Mayer 1981; Santhanam and Sein 1994; Davis and Bostrom 1993). Much of the IS literature involving the development of mental models applies to new manufacturing technologies as well. With the increased automation of manual tasks, the workforce has shifted toward cognitive tasks involving interactions with AMTs (Nembhard and Uzumeri 2000; Arzi and Shtub 1997). AMTs combine production and information-processing capabilities and call for the use of cognitive rather than physical skills. The narrowing skill divide between office and shop floor jobs thus warrants extending IS research findings to cover production.

When learning new information systems, trainees may develop models based on prior experience with similar technologies (Young 1983). In this case, they learn by analogy. With new information technologies, however, it is not rare for trainees to have little or no a priori knowledge. This is why the use of external aids or "advance organizers" (Ausubel 1968) is necessary to provide a framework for the construction of new knowledge structures. Conceptual models of new systems refer to the external aids that are used in training (Mayer 1983) and help users form internal mental models of the system in question (Santhanam and Sein 1994).

The quality of external aids that are used in training affects the quality of mental model development which, in turn, affects performance (Sauer et al. 2000). In the

IS field, accurate mental models have indeed been associated both with efficiency (Staggers and Norcio 1993) and effectiveness (Sein and Bostrom 1989; Karuppan and Karuppan 2008). Accurate mental models are also more resilient to the passage of time as better comprehension greatly influences one's ability to recall information (Thorndyke 1977). Using the suitcase metaphor, Towse et al. (2005) explained that the amount of learning that could be "packed" or stored in memory depended on packing efficiency and on the arrangement of contents. In other words, greater amounts of knowledge can be stored in one's memory if it is well organized. Therefore, when flexible workers are assigned to new jobs or tasks, a hiatus between successive assignments will cause workers to focus on other cognitive demands, and performance will suffer (e.g., Peterson and Peterson 1959) unless they have developed robust mental models of their tasks/jobs.

The development of mental models and the preservation of their integrity over time both depend on two categories of factors: (i) those that cannot be controlled by the trainer (individual characteristics, task complexity, and length of interruption intervals), and (ii) the methodological factors that can be manipulated during training (amount and type of training, and test contents).

UNCONTROLLABLE FACTORS

INDIVIDUAL CHARACTERISTICS

There are multiple indications in various learning theories that individual characteristics shape the learning process. Skinner (1969) underscored the impact of a learner's past history on behavior in the theory of operant conditioning. Traits such as prior knowledge, skills, and previous experience affect the learning process (Cronbach and Snow 1977). Social learning theory (Bandura 1977, 1986, 1997) further buttresses these arguments by claiming that different individuals need different conditions to model others' behaviors and, by extension, to learn. Demographic/situational variables, personality variables, and cognitive/learning style shape the learning process and its effectiveness (Zmud 1979).

Demographic/situational variables include—but are not limited to—age, gender, education, intellectual abilities, prior experience, personality, and learning styles. In this chapter, the focus is on the last four characteristics. Higher-ability individuals have consistently been found to process information and make decisions faster (Taylor and Dunnette 1974), develop more coherent mental models (Hunt and Lansman 1975), and therefore retain more knowledge and skills (especially abstract and theoretical) over periods of non-use than do lower-ability individuals (Arthur et al. 1998; Farr 1987). Individuals with higher working memory capacity also perform better because they retrieve information more efficiently in the presence of interference (e.g., Bunting et al. 2004; Cantor and Engle 1993). Prior experience and task knowledge facilitate the integration of new knowledge. They are related to more gradual learning, but also to slower forgetting, leading Nembhard and Uzumeri (2000) to conclude that individuals with prior experience are good candidates for cross-training.

Personality traits are fully covered by the "big five" factor system (De Raad and Schouwenburg 1996). The five traits include extraversion, agreeableness, conscientiousness, emotional stability, and openness to experience (also known as intellect/creativity/imagination) (De Raad and Van Heck 1994). Conscientiousness captures the traits of a good learner: well organized, efficient, systematic, and practical (Goldberg 1992). It is therefore not surprising that conscientiousness is a significant predictor of school (Schuerger and Kuna 1987), college (Wolfe and Johnson 1995), and job (Barrick and Mount 1993) performance. Although extraversion has been negatively related to undergraduate and high school grade point average (Goff and Ackerman 1992), it is positively related to job performance (e.g., Barrick and Mount 1991). Emotional stability and intellect are also prominent traits associated with effective learning (De Raad and Schouwenburg 1996). Moreover, the last trait, intellect, reflects the close alliance between intelligence and personality (Ackerman 1988). Personality indeed influences information processing or cognitive strategies. This connection has been discussed in terms of cognitive styles, thinking styles, learning styles, and intellectual engagement (De Raad and Schouwenburg 1996).

Learning styles refer to an individual's preferred way of learning (Anis et al. 2004). This choice is determined by the individual's goals and objectives. Abstract learners (Sein and Bostrom 1989; Bostrom et al. 1990) and active experimenters (Bostrom et al. 1990; Frese et al. 1988; Karuppan and Karuppan 2008) have been found to develop more accurate mental models of new technologies. The rationale is that abstract thinkers are more predisposed to decode and understand basic system configurations. Similarly, actual use of and experimentation with a system or technology is a requirement for effective learning (Brown and Newman 1985; Carroll and Mack 1984). Further influencing learning and forgetting phenomena is the type of tasks that are performed (Farr 1987).

TASK COMPLEXITY

A great deal of literature has dealt with the categorization of tasks into various types. A popular dichotomy is simple versus complex. The complexity of a task affects the learning rate. Bishu and Chen (1989) found that simple tasks are learned more quickly than difficult tasks. They also found that performance improvements were greater when tasks were learned from difficult to simple rather than the other way around. However, such performance improvements seem to occur only in the short run (Bohlen and Barany 1976). There is strong evidence to suggest that complex knowledge tends to be quickly forgotten (Hall et al. 1983; McCreery and Krajewski 1999; McCreery et al. 2004; Nembhard and Osothsilp 2002; Shields et al. 1979). Nembhard and Uzumeri's (2000) findings also indicated that complexity—measured in their study in terms of sewing assembly times—affected learning and forgetting, but prior experience was a moderator. Experienced workers learned (in essence, they "relearned") complex tasks faster than they did simple tasks; they also forgot these complex tasks more rapidly than inexperienced workers did, which the authors recommended be studied further. Arzi and Shtub (1997) discovered that the forgetting effect was stronger for tasks requiring a higher degree of decision making, but that individual capabilities and motivation played an important role in this process. These results suggest that personal and task characteristics interact to predict learning and forgetting.

These considerations are especially relevant in AMT environments. The sophisti-cated technologies implemented in many of today's production environments have con-tributed to increasing job complexity in multiple ways. Not only have they placed greater demands on attention and concentration (Aronsson 1989), but they have also increased cognitive requirements. Information technologies are believed to promote innova-tion and problem solving through learning, creating a "fusion" of work and learning (Torzadeh and Doll 1999). When complex technologies fail on the shop floor, problem referrals to supervisors are no longer considered as viable alternatives because the delays they engender threaten the fluidity of the production system (Kathuria and Partovi 1999; Cagliano and Spina 2000; Karuppan and Kepes 2006). Consequently, operators must utilize cognitive skills under time pressure to bring back system normalcy and prevent further problem occurrences (Pal and Saleh 1993). As mentioned earlier, high levels of complexity and the flexibility demands placed on the workforce contribute to excessive stimulation. Although job simplification through redesign is an attractive solution, it is a partial and temporary one in terms of overstimulation, as work demands always shift. Process redesigns are aimed at boosting productivity. The elimination of steps or tasks in a process frees up resources, which are then re-allocated to new tasks and jobs. In parallel, the constant push for product and process innovation will continue to intensify learning requirements, casting overstimulation as a permanent threat.

Unfortunately, task complexity has been operationalized very loosely in the liter-ature, making comparisons across studies difficult to make. Study-specific examples include: slower learning rate, total proportion of learning possible, speed of forget-ting, and product learning (e.g., McCreery and Krajewski 1999; McCreery et al. 2004), which are actually the outcomes of complexity; method, machine, and mate-rial attributes (Nembhard 2000); and fault states (Sauer et al. 2000). Wood (1986) pointed out the failure of the empirical approach to produce definitions of tasks with sufficient construct validity. Through a review of the various frameworks used to define task complexity, he proposed a theoretical approach that describes task complexity along task inputs (acts and information cues) and outputs (products). *Products* are the final results of the task (Naylor et al. 1980). Examples include a part, a report, or a treated patient. *Acts* are the patterns of behaviors or activities required for the creation of the product. *Information cues* are the pieces of informa-tion that the individual uses to make judgments while performing the task. Based on these elements, Wood (1986) proposed three types of task complexity, two of which—component and dynamic—are especially relevant to the study of work-force flexibility. Component complexity is determined by the number of distinct acts required to complete the task and the number of distinct information cues that must be processed. Component redundancy (i.e., the degree of overlap among demand requirements for the task), reduces component complexity, whereas the number of subtasks required to do a particular task increases it. Component redundancy is not a measure of task complexity but it influences it. It evokes the notion of task similar-ity. The similarity between old and new tasks eases the learning process as well as the development (and even reinforcement) of the mental model. By the same token, it counters forgetting. Jaber et al.'s (2003) operationalization of task similarity exem-plifies the relationship between task complexity and similarity. If task 1 comprises acts A, B, and C, and task 2 includes A, C, H, I, and J, the complexity (in terms of the number of acts) of either task will be reduced by the redundancy of acts A and C in both tasks. Dynamic complexity is the result of external changes affecting the relationships between task inputs and outputs. Non-stationary acts and information cues—e.g., as the result of the process changes required for the task—place new demands on the worker's skills and knowledge. The above considerations have important implications for flexibility deployments, since they hint at the degree of additional stimulation that complexity may produce.

LENGTH OF INTERRUPTION INTERVALS

Once workers are trained for a task/job, the length of the interruption interval between tasks/jobs is one of the most important contributors of forgetting (Anderson et al. 1999; Badiru 1995; Bailey 1989; Jaber and Bonney 1996; Jaber and Kher 2004). The period of skill non-use has, without fail, been negatively related to performance in the literature. Skill decay is commensurate with the length of these non-practice intervals. However, its effect on skill decay is moderated by a host of factors, some of which are mentioned above. For example, in their meta-analysis of factors influencing skill decay and retention, Arthur et al. (1998) found that the effect of interruption intervals was exacerbated when the tasks involved were more demanding. Adding to the difficulty of gauging the impact of this variable on skill decay is the fact that the interruption interval may not be clear-cut. For example, workers may be assigned to new tasks that share some common components with old tasks. This component redundancy not only facilitates the learning of new tasks, but also preserves some knowledge of the old ones through continued practice. Finally, some individuals may mentally practice some tasks even when they are not performing them. Consequently, there are claims that cognitive tasks should be more resilient to the passage of time than physical tasks because they naturally lend themselves (e.g., Ryan and Simon 1981) to mental practice. Weakening these claims are the facts that the opposite has been observed in the absence of practice (Arthur et al. 1998) and that the mental practice of unneeded tasks is hardly realistic in actual work settings (Farr 1987). In summary, interactions with other factors complicate the relationships between the length of interruption interval and retention. However, there seems to be a consensus that the duration of the interruption and task complexity exacerbate forgetting.

The learning and forgetting research relating to uncontrollable factors is certainly intellectually appealing, but its practical value may be limited. Their interactions allow for a multitude of task-individual-interval combinations, which are daunting for trainers and production managers alike. Interventions known to influence the accuracy of mental models—method of training, over-learning, and test contents—are more easily manageable.

CONTROLLABLE FACTORS

TRAINING METHODS

The following discussion does not address training delivery methods (e.g., online versus face-to-face) but rather focuses on the development of mental models via conceptual models or procedural training. Such methods are most suitable to learn how to interact with information technologies and systems (e.g., Santhanam and Sein 1994). Conceptual models refer to the external aids given to new users of a system. They can be analogical or abstract. Analogical models provide analogies between the system on which the worker is being trained and a system with which the worker is familiar. For example, computer numerical control (CNC) compensation may be explained in terms of a marksman compensating for the distance to the target before firing a shot. Analogies are powerful tools to help trainees connect new knowledge to existing knowledge (Lakoff and Johnson 1980). They are rather concrete compared to abstract conceptual models that use schema, structure diagrams, hierarchical charts, or mathematical expressions to represent the new system being learned (Gentner 1983). The contribution of either type of conceptual model to the development of an accurate mental model depends on personal characteristics, such as cognitive ability and learning style. It seems that abstract conceptual models benefit abstract and active learners, whereas analogical models are better suited for concrete learners (Bostrom et al. 1990; Sein and Bostrom 1989). The effectiveness of either model may also be subject to the degree of component redundancy between old and new assignments, which would naturally lend itself to meaningful analogies.

As its name suggests, procedural training involves the instruction of step-by-step procedures or predetermined sequences that guide the user's interaction with a system. No relationship between the system and its components is provided, and the user mentally establishes the linkages between the components and confirms or modifies them as learning progresses. This is consistent with knowledge assembly theory (Hayes-Roth 1977). The concept is similar to using the step-by-step instructions of a GPS system to drive to various destinations. If the driver routinely uses the GPS system, he/she will eventually develop a mental model of the city. On the other hand, a map would provide a conceptual model. Some have argued that procedural training is more appropriate than a conceptual model for the instruction of simple systems (Kieras 1988). In their experiment involving an e-mail system, Santhanam and Sein (1994) did not find any difference between the training methods, but they noted that users who had formed a conceptual mental model of the system outperformed those who had developed a more procedural representation of the system.

OVER-LEARNING VIA TRAINING AND USAGE

Additional training beyond that required for initial mastery is referred to as over-learning (e.g., Farr 1987). Consistent with early claims that the longer a person studies, the longer he/she will be able to retain knowledge (Ebbinghaus 1885), over-learning is considered to be the dominant factor of knowledge retention. According to Hurlock and Montague (1982, 5), "[...] any variable that leads to high initial levels of learning, such as high ability or frequent practice, will facilitate skill retention." Frequent practice connotes over-learning. With subsequent task performance, initial assumptions about the system can be rectified or solidified, leading to the enrichment of the mental model (de Kleer and Brown 1983; Norman 1983). Over-learning also increases the automaticity of responses and greater self-confidence, which reduce overstimulation and subsequent stress (Arthur et al. 1998; Martens 1974). Since stress elicits the secretion of cortisol, which compromises memory and cognitive functions such as knowledge retrieval (Bremner and Narayan 1998), over-learning shields memory from decay.

The implementation of over-learning seems to call for longer training sessions. This is especially true for complex tasks for which learning is slower. In light of difficult economic times, many organizations will be tempted to rush training even though it will probably result in performance decrements. Malhotra et al. (1993) found that workers who were not fully trained on a particular task were less productive. They advocated speeding up the learning process. However, since rapid learning is associated with rapid forgetting, their recommendations may be risky. Indeed, one should ensure that trainees have developed adequate mental models prior to the actual re-assignment of work duties. This can be accomplished through an assessment of mental model quality via testing.

TEST CONTENTS

Performance is the only measure of mental model quality. However, performance for a simple task is not comparable to performance for complex tasks. Reminders of this argument are the consistent reports of overstimulation effects for complex tasks. As a result, the adequacy of a mental model is best judged by performance in creative, far-transfer tasks that require the completion of complex command sequences, which are not presented in training manuals (Mayer 1981; Sein and Bostrom 1989). These tasks differ from simple, near-transfer tasks to which trainees are usually exposed during training and whose performance demonstrates the ability to repeat commands, rather than the acquisition of a robust mental model (Weimin 2001). Far-transfer tasks force users to adapt the knowledge learned in training to novel situations. High performance on these tasks thus demonstrates better comprehension and an individual's likelihood to integrate new knowledge in a logical fashion. Therefore, gauging the necessary amount of training or over-learning can be easily accomplished by setting benchmarks on performance tests of far-transfer tasks.

CONCLUSION

To summarize, the learning of new jobs or tasks occurs via training and/or usage. The degree of mastery acquired through training and usage is reflected in the accuracy of the trainee's mental model of the job/task. Individual and task/job characteristics influence this process, thereby moderating the relationship between training/usage and the accuracy of the mental model. Accurate mental models lead to better performance on the job, operationalized here as workforce flexibility (RN, RH) with minimal losses of efficiency and effectiveness (MOB, UN). Accurate mental models also provide insulation against memory seepage following the assignment of new tasks. In other words, a solid mental model of a particular set of tasks would influence the quality of the mental model after interruption. Further affecting retention and therefore the quality of the mental model after interruption are the length of the interruption interval, individual characteristics, and task complexity. As Arthur et al. (1998) demonstrated, individual characteristics and task complexity also moderate the relationship between the length of the interruption interval and retention. All the linkages between variables are depicted in the framework in Figure 10.1. Although not exhaustive, it extends the learning process model proposed by Santhanam and Sein (1994) and captures the elements most prevalent in the learning and forgetting literature.

Much of the empirical research supporting the theory leading to the development of this framework relies on controlled laboratory and simulated experiments, which enable a "clean" manipulation of variables and are well suited to explore causality. This high degree of "purity" in the research methods has generated numerous insights. Nevertheless, the lack of generalizability of experiments to "noisier" environments is a weakness that needs to be addressed. Paradoxically, the popularity of the research stream in multiple disciplines has contributed to its richness, but it has also led to fragmentation. In the absence of a unified approach, the operationalization of variables (e.g., task characteristics) has been study-specific and even questionable in terms of construct validity. The above considerations suggest that validations of the vast experimental research in field settings with diverse individuals, truly complex systems, and well-defined constructs, constitute the next logical step. There have been some isolated attempts to do so (e.g., Karuppan and Karuppan 2008; Nembhard 2000; Nembhard and Osothsilp 2002); they just need to be more widespread and more interdisciplinary.

The primary difficulty emerging in field research is the interference of a host of individual, task, and site characteristics with the relationships that are under investigation. A serious problem in research, it is even more acute in practice. Besides the controversy that a distinction between low- and high-ability individuals would create, the intricacy of the relationships among various characteristics poses challenges for the training decision makers. Regarding the methods of instruction, a great deal

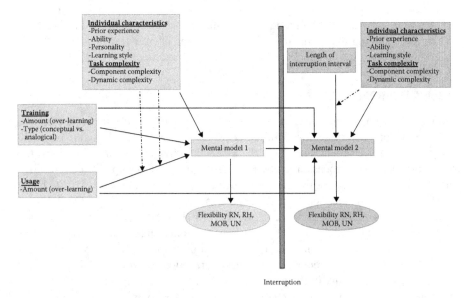

FIGURE 10.1   Proposed integrative framework of training for flexibility.

of research has identified the interactions between the methods of instruction and the individual and task characteristics, and has also suggested that training be customized accordingly. Unfortunately, and at least for the time being, the pressures for cost control and resource utilization in real-world settings limit the applicability of this research. Loo (2004) had recognized the burden of instruction customization and has instead advocated the use of a variety of instructional methods to appeal to a variety of individuals. Consequently, both conceptual and procedural models of AMT jobs/tasks should be made available to learners.

Construing task complexity in terms of the "amount of learning possible" (e.g., McCreery et al. 2004) raises an interesting question. What does "fully trained" really mean? Does getting a passing grade on a test constitute complete training? One could easily argue that one learns something on the job every single day. In such cases, over-learning is not confined to the actual training session format; it extends to daily practice and becomes confounded with experience. This observation fits with existing evidence regarding flexibility deployments. The rules suggest that flexibility deployments should favor workers with more extensive experience. Experience gives workers the time to build and refine a robust mental model, which is more resilient to decay. In order to account for task complexity in a pragmatic fashion, the extent of flexibility RN and RH should also be determined by the degree of component redundancy among tasks. When component redundancy is limited, so should deployments. Component redundancy is also confounded with experience since it assumes that workers have already learned and practiced elements of the new assignments. In the presence of dynamic complexity, extra time for relearning should be granted in order to avoid significant performance decrements following the return to a job that has been altered by external events. These rules are simple. Yet personal communication with industrial partners suggests that workforce flexibility is not measured and is therefore not examined rigorously like cost and quality are. The old adage of "more is better" seems to prevail until major performance decrements and employee resistance reach unacceptable levels. Clearly, academic research in the field has not filtered through to other professional communication vehicles. They may not be as prestigious as academic outlets, but professional publications have the advantage of increasing the visibility of a topic and ultimately the opportunities for external research funding.

REFERENCES

Ackerman, P.L., 1988. Determinants of individual differences during skill acquisition: Cognitive abilities and information processing. *Journal of Experimental Psychology* 117(3): 288–318.

Allen, M., 1963. Efficient utilization of labor under conditions of fluctuating demand. In *Industrial scheduling*, eds. J. Muth and G. Thompson, 252–276. Englewood Cliffs, NJ: Prentice-Hall.

Anderson, J.R., Fincham, J.M., and Douglass, S., 1999. Practice and retention: A unifying analysis. *Journal of Experimental Psychology* 25(5): 1120–1136.

Anis, M., Armstrong, S.J., and Zhu, Z., 2004. The influence of learning styles on knowledge acquisition in public sector management. *Educational Psychology* 24(4): 549–571.

Aronsson, G., 1989. Changed qualifications in computer-mediated work. *Applied Psychology: An International Review* 38(1): 57–71.

Arthur, W., Jr., Bennett, W., Jr., Stanush, P.L., and McNelly, T.L., 1998. Factors that influence skill decay and retention: A quantitative view and analysis. *Human Performance* 11(1): 57–101.

Arzi, Y., and Shtub, A., 1997. Learning and forgetting in mental and mechanical tasks: A comparative study. *IIE Transactions* 29(9): 759–768.

Ausubel, D.P., 1968. *Educational psychology: A cognitive view*. Holt, Reinhart, and Winston: New York.

Badiru, A.B., 1995. Multivariate analysis of the effect of learning and forgetting on product quality. *International Journal of Production Research* 33(3): 777–794.

Bailey, C.D., 1989. Forgetting and the learning curve: A laboratory study. *Management Science* 35(3): 340–352.

Bandura, A., 1977. *Social learning theory*. Englewood Cliffs, NJ: Prentice-Hall.

Bandura, A., 1986. *Social foundations of thought and action: A social cognitive theory*. Upper Saddle River, NJ: Prentice-Hall.

Bandura, A., 1997. *A self-efficacy—The exercise of control*. New York, NY: W.H. Freeman and Company.

Barrick, M.R., and Mount, M.K., 1991. The big five personality dimensions and job performance: A meta-analysis. *Personnel Psychology* 44(1): 1–26.

Barrick, M.R., and Mount, M.K., 1993. Autonomy as a moderator of the relationships between the big five personality dimensions and job performance. *Journal of Applied Psychology* 78(1): 111–118.

Bishu, R.R., and Chen, Y., 1989. Learning and transfer effects in simulated industrial information processing tasks. *International Journal of Industrial Ergonomics* 4(3): 237–243.

Bohlen, G.A., and Barany, J.W., 1976. A learning curve prediction model for operators performing industrial bench assembly operations. *International Journal of Production Research* 14(2): 295–302.

Bolino, M.C., and Turnley, W.H., 2005. The personal costs of citizenship behavior: The relationship between individual initiative and role overload, job stress, and work-family conflict. *Journal of Applied Psychology* 90(4): 740–748.

Bostrom, R.P., Olfman, L., and Sein, M.K., 1990. The importance of learning style in end-user training. *MIS Quarterly* 14(1): 101–119.

Bransford, J.D., 1979. *Human cognition*. Monterey: Wadsworth.

Bremner, J.D., and Narayan, M., 1998. The effects of stress on memory and the hippocampus throughout the life cycle: Implications for childhood development and aging. *Development and Psychopathology* 10:871–885.

Brown, J.S., and Newman, S.E., 1985. Issues in cognitive and social ergonomics: From our house to Bauhaus. *Human-Computer Interaction* 1(4): 359–391.

Bunting, M.F., Conway, A.R.A., and Heitz, R.P., 2004. Individual differences in the fan effect and working memory capacity. *Journal of Memory and Language* 51(4): 604–622.

Cagliano, R., and Spina, G., 2000. Advanced manufacturing technologies and strategically flexible production. *Journal of Operations Management* 18(2): 169–190.

Cantor, J., and Engle, R.W., 1993. Working memory capacity as long-term memory activation: An individual differences approach. *Journal of Experimental Psychology: Learning, Memory, and Cognition* 19(5): 1101–1114.

Carroll, J.M., and Mack, R.L., 1984. Learning to use a wordprocessor: By doing, by thinking and by knowing. In *Human Factors in Computing Systems*, eds. J.C. Thomas and M.L. Schneider, 13–51. Norwood: Ablex Publishing Company.

Champoux, J.E., 1978. A preliminary examination of some complex job scope-growth need strength interactions. Paper read at Academy of Management Proceedings.

Champoux, J.E., 1980. A three sample test of some extensions to the job characteristics model of work motivation. *Academy of Management Journal* 23(3): 466–478.

Champoux, J.E., 1992. A multivariate analysis of curvilinear relationships among job scope work context satisfaction and affective outcomes. *Human Relations* 45(1): 87–111.

Craik, F.I.M., Govoni, R., Naveh-Benjamin, M., and Anderson, N.D., 1996. The effects of divided attention on encoding and retrieval processes in human memory. *Journal of Experimental Psychology: General* 25(2): 159–180.

Cronbach, L.J., and Snow, R.E., 1977. *Aptitudes and instructional methods: A handbook for research on interactions.* New York: Irvington.

Davis, S.A., and Bostrom, R.P., 1993. Training end-users: An experimental investigation of the roles of the computer interface and training methods. *MIS Quarterly* 17(1): 61–81.

de Kleer, J., and Brown, J.S., 1983. Assumptions and ambiguities in mechanistic mental models. In *Mental models*, eds. D. Gentner and A.L. Stevens, 15–34. Hillsdale: Lawrence Erlbaum.

De Raad, B., and Schouwenburg, H.C., 1996. Personality in learning and education: A review. *European Journal of Personality* 10(5): 303–336.

De Raad, B., and Van Heck, G.L., 1994. The fifth of the big five. *European Journal of Personality* 8(Special Issue): 225–356.

Easterbrook, A., 1959. The effect of emotion on cue utilization and the organization of behavior. *Psychological Review* 66(3): 187–201.

Ebbinghaus, H., 1885. *Memory: A contribution to experimental psychology.* New York: Dover (translated in 1962).

Farr, M.J., 1987. *Long-term retention of knowledge and skills: A cognitive and instructional perspective.* New York: Springer-Verlag.

Felan, J.T., III, Fry, T.D., and Philipoom, P.R., 1993. Labour flexibility and staffing levels in a dual-resource constrained job shop. *International Journal of Production Research* 31(10): 2487–2506.

French, J.R.P., and Caplan, R.D., 1973. Organizational stress and individual strain. In *The failure of success*, ed. A.J. Murrow, 30–66. New York: AMACOM.

French, J.R.P., Jr., Caplan, R.D., and Van Harrison, R., 1982. *The mechanisms of job stress and strain.* Chichester: Wiley.

Frese, M., Albrecht, K., Altmann, A., Lang, J., Papstein, P.V., Peyerl, R., Prumper, J., Schulte-Gocking, H., Wankmuller, I., and Wendel, R., 1988. The effects of an active development of the mental model in the training process: Experimental results in a word processing system. *Behavior and Information Technology* 7(3): 295–304.

Fryer, J.S., 1976. Organizational segmentation and labor transfer policies in labor and machine limited production systems. *Decision Sciences* 7(4): 725–738.

Gardner, D.G., 1986. Activation theory and task design: An empirical test of several new predictions. *Journal of Applied Psychology* 71(3): 411–418.

Gardner, D.G., and Cummings, L.L., 1988. Activation theory and job design: Review and reconceptualization. *Research in Organizational Behavior* 10: 81–122.

Gentner, D., 1983. Structure-mapping: A theoretical framework for analogy. *Cognitive Science* 7(2): 155–170.

Gerwin, D., 1987. An agenda for research on the flexibility of manufacturing processes. *International Journal of Operations and Production Management* 7(1): 38–49.

Goff, M., and Ackerman, P.L., 1992. Personality-intelligence relations: Assessment of typical intellectual engagement. *Journal of Educational Psychology* 84(4): 532–552.

Goldberg, L.R., 1992. The development of markers of the big five factor structure. *Psychological Assessment* 4(1): 26–42.

Hackman, J.R., and Oldham, G.R., 1975. Development of the job diagnostic survey. *Journal of Applied Psychology* 60(2): 159–170.

Hackman, J.R., and Oldham, G.R., 1976. Motivation through the design of work: Test of a theory. *Organizational Behavior and Human Performance* 16(2): 250–279.

Hall, E.R., Ford, L.H., Whitten, T.C., and Plyant, L.R., 1983. *Knowledge retention among graduates of basic electricity and electronics schools*. Training Analysis and Evaluation Group, Department of the Navy (AD-A131855): Orlando.

Hayes-Roth, B., 1977. Evolution of cognitive structures and processes. *Psychological Review* 84(3): 260–278.

Hogg, G.L., Phillips, D.T., and Maggard, M.J.,1977. Parallel-channel dual-resource-constrained queuing systems with heterogeneous resources. *AIIE Transactions* 9(4): 352–362.

Hulin, C.L., and Blood, M.R., 1968. Job enlargement, individual differences, and worker responses. *Psychological Bulletin* 69(1): 41–55.

Humphreys, M.S., and Revelle, W., 1984. Personality, motivation, and performance: A theory of the relationship between individual differences and information processing. *Psychological Review* 91(2): 153–184.

Hunt, E., and Lansman, M., 1975. Cognitive theory applied to individual differences. In *Handbook of learning and cognitive processes*, ed. W.K. Estes, 81–107. Hillsdale, NJ: Lawrence Erlbaum.

Hurlock, R.E., and Montague,W.E., 1982. *Skill retention and its implications for Navy tasks: An analytical review*. San Diego: Navy Personnel Research and Development Center.

Jaber, M.Y., and Bonney, M., 1996. Production breaks and the learning curve: The forgetting phenomenon. *Applied Mathematics Modeling* 20(2): 162–169.

Jaber, M.Y., and Kher, H.V., 2002. The dual-phase learning-forgetting model. *International Journal of Production Economics* 76(3): 229–242.

Jaber, M.Y., and Kher, H.V., 2004. Variant versus invariant time to total forgetting: The learn-forget curve model revisited. *Computers & Industrial Engineering* 46(4): 697–705.

Jaber, M.Y., Kher, H.V., and Davis, D., 2003. Countering forgetting through training and deployment. *International Journal of Production Economics* 85(1): 33–46.

Johnson-Laird, P.N., 1983. *Mental models: Towards a cognitive science of language, inference, and consciousness*. Cambridge: Cambridge University Press.

Kahn, R.L., and Byosiere, P., 1992. Stress in organizations. In *Handbook of industrial and organizational psychology*, eds. M.D. Dunnette and L.M. Hough, 571–650. Palo Alto: Consulting Psychologists Press.

Karuppan, C., 2008. Labor flexibility: Rethinking deployment. *International Journal of Business Strategy* 8(2): 108–113.

Karuppan, C.M., and Ganster, D.C., 2004. The labor-machine dyad and its influence on mix flexibility. *Journal of Operations Management* 22(4): 533–556.

Karuppan, C.M., and Karuppan, M., 2008. Resilience of super users' mental models of enterprise-wide systems. *European Journal of Information Systems* 17(1): 29–46.

Karuppan, C.M., and Kepes, S., 2006. The strategic pursuit of flexibility through operators' involvement in decision making. *International Journal of Operations & Production Management* 26(9): 1039–1064.

Kathuria, R., and Partovi, F.Y., 1999. Workforce management practices for manufacturing flexibility. *Journal of Operations Management* 18(1): 21–39.

Kieras, D.E., 1988. What mental models should be taught: Choosing instructional content for complex engineering systems. In *Intelligent tutoring systems: Lessons learned*, ed. J. Psotka, L.D. Massey, and S.A. Mutter, 85–111. Hillsdale: Lawrence Erlbaum.

Koste, L.L., and Malhotra, M., 1999. A theoretical framework for analyzing the dimensions of manufacturing flexibility. *Journal of Operations Management* 18(1): 75–93.

Lakoff, G., and Johnson, M., 1980. *Metaphors we live by*. Chicago: The University of Chicago Press.

Lindsley, D.B., 1951. Emotion. In *Handbook of experimental psychology*, ed. S.S. Stevens. New York: Wiley. 473–516.

Loo, R., 2004. Kolb's learning styles and learning preferences: Is there a linkage? *Educational Psychology* 24(1): 99–108.

Malhotra, M.K., Fry, T.D., Kher, H.V., and Donahue, J.M., 1993. The impact of learning and labor attrition on worker flexibility in dual esource constrained job shops. *Decision Sciences* 24(3): 641–663.

Malmo, R.B., 1959. Activation: A neuropsychological dimension. *Psychological Review* 66(6): 367–386.

Martens, R., 1974. Arousal and motor performance. In *Exercise and sport sciences review*, ed. J.H. Wilmore, 155–188. New York: Academic Press.

Martin, R., and Wall, T., 1989. Attentional demand and cost responsibility as stressors. *Academy of Management Journal* 32(1): 69–86.

Mayer, R.E., 1981. The psychology of how novices learn computer programming. *Computing Surveys* 13(1): 121–141.

Mayer, R.E., 1983. Can you repeat that? Qualitative effects of repetition and advance organizers on learning from scientific prose. *Journal of Educational Psychology* 75(1): 40–49.

McCreery, J.K., and Krajewski, L.J., 1999. Improving performance using workforce flexibility in an assembly environment with learning and forgetting effects. *International Journal of Production Research* 37(9): 2031–2058.

McCreery, J.K., Krajewski, L.J., Leong, G.K., and Ward, P.T., 2004. Performance implications of assembly work teams. *Journal of Operations Management* 22(4): 387–412.

Naveh-Benjamin, M., 2001. and H.L. Roediger, III, 193–207. New York: Psychology Press.

Naylor, J.C., Pritchard, R.D., and Ilgen, D.R., 1980. *A theory of behavior in organizations.* New York: Academic Press.

Nembhard, D.A., 2000. The effects of task complexity and experience on learning and forgetting: A field study. *Human Factors* 42(2): 272–286.

Nembhard, D.A., and Osothsilp, N., 2002. Task complexity effects on between-individual learning/forgetting variability. *International Journal of Industrial Ergonomics* 29(5): 297–306.

Nembhard, D.A., and Uzumeri, M.V., 2000. Experiential learning and forgetting for manual and cognitive tasks. *International Journal of Industrial Ergonomics* 25(3): 315–326.

Norman, D.A., 1983. Some observations on mental models. In *Mental models*, eds. D. Gentner and A. L. Stevens, 7–14. Hillsdale: Lawrence Erlbaum.

Pal, S.P., and Saleh, S., 1993. Tactical flexibility of manufacturing technologies. *IEEE Transactions on Engineering Management* 40(4): 373–380.

Parthasarthy, R., and Sethi, S.P., 1992. The impact of flexible automation on business strategy and organizational structure. *Academy of Management Review* 17(1): 86–111.

Peterson, L.R., and Peterson, M.J., 1959. Short-term retention of individual verbal items. *Journal of Experimental Psychology* 58:193–198.

Rouse, W.B., and Morris, N.M., 1986. On looking into the black box: Prospects and limits in the search for mental models. *Psychological Bulletin* 100(3): 349–363.

Ryan, E.D., and Simons, J., 1981. Cognitive demand, imagery, and the frequency of mental rehearsal as factors influencing acquisition of motor skills. *Journal of Sports Psychology* 3(1): 35–45.

Santhanam, R., and Sein, M.K. 1994. Improving end-user proficiency: Effects of conceptual training and nature of interaction. *Information Systems Research* 5(4): 378–399.

Sauer, J., Hockey, G.R.J., and Wastell, D.G., 2000. Effects of training on short- and long-term skill retention in a complex multi-task environment. *Ergonomics* 43(12): 2043–2064.

Schaubroeck, J., and Ganster, D.C., 1993. Chronic demands and responsivity to challenge. *Journal of Applied Psychology* 78(1): 73–85.

Schuerger, J.M., and Kuna, D.L., 1987. Adolescent personality and school performance: A follow-up study. *Psychology in the Schools* 24(3): 281–285.

Schultz, P., Kirschbaum, C., Prüßner, J., and Hellhammer, D., 1998. Increased free cortisol secretion after awakening stressed individuals due to work overload. *Stress and Health* 14(2): 91–97.

Sein, M.K., and Bostrom, R.P. 1989. Individual differences and conceptual models in training novice users. *Human-Computer Interaction* 4(3): 197–229.

Shields, J.L., Goldberg, S.I., and Dressel, J.D., 1979. *Retention of basic soldering skills.* Alexandria, VA: US Army Research Institute for the Behavioral and Social Sciences.

Singh, J., 1998. Striking a balance in boundary-spanning positions: An investigation of some unconventional influences of role stressors and job characteristics on job outcomes of salespeople. *Journal of Marketing* 62(3): 69–86.

Skinner, B.F., 1969. *Contingencies of reinforcement: A theoretical analysis.* Englewood Cliffs, NJ: Prentice-Hall.

Staggers, N., and Norcio, A.F., 1993. Mental models: Concepts for human-computer interaction research. *International Journal of Man-Machine Studies* 38(4): 587–605.

Sünram-Lea, S.I., Foster, J.K., Durlach, P., and Perez, C., 2002. Investigation into the significance of task difficulty and divided allocation of resources on the glucose memory facilitation effect. *Psychopharmacology* 160(4): 387–397.

Taylor, R.N., and Dunnette, M.D., 1974. Relative contribution of decision-maker attributes to decision process. *Organization Behavior and Human Performance 12*(2): 286–298.

Thorndyke, P., 1977. Cognitive structures in comprehension and memory for narrative discourse. *Cognitive Psychology* 9(1): 77–110.

Torzadeh, G., and Doll, W.J., 1999. The development of a tool for measuring the perceived impact of information technology on work. *Omega* 27(3): 327–339.

Towse, J.N., Hitch, G.J., Hamilton, Z., Peacock, K., and Hutton, U.M.Z., 2005. Working memory period: The endurance of mental representations. *The Quarterly Journal of Experimental Psychology* 58A(3): 547–571.

Treleven, M., 1989. A review of the dual resource constrained system research. *IIE Transactions* 21(3): 279–287.

Upton, D.M., 1997. Process range in manufacturing: An empirical study of flexibility. *Management Science* 4(8): 1079–1092.

Weimin, W., 2001. The relative effectiveness of structured questions and summarizing on near and far transfer tasks. Paper read at 24th National Convention of the Association for Educational Communications and Technology, Atlanta, GA.

Wolfe, R.N., and Johnson, S.D., 1995. Personality is a predictor of college performance. *Educational and Psychological Measurement* 55(2): 177–185.

Wood, R.E., 1986. Task complexity: Definition of the construct. *Organizational Behavior and Human Decision Processes* 37(1): 60–82.

Xie, J.L., and Johns, G., 1995. Job scope and stress: Can job scope be too high? *Academy of Management Journal* 3(5): 1288–1309.

Young, R.M., 1983. Surrogates and mappings: Two kinds of conceptual models for interactive devices. In *Mental models*, eds. D. Gentner and A.L. Stevens, 32–52 Hillsdale: Lawrence Erlbaum.

Zmud, R.W., 1979. Individual differences and MIS success: A review of the empirical literature. *Management Science* 25(10): 966–979.

11 Accelerated Learning by Experimentation

Roger Bohn and Michael A. Lapré

CONTENTS

DELIBERATE LEARNING

Experimentation was a key part of the Scientific Revolution. Galileo (1564–1642) is often credited with being the first to use experiments for both discovery and proof of scientific relationships, although this could also be said of Ibn al-Haytham (965–1040) who lived some 600 years earlier. Experimentation as a core concept in management was introduced by Herbert Simon, who discussed both management and engineering as processes of systematic search over a field of alternatives (March and Simon 1958; Newell and Simon 1972). Deliberate systematic experimentation to improve manufacturing probably goes back to chemical engineering in the late nineteenth century. Frederick Taylor ran thousands of metal-cutting experiments over several decades, and in many ways was a pioneer in systematic learning, a decade before his controversial research on managing people (Bohn 2005). Systematic experimentation in marketing began around 1940 (Applebaum and Spears 1950).

A general definition of an experiment is: "A deliberate comparison of outcomes from a varied but repeatable set of conditions, with an effort to explain different outcomes by differences in conditions." This definition includes controlled experiments (in which possible causal conditions are deliberately manipulated), and natural experiments (in which they are measured but not deliberately altered). The definition excludes purely descriptive investigations such as surveys or satellite photographs in which the results are solely tabulated or displayed. For example, a control chart by itself is not an experiment, although control charts can provide data for natural experiments.

Experiments are a key mechanism for industrial learning, and are therefore an important managerial lever to accelerate learning curves (Adler and Clark 1991; Dutton and Thomas 1994; Lapré et al. 2000). The learning curve phenomenon has been observed frequently for measures of organizational performance such as quality and productivity. The rate of improvement is called the "learning rate." Learning rates show tremendous variation across industries, organizations, and organizational units (Dutton and Thomas 1984; Lapré and Van Wassenhove 2001). Bohn (1994) and Lapré (2011) discuss the model inside the learning curve. Both experience and deliberate activities can be sources for learning; learning can yield better organizational knowledge, which in turn can lead to changed behavior, and subsequently to improved performance. None of these steps are automatic. Dutton and Thomas (1984) call learning from experience "autonomous learning" and learning from deliberate activities "induced learning." The typical examples of deliberate activities are quality improvement projects and productivity improvement projects. Such projects often rely on a series of experiments. Other induced learning methods at the level of the firm are deliberate knowledge transfers from outside the organization, and the training of workers—but these are available only for knowledge that already exists and is accessible. Even when firms transfer knowledge from the outside, some adaptation to local circumstances—and thus experimentation—is almost always required (Leonard-Barton 1988). Hence, sound management of experimentation is important in order to attain the effective management of the learning rate.

In an extreme example, experimentation can be the sole driver of a learning curve. Lapré and Van Wassenhove (2001) studied productivity learning curves at four production lines at Bekaert, the world's largest independent producer of steel wire. One production line was run as a learning laboratory in a factory. The productivity learning curve was explained by the cumulative number of productivity improvement projects, which consisted of a series of natural and controlled experiments. The other three production lines were set up to replicate the induced learning. Interestingly, the other three lines struggled with learning from experimentation and relied on autonomous learning instead. Even within the same organization, it can be difficult to manage the required scientific understanding and human creativity (we will later refer to this as the "value of the underlying ideas").

Experiments are used in a variety of settings. Generally, the experiments themselves are embedded into broader processes for deliberate learning, such as line start-ups, quality programs, product development, or market research. Examples of situations dealt with by experimentation include:

- Diagnosing and solving a newly observed problem in a complex machine
- Improving the layout or contents of web pages that are displayed to consumers
- Breeding new varieties of plants
- Developing a new product by building and testing prototypes
- Conducting a clinical trial of a new medication on a population of patients
- Scientific modeling via simulation, such as global climate change models

Experimentation is not the only method of performance improvement in industry. Other approaches start from existing knowledge and attempt to employ it more widely or more effectively. These include training workers, installing improved machinery (with the knowledge embedded in the machines), and improving process monitoring to respond faster to deviations. In such approaches, experiments are still needed to validate changes, but they do not play a central role.

SEQUENTIAL EXPERIMENTATION

Experiments are generally conducted in a series, rather than individually. Each cycle in the series consists of planning the experiment, setting up the experimental apparatus, actually running the trial and collecting data, and analyzing the results (plan, set-up, run, analyze). For example, in product development, experiments focus on prototypes, and the prototyping cycle consists of designing (the next prototype), building, testing (running trials with the prototype), and analyzing (e.g., see Thomke 1998). The "planning" stage includes deciding what specific topic to investigate, deciding where and how to experiment (discussed below), and the detailed design of the experiment. The analysis of results from each experiment in the series suggests further ideas to explore.

Experiments can be designed for the general exploration of a situation, to compare discrete alternative actions, to estimate the coefficients of a mathematical model that will be used for optimization or decision making, or to test a hypothesis about causal relationships in a complex system. The goals of experiments in a series usually evolve over time, such as moving from general to specific knowledge targets. For example, to fix a novel problem in a multistage manufacturing process, engineers may first isolate the location of the cause (general search), and then test hypotheses about the specific causal mechanism at work. They may then test possible interventions to find out whether or not they fix the problem and what, if any, are the side effects. Finally, they may run a series of experiments to quantitatively optimize the solution.

The goals of experiments depend on how much is already known. Knowledge about a particular phenomenon is not a continuous variable, but rather it passes through a series of discrete stages. The current stage of knowledge determines the kinds of experiments needed in order to move to the next stage (Table 11.1). For example, to reach Stage 3 of causal knowledge, one must learn the direction of an effect, which may only require a simple two-level experiment, while at Stage 4 the full effects are understood quantitatively. This requires an elaborate design, often

TABLE 11.1

Stages of Causal Knowledge and the Types of Experiments Needed to Progress

Causal Knowledge Stage: How Cause x_i Affects Outcome Y	Experimental Design Needed to Get to this Stage
6. Integrated multivariate causal system	Multivariate experiment with initial, intermediate, and final variables all measured
5. Scientific model (formulaic relationship derived from scientific theory)	Test fit to an equation derived from first principles
4. Magnitude and shape of effect (empirical relationship)	Compare range of levels of x_i (response surface estimation)
3. Direction of effect on Y	Compare two selected alternatives
2. Awareness of variable x_i	Screening experiment on multiple possible causes; general exploration
1. Ignorance	(Starting condition)

involving the changing of multiple variables. Whether the additional knowledge is worth the additional effort depends on the economics of the particular situation. Moving from rudimentary to complete knowledge often takes years because advanced technologies have complex networks of relationships, with hundreds of variables to consider.

LEARNING IN SEMICONDUCTOR MANUFACTURING

Deliberate learning takes place almost constantly in semiconductor fabrication. Fixed costs of fabrication are very high, so the production rate (throughput) and yield (fraction of output that is good) are critical. Yields of new processes sometimes start well below 50%, and a percentage point of yield improvement is worth millions of dollars per month (Weber 2004).

Because of the very high complexity of the fabrication process, no changes are made unless they have been tested experimentally. Hundreds of engineers in each wafer fabrication facility (called "fab") engage in constant learning cycles. Some seek changes that will permanently improve yields. Learning targets can include the product design at several levels, from circuit design to specific masks, changes in the process recipe, changes in the tooling recipe (such as a time/intensity profile for a deposition process), or alterations in a $10 million tool. Other problem-solving activities can locate and diagnose temporary problems, such as contamination. Natural experiments (discussed below) may be adequate to find such problems, but engineers always test the solution by using a controlled experiment before putting the tool back into production. Other experiments are used to test possible product improvements. Finally, some experiments are run for development purposes, either of new products or of new processes.

Learning curves for yield improvement can differ considerably, even within the same company. In the case of one semiconductor company for which we had data, the cumulative production that was required in order to raise yields by 10 percentage points from their initial level ranged from below 10 units, to more than 200 units.[*] There were many reasons behind this large range, including better systems put in place for experimentation and observation (natural experiments), transfer of knowledge across products, and more resources for experiments at some times than at others.

METHODS OF ACCELERATING LEARNING

Deliberate (induced) learning is a managed process. This section discusses the drivers of effective learning by experimentation. There is no perfect way to experiment, and choosing methods involves multiple tradeoffs. In some situations, a better strategy can have a 10-fold effect on the rate of learning. Even when using a single strategy, effectiveness can vary dramatically.

We divide the analysis into three sections. First we show that there are four basic types of experiments. Second, we present criteria for predicting the rate of learning from experiments. These include cost per experiment, speed, and specific statistical properties. Third, we discuss the choice of where to experiment (location). At one extreme are full-scale experiments in the real world, while more controlled locations are usally superior. Different combinations of experiment type and location can improve some of the criteria, while worsening others.

This three-part framework (types, criteria, and locations) was first developed for manufacturing (Bohn 1987) and was then applied to product development (Thomke 1998). The framework also fits market research and other industrial learning about behavior. It can also be used in clinical trials.

TYPES OF EXPERIMENTS AND THEIR CHARACTERISTICS

There are four main types of experiments, each with different ways of manipulating the causal variables. (1) *Controlled experiments* make deliberate changes to treatments for several groups of subjects, and compare their properties. For example, medical *clinical trials* treat separate groups of patients with different drugs or doses. The "control group" captures the effects of unobserved variables; the difference in outcomes between the control and treated groups is the estimated effect of the treatment. Treatments in controlled experiments can be elaborated indefinitely. A classic reference on this subject is Box et al. (1978).

(2) *Natural experiments* use normal production as the data source.[†] The natural variation in causal variables is measured carefully, and is related to the natural variation in outcomes using regression or related techniques. Natural experiments are generally inexpensive; the main costs involved are for analysis, and, if necessary, special measurements. As a result, very large sample sizes are possible. On the other hand, natural experiments can only measure the impact of changes that occur due to

---

[*] Arbitrary production units used for disguise.

[†] The term "natural experiment" has apparently never been formally defined, although various authors have used it, or have discussed a similar concept using a different name. Murnane and Nelson (1984) refer to natural experiments but without defining them. Box et al. (1978) referred to them as "happenstance data."

natural variations in the process. They cannot predict the effects of radical changes, such as a new type of equipment, or a completely new procedure.

A fundamental problem with natural experiments is confounding. If A and B vary together, does A cause B, or does B cause A, or are both caused by an invisible third variable (Box et al. 1978)? There are also tricky questions about causality. Suppose that the length of time spent by customers in a grocery store is measured and turns out to be positively correlated to the amount of money spent. Do the customers: (a) spend more because they had more time to look at merchandise, or (b) spend more time shopping because they intended to buy more at the time they entered the store? Gathering additional data via a questionnaire might resolve this, while still remaining a natural experiment, but even if the first case is correct, an intervention that increases the time spent in the store will not necessarily lead to increased spending.[*] This simple example highlights the importance of an explicit *causal model* for learning from experiments; an appropriately complex causal model is needed in order to understand the effect of interventions that can change A, compared with just statistically establishing and concluding that A increases B (Pearl 2001).[†] The causal model can be determined from outside knowledge, or by appropriate controlled experiments, but it cannot generally be tested purely by natural experiments.

(3) *Ad hoc experiments*, like controlled experiments, use deliberate changes. However, the changes are made without a careful control group or experimental design. A simple "before-and-after" comparison is used to estimate the impact of the treatment. Because many unobserved variables can also change over time, ad hoc experiments can be very misleading and can have a poor reputation. However, young children playing with blocks learn very effectively in this way, and quickly learn basic cause-and-effect relationships. This form of learning is sometimes called *trial and error* (Nelson 2008).

(4) *Evolutionary operation (EVOP) experiments* are a hybrid between controlled and natural experiments. Slight changes in the production process are made deliberately, and the resulting changes are measured and statistically associated with the process shifts. The changes are small enough so that the process still works and the output is still good. Subsequent changes can move further on in whichever directions yield the most improved results—this is the "evolution."

EVOP was proposed decades ago for factories, but as far as we know, it was little used (Hunter and Kittrell 1966).[‡] Recently, however, it has become a very common approach to learning on the Internet. As is discussed below, Amazon and Google both use multiple EVOP experiments to tune their user-interface (web page) design, search algorithms for responding to queries, selection and placement of ads on the page, and so forth. Seemingly minor design issues such as colors and the precise location of "hot spots" on the page can be experimented on very easily, quickly, and cheaply.

---

[*] For example, management could slow down the checkout process, or show a free movie. These might seem far fetched in a grocery store, but similar problems would cloud the results of a natural experiment in a casino, a theme park, or a bookstore.

[†] The standard statistical and mathematical notations are not even capable of distinguishing among the different types of causes (see Pearl 2001).

[‡] It was also proposed for marketing, as a form of "adaptive control" (Little 1966).

All four types of experiments are generally feasible for learning in ongoing operations, but only controlled and ad hoc experiments are available for novel situations. Deciding which type to use depends on the interactions among several criteria, which we turn to next.

CRITERIA FOR EVALUATING EXPERIMENTATION METHODS

There are advantages and drawbacks to the different types of experiments and the different designs within a type. There is, as yet, no "theory of optimal experimentation," except within very stylized models. However, a small number of high-level properties are essential to predicting the overall effectiveness of the different approaches. Key characteristics include speed, signal-to-noise (S/N) ratio, cost per cycle, value and variety of ideas, and fidelity.

- *Speed* can be defined as the inverse of the *information cycle time* from the beginning to the end of each plan–set-up–run–analyze cycle. Shorter cycle times directly accelerate learning by requiring less time to run a series of experiments. Less directly, a faster cycle helps the experimenters "keep track of" the experiment, the reasoning behind it, and any unrecorded subtleties in the experimental conditions, along with any results that may be important only in retrospect.
- *Signal-to-noise ratio* (S/N) can be defined as the ratio of the true (unknown) effect of the experimental change to the standard deviation of measured outcomes. The S/N drives the common significance test such as the "t test," which measures the probability of a false positive result. It also drives the statistical *power*, which measures the probability of a false negative (overlooking a genuine improvement) (Bohn 1995a). The S/N can generally be improved by increasing the sample size, but more subtle methods are usually available and are often less expensive.*
- *Cost* per cycle. The lower the cost, the more experimental cycles can be run (or the more conditions can be tested in parallel). The financial costs of controlled experiments usually include the cost of the materials used in the experiment, but the most important costs are often non-financial—notably, the opportunity costs of busy engineers, the production capacity, computer time (for simulation), or the laboratory-quality metrology equipment. Controlled experiments carried out on production equipment often require elaborate set-ups, which increases the opportunity costs. One of the great benefits of natural experiments and EVOP is that the only costs are for special measurements, if any, and the analysis of the results. Costs are divisible into variable costs (meaning proportional to the sample size) and fixed costs (which depend on the complexity of the experimental set-up, but not the sample size). A third type of cost is capital costs for expenditures to create the experimental system itself. The cost of a pilot line can be considered a capital cost to enable more/better experiments.

---

* S/N is a core concept in communications engineering.

In semiconductor fabs, experimental lots are interspersed with normal production lots and the cost of experimentation is managed by quotas, rather than a financial budget. So-called "hot" lots are accelerated through the process by putting them at the front of the queue at each process step, giving roughly a two-fold reduction in the information cycle time, but increasing the cycle time for normal lots. For example, a development project could be given a quota of five normal lots and two hot lots per week. It is up to the engineers to assign these lots and their wafers to different questions within the overall development effort. Even for hot lots the information cycle is generally more than a month, so one fab could have 50 experiments in progress at one time.

- *Value and variety of the underlying ideas* being tested. Ideas for experiments come from a variety of sources, including prior experiments, outside organizations, scientific understanding of the problem, and human creativity. Strong ideas convey a double benefit: they improve the S/N of the experiment and, if the experiment reaches the correct conclusion, they increase the benefit derived from the new knowledge.

  In mature production processes and markets, most ideas will have a negative expected value—they make the situation worse. A higher variety of underlying ideas raises the total value of experimentation. This follows from thinking of experiments as real options, where the cost of the experiment is the price of buying the option, the current value of "best practice" is its exercise price, and the revealed value of the new method is the value of the underlying asset.[*] According to the Black-Scholes formula and its variants, higher variance of the asset increases the value of options. There is a further benefit from the higher S/N that is not captured in the standard formulas, namely, the reduced probabilities of statistical errors.

- *Fidelity* of the experiment can be defined as the degree to which the experimental conditions emulate the world in which the results will be applied. Fidelity is a major concern in choosing both the type and the location of experiments, which we discuss later.

Ideal experimentation strategies and tactics would increase the value of all five criteria. More commonly though, the choices of how to experiment will involve tradeoffs between the criteria. For example, S/N ratio increases with sample size, but so does the cost (except for natural experiments). Speed can usually be increased by paying a higher price. The degree to which this is worthwhile depends on the opportunity cost of slower learning. This depends on both business and technical factors. Terwiesch and Bohn (2001) show that under some conditions of factory ramp-up, the optimal policy is bang-bang: at first, carry out zero production and devote all resources to experiments; later carry out no experiments and devote all resources to production. At least in semiconductors, the normal pattern for debugging a new process is to start with 100%

---

[*] Terwiesch and Ulrich (2009) show how higher variance is good in *tournaments*. Tournaments are a highly structured form of experimentation, in which only one proposal will be selected out of many.

experiments, then shift to a quota, such as 5% experiments, and go to zero experiments near the end of life.

The recent popularity of "data mining" reflects the power of natural experiments under the right circumstances. For example, Harrah's Casino uses natural experiments for insights that drive a variety of controlled experiments (Lee et al. 2003). In data mining, a company has invested in a database of historical data and analytic hardware, software, and expertise. Once this large investment is operational, each experiment costs only a few hours, or days, of people's time and server processing time. The sample size can be in the millions, so the S/N ratio may be excellent, even if the environment has high variability and the data has measurement errors. So, three of the five criteria are extremely good: S/N, speed, and cost. On the other hand, fidelity is unclear, since much of the data are old and subject to the causal ambiguity problems discussed earlier.

APPROXIMATE REALITY: THE LOCATION OF EXPERIMENTS

The goal of experimentation in industry is to develop knowledge that can be applied in real environments, such as a high-volume manufacturing plant, a network of service delivery points (automated teller machines, web servers, stores, etc.), the end-use environments of a product, a population of customers, or sufferers from a particular disease. Yet it is usually preferable to do experiments elsewhere—in an environment that emulates key characteristics of the real world, but suppresses other characteristics in order to improve the S/N, cost, or information cycle time. Scientists have experimented on models of the real world for centuries, and new product developers have used models at least since the Wright brothers' wind tunnels. The literature on product development identifies two orthogonal choices for how to model: *analytical to physical* and *focused to comprehensive* (Ulrich and Eppinger 2008). Both of these axes apply to all learning domains.[*]

Organizations that do a lot of learning by experimentation often designate special facilities for the purpose, referred to as *pilot lines, model stores, laboratories, test chambers, beta software versions,* and other names. These facilities usually have several special characteristics. First, they are more carefully controlled environments than the real world.[†] For example, clinical trials use prescreened subjects, with clear clinical symptoms, no other diseases, and other characteristics that increase the likely signal size and decrease the noise level. Second, they are usually better instrumented, with more variables measured, more accurately and more often. For example, test stores may use video cameras to study shopper behavior. Third, these facilities are more flexible, with more skilled workers and different tools.

Moving even further away from the real world are virtual environments, such as simulation models and mathematical equations. Finite element models have revolutionized experimentation in a number of engineering disciplines, because they allow science-derived first principles to be applied to complex systems.

---

[*] The original literature on manufacturing experimentation conflated them into a single axis, "location" (Bohn 1987).

[†] But see the Taguchi argument for using normal quality materials, discussed in the conclusion.

Ulrich and Eppinger (2008) call the degree of reality or abstraction the "analytical/physical" dimension. They apply it to prototypes in product development, but their spectrum from physical to analytical also applies in manufacturing, medicine, and other domains, as illustrated in Table 11.2. For example, a steel company found that water and molten steel had about the same flow characteristics, and therefore prototyped new equipment configurations using scale models and water (Leonard-Barton 1995). In drug development, test tube and animal experiments precede human trials.

There are many benefits of moving away from full reality toward analytical approximations (top to bottom in Table 11.2). Cost per experiment, information cycle time, and the S/N ratio all generally improve. The S/N ratio can become very high in deterministic mathematical models. The disadvantage of moving toward analytical models is the loss of fidelity: the results of the experiment will be only partly relevant to the real world. In principle, the best approach is to run an experiment in the simplest possible (most analytical) environment that will still capture the essential elements of the problem under study. In practice, there is no single level of abstraction that is correct for every part of a problem, and learning therefore requires a variety of locations, with preliminary results developed at more analytical levels and then checked in more physical environments.

Full-scale manufacturing can be very complex. Lapré et al. (2000) studied organizational learning efforts at Bekaert, the world's largest independent producer of steel wire. Bekaert's production process can be characterized by detail complexity

TABLE 11.2
The Spectrum of Locations from Physical to Analytical

Locations of Experiments: Physical/Analytic Range	Manufacturing	Aviation Product Development Example	Drug Testing Example	Retail Behavior Example
Full-scale reality (most physical)	Manufacturing line	Flight test	Human trial	Full-scale trial
Scale model	Pilot line	Wind tunnel	Animal model	Test market
Laboratory	Lab		In vitro test	Focus group
Complex mathematical model	Finite-element simulation	CAD (finite element) simulation with graphical output	Rational drug design model	
Simple mathematical model (most analytical)	Simultaneous equation model for annealing in a furnace	Strength of materials model	Half-life model of drug metabolism	Advertising response model

Note: Example locations shown for four learning situations. Fidelity is highest at the top (full scale), but information cycle times, signal-to-noise ratio, and cost per experiment get better as the domain moves toward analytical (at the bottom).

(hundreds of machines, and hundreds of process settings), dynamic complexity (lots of dependencies between the production stages), and incomplete knowledge concerning the relevant process variables and their interactions. They found that Bekaert factories sometimes used the results from experiments run in laboratories at a central research and development (R&D) facility. However, on average, these laboratory insights actually slowed down the rate of learning in full-scale manufacturing. Small-scale laboratories at the central R&D facility were too different from the reality of full-scale manufacturing. Ignoring the complexity of full-scale manufacturing (such as equipment configurations) actually caused deterioration in performance. Thus, in manufacturing environments such as Bekaert's, fidelity issues mean that the locations used for experiments need to be more physical and less analytical. Bekaert, therefore, did most of its experiments in a special "learning line" set up inside its home factory.

FOCUSED/COMPREHENSIVE SPECTRUM

The second aspect of experimental "location" is what Ulrich and Eppinger (2008) call "focused/comprehensive" dimension. An experiment can be run on a subsystem of the entire process or product rather than the entire system. It is easier to try out different automobile door designs by experimenting on a door than it is on an entire automobile, and easier still to experiment on a door latch. The effects of focused studies are similar to the effect of moving from physical to analytical locations. Once again, the tradeoff is loss of fidelity: subsystems have interactions with the rest of the system that will be missed. So this technique is more applicable in highly modular systems. At Bekaert, there were high interactions among the four main process stages, so single-stage trials were risky.

Experimenting on a subsystem has always been common in product development. In manufacturing and other complex processes, it is more difficult and more subtle, yet can have an order-of-magnitude effect on the speed of learning. The key is to understand the process well enough to measure variables that *predict* the final outcome before it actually happens. The case study "short-loop experiments for AIDS" in the next section gives a dramatic example where the learning cycle time was reduced from years to months.

In semiconductor fabrication, experiments that only go through part of the process, such as a single tool or single layer, are sometimes called *short-loop experiments* (Bohn 1995b).[*] Suppose that a megahertz-reducing mechanism has been tentatively diagnosed as occurring in layer 10 out of 30, in a process with an average cycle time of two days per layer. Mathematical models of the product/process interaction suggest a solution, which will be tested by a split-lot experiment in which 14 wafers are treated by the proposed new method and 10 by the old method. Clearly, 18 days can be saved by splitting a previously routine production lot at the end of layer 9, rather than at the beginning of the line. However, must the test lot be processed all the way to the end before measuring the yields of each wafer? If so, the information cycle time will be 42 days, plus a few more for setting up the experiment and testing.

---

[*] Juran referred to such experiments as "cutting a new window" into the process (Juran and Gryna 1988).

Furthermore, megahertz differences among the 24 wafers will be due to the effect of the process change plus all the normal wafer-to-wafer variation in all 30 layers, leading to a poor S/N ratio (Bohn 1995b).

The information cycle time and S/N will be far better if the effect of the change can be measured directly after the tenth layer, without processing the wafers further. This requires a good understanding of the intermediate cause of the problem, and the ability to measure it accurately in a single layer. Furthermore, it requires confidence that the proposed change will not interact with any later process steps.

Good (fast and accurate) measurement of key properties is critical to focused experiments. To allow better short-loop experiments, most semiconductor designs include special test structures for measuring the electrical properties of key layers and features. The electrical properties of these test structures can be measured and compared across all wafers. Running the trial and measuring the test structure properties can be done in a few days. Depending on how well the problem and solution are understood, this may be sufficient time to go ahead and make a process change. Even then, engineers will need to pay special attention to the megahertz tests of the first production lots using the new method. This checks the fidelity of the early short-loop results against the full effect of the change. If there is a discrepancy, it means that the model relating the test structures to performance of the full device needs to be revised.

An analogous measurement for focused consumer behavior experiments is the ability to measure the emotional effect of advertisements on consumers in real time, and the ability to prove that the measured emotional state predicts their later purchasing behavior. Once these conditions exist, consumers can be shown a variety of advertisements quickly, with little need to measure their actual purchasing behavior. Measuring emotional response using functional magnetic resonance imaging (fMRI) is still in its infancy, but as it becomes easier it will have a big effect on the rate of progress in consumer marketing situations (for a review on fMRI, see Logothetis 2008).

CASE STUDIES

This section illustrates how learning has been accelerated through more effective experimentation in a variety of situations.

EXPERIMENTATION ON THE INTERNET

Amazon, Google, and other dot-com companies have exploited the favorable properties of the web for relentless experimentation on the interaction between their users and websites. Google has set up a substantial infrastructure to run experiments quickly and cheaply on its search site.[*] For example, two search algorithms can be interleaved to compare results. The dependent (outcome) variables include the number of results clicked, how long the user continues to search, and other measures of

---

[*] Presentation by Hal Varian, Google chief economist, UCSD May 2007. See also Shankland (2008) and Varian (2006).

user satisfaction. Multiple page layout variables such as colors, white space, and the number of items per page are also tested. Through such tests, Google discovers the "real estate value" of different parts of the search page. Google tested 30 candidate logos for the Google checkout service, and, overall, it now runs up to several hundred experiments per day.[*]

Such experiments have excellent properties with respect to most of the criteria we have discussed. They have very fast information cycle times (a few days), very low variable cost because they use EVOP, and high fidelity because they are run in the real world. Even if individual improvements are very small, or more precisely, have low standard deviation, the S/N ratio can still be excellent because of very large sample sizes. Google runs experiments on approximately 1% of the relevant queries, which in some cases can give a sample of one million in a single day. This can reliably identify a performance improvement of only 1 part in 10,000.

However, Internet-based experiments still face potential problems on the focused/ comprehensive dimension. In many experiments the goal is to affect the *long*-term behavior of customers; but in a single user session only their *immediate* behavior is visible. A change could have different effects in the short and long run, and sometimes even opposite effects. Thus, experiments that measure immediate behavioral effects are essentially short-loop experiments, with a corresponding loss of fidelity. A company like Amazon can instead track the long-term buying behavior of customers in response to minor changes, but such experiments are slower and noisier. It is also difficult to measure the long-term effects on customers who do not log in to sites.

APPLE BREEDING: FIVE-FOLD REDUCTION OF INFORMATION CYCLE TIME

Breeding plants for better properties is an activity that is probably millennia old. Norman Borlaug's "Green Revolution" bred varieties of rice and other traditional crops that were faster growing, higher yielding, and more resistant to drought and disease. However, some plants, including apple trees and related fruits, have poor properties for breeding (Kean 2010). One difficulty is that apple trees take about five years to mature and begin to bear fruit, leading to very long experimental cycles. Finding out whether a seedling is a dwarf or full-sized tree can take 15 years. A second difficulty is that, because of Mendelian inheritance, traditional breeding produces a high percentage of undesirable genetic combinations, which cannot be identified until the trees begin bearing. Finally, the underlying genetic variation in American apples is small, as virtually all domestic plants are cloned from a small number of ancestors, which themselves originated from a narrow stock of seeds promulgated by Johnny Appleseed and others 200 years ago. As a result, apple-breeding experiments have been slow and expensive, with low variation in the underlying "ideas."

Botanists are now using four techniques to accelerate the rate of learning about apples. First, a USDA scientist collected 949 wild variants of the ancestor to domestic apples from Central Asia. This approximately doubled the stock of

---

[*] Erik Brynjolfsson, quoted in Hopkins (2010, 52). The number 50 to 200 experiments at once is given in Gomes (2008).

apple-specific genes available as raw material for learning (idea variety). Second, the time from planting to first apples is being shortened from five to about one year by inserting "fast flowering" genes from a poplar tree (information cycle time). Third, when two trees with individual good traits are crossed to create a combination of the two traits, DNA screening will be able to select the seedlings that combine both of the favored traits, without waiting until they mature. This reduces the number that must be grown by 75% (short-loop experiment to reduce cost). Finally, some of the wild trees were raised in a special greenhouse deliberately full of pathogens (specialized learning location, to improve cycle time and S/N).

SHORT-LOOP EXPERIMENTS FOR AIDS: THE CRITICAL ROLE OF MEASUREMENT

The human immunodeficiency virus (HIV) is the virus that eventually leads to acquired immunodeficiency syndrome (AIDS), but the delay between the initial HIV infection and AIDS onset can be many years, even in untreated individuals. This delay made experiments on both prevention and AIDS treatment quite slow. It was not even possible to be sure someone was infected until they developed AIDS symptoms. Both controlled and natural experiments were very slow in consequence. In the early 1990s, new tests allowed a quantitative measurement of HIV viral loads in the bloodstream. Researchers hypothesized that viral load might be a proxy measurement for the occurrence and severity of infection, potentially permitting short-loop experiments that would be years faster than waiting for symptoms to develop. There is, however, always the question of the fidelity of short-loop experiments: is viral load truly a good proxy for severity of infection?

Using a controlled experiment to validate the viral load measure would be impossible for ethical and other reasons. As an alternative, a natural experiment consists of measuring viral loads in a number of individuals who may have been exposed, and then waiting to see what happens to them. Such an experiment would take years, and it would require a large sample to get a decent S/N ratio. Fortunately, it was possible to run a quick natural experiment using historical data (Mellors 1998). Mellors and colleagues measured viral loads in stored blood plasma samples from 1600 patients. These samples had been taken 10 years earlier, before the measurement technique existed. From the samples plus mortality data on the individuals, they calculated the survival and life expectancy rates as a function of the initial viral load. They found very strong relationships. For example, if the viral load was below 500, the six-year survival rate was over 99% versus 30% if the load was over 30,000.

This natural experiment on historical data achieved a high S/N ratio in a relatively short period. However, even with strong results, "correlation does not prove causation." In medicine the use of historical data is called a "retrospective trial." Fortunately, several large-scale prospective and controlled trials were able to demonstrate that if a treatment reduced viral load, it also reduced the likelihood of the disease progressing from infection to full AIDS. Therefore, viral load became a useful proxy for short-loop experiments on possible treatments, as well as for treating individual patients.

An ongoing example of developing a new measurement to permit shorter loop experiments is the recent discovery of seven biomarkers for kidney damage (Borrell 2010). These biomarkers permit faster detection and are sensitive to lower levels of damage in animal studies, and presumably for humans as well. This will allow for the better assessment of toxicity at an earlier stage of drug trials.

Both examples highlight how identifying new variables and developing practical measurements for them can accelerate learning rates several-fold. Especially good variables can predict final outcomes early, thus allowing for more focused experiments.

FIDELITY PROBLEMS DUE TO LOCATION IN CLINICAL TRIALS

Sadly, searches for short-loop measurements that predict the course of illness are not always so successful. A particularly insidious example of the problem with short-loop experiments is the use of selective health effects as the outcome measurement in controlled clinical drug trials. If a drug is designed to help with disease X, in a large clinical trial should the outcome measure be symptoms related to X, clinical markers for X, mortality due to X, or measures of total health? Should immediate effects be measured, or should patients be followed over multiple years? The assumed causal chain is: new medication → clinical markers for disease X → symptoms for X → outcomes for X → total health of those taking the medication. The target outcome is the *total* health effect of the new medication, and if the disease is a chronic one that requires long-term treatment, this can take years to measure. Looking at just the markers for disease X, or even at the outcomes of disease X alone, are forms of short-loop experiments.

However, the problem with short-loop experiments is fidelity: Is the result for disease X a reliable predictor of overall health effects? In the human body, with its variety of interlocking homeostatic (feedback) systems, multiple complex effects are common. Even if X gets better, overall health may not. Although clinical trials are supposed to look for side effects, side effect searches generally have a low S/N. They are also subject to considerable conflict of interest, as the drug trials are usually paid for by the pharmaceutical company. There is ample evidence that this sometimes affects the interpretation and publication of study results (see, e.g., DeAngelis and Fontanarosa 2008).

A tragic example of this problem was the anti-diabetes drug, Avandia. It was one of the world's highest-selling drugs, with sales of $3.2 billion in 2006 (Harris 2010). Untreated diabetes raises the mean and variance of blood glucose levels, and better control of the glucose level is viewed as a good indicator of effectiveness in treating diabetes. However, once Avandia was put on the market the drug caused serious heart problems. In fact, one estimate was that it raised the odds of death from cardiovascular causes by 64% (Nissen and Wolski 2007). Given the already high risk of heart attacks for diabetics, this indicates that it decreased overall survival. Unfortunately, many drug trial reports do not even list overall mortality for the treatment and placebo populations, making it impossible to calculate the overall effects. Doctors are essentially forced to assume that the short-loop experiment has adequate fidelity for overall patient health.

Using intermediate disease markers for a focused clinical trial is not the only problem with the fidelity of clinical trials. Trial sample populations are carefully screened to improve the experimental properties for the primary (desired) effect. This is a move along the analytical/physical spectrum, because the trial population is not the same as the general population, which will take the drug if it is approved. For example, trials often reject patients who are suffering from multiple ailments or are already taking another medication for the targeted disease. For ethical reasons, they almost always avoid juvenile patients. These changes potentially reduce the fidelity of the experiment.

An example is the use of statins to reduce cholesterol and heart disease. While multiple clinical studies show that statins lower the risk of heart attack, they also have side effects that include muscle problems, impotence, and cognitive impairment (Golomb and Evans 2008). Few of the clinical trials of statins measure *overall* mortality, and the available evidence suggests that they improve it only in men with pre-existing heart disease. Most of the clinical trials have been conducted on this population, yet statins are now widely prescribed for women and for men with high cholesterol but without heart disease. The evidence suggests that for these populations, they do not actually improve overall mortality (Parker-Pope 2008). Since they also have side effects, this suggests that they are being widely over-prescribed.

ASTRONOMY AND OTHER OBSERVATIONAL FIELDS

In some situations, causal variables cannot be altered and so only natural experiments are possible. However, when observations are expensive and difficult, the learning process is very similar to a series of controlled experiments. Often the experiments require elaborate data collection, and the choice of where to observe and what to collect raises the same issues as setting up controlled experiments. The only difference is that there is usually no choice of location, which is at the "real world" end of the physical-analytical spectrum.[*]

The classic observational science is astronomy, since extra-planetary events cannot be altered, though the observation time on instruments is scarce and rationed, especially for wavelengths blocked by Earth's atmosphere. One approach to finding extra-solar planets is to look for periodic dimming in the light from a star. The magnitude of the dimming gives an indication of planetary size. The S/N ratio of the measurements is critical, as small planets will have little effect. Choosing stars to observe that are more likely to have planets and to have high S/N ratios is therefore crucial. Even so, a large sample is needed because if the planet's orbit is not on the same plane as our sun, no occlusion will be visible. A special satellite mission, "Kepler," has been launched for this purpose (Basri et al. 2005). A recent paper posits that the Kepler mission can also detect Earth-sized moons in habitable zones (Kipping et al. 2009). The authors used mathematical models of hypothetical moons to evaluate different detection strategies and their statistical efficacy.

---

[*] Even in observational fields like astronomy, controlled experiments are possible using mathematical models, though the results cannot be validated by controlled experiments in the real world.

CONCLUSIONS

This chapter shows how the rate of learning by experimentation—and, by extension, the slope of many learning curves—is heavily determined by the ways in which experiments are conducted. We identified five criteria that measure experimental effectiveness. They are: information cycle time (speed), S/N ratio, cost, value and variety of ideas, and fidelity. The statistical literature on experimentation deals formally with only one of these, the S/N ratio, but it offers insights that are helpful in dealing with the others. For example, the powerful method of *fractional factorial* experiments looks at the effects of multiple variables simultaneously rather than one at a time, thereby improving both speed and cost.

These five criteria are, in turn, largely determined by how the experiment is designed, and we discuss two high-level design decisions: location and type of experiment. "Location" has two orthogonal axes, referred to as analytical to physical, and focused to comprehensive. Generally, more analytical or more focused locations reduce the fidelity of the experiment, but improve other criteria such as information cycle time and cost. We discuss four types of experiments: controlled, natural, evolutionary, and ad hoc. Although controlled experiments are often viewed as inherently superior, for some types of problems they are impossible, and in other cases they are dominated by other types of experiments.

We have not discussed the meta-problem of designing good learning environments. For example, "just in time" manufacturing principles encourage just the key properties of fast information cycle time and good S/N ratio, while providing perfect fidelity (Bohn 1987). We have also said little about *what* is being learned, since it is heavily situation specific. However, G. Taguchi claimed that the knowledge being sought in experiments was often too narrow. He pointed out that in many situations the *variability* in outcomes is just as important as the mean level of the outcome. He proposed specific experimental methods for simultaneously measuring the effects of process changes on both mean and variation. For example, he suggested that pilot lines and prototypes should be built with material of normal quality, rather than using high-quality material to improve the S/N ratio of experiments. Taguchi's insight reminds us that "the first step in effective problem solving is choosing the right problem to solve."

ACKNOWLEDGMENTS

A multitude of colleagues, engineers, and managers have helped with these ideas over many years. Roger Bohn's research on this topic was supported by the Harvard Business School Division of Research, and the Alfred P. Sloan Foundation Industry Studies Program. His friends at HBS were especially influential, including Oscar Hauptman, Jai Jaikumar, Therese Flaherty, and Roy Shapiro. Gene Meiran and Rick Dehmel taught him about semiconductors. Jim Cook provided a number of comments and ideas. Michael Lapré's work was supported by the Dean's Fund for Faculty Research from the Owen School at Vanderbilt.

REFERENCES

Adler, P.S., and Clark, K.B., 1991. Behind the learning curve: A sketch of the learning process. *Management Science* 37(3): 267–281.

Applebaum, W., and Spears, R.F., 1950. Controlled experimentation in marketing research. *Journal of Marketing* 14(4): 505–517.

Basri, G., Borucki, W.J., and Koch, D., 2005. The Kepler mission: A wide-field transit search for terrestrial planets. *New Astronomy Reviews* 49(7–9): 478–485.

Bohn, R.E., 1987. *Learning by experimentation in manufacturing.* Harvard Business School Working Paper 88–001.

Bohn, R.E., 1994. Measuring and managing technological knowledge. *Sloan Management Review* 36(1): 61–73.

Bohn, R.E., 1995a. Noise and learning in semiconductor manufacturing. *Management Science* 41(1): 31–42.

Bohn, R.E., 1995b. The impact of process noise on VLSI process improvement. *IEEE Transactions on Semiconductor Manufacturing* 8(3): 228–238.

Bohn, R.E., 2005. From art to science in manufacturing: The evolution of technological knowledge. *Foundations and Trends in Technology, Information and Operations Management* 1(2): 129–212.

Borrell, B., 2010. Biomarkers for kidney damage should speed drug development. *Nature,* May 10. http://www.nature.com/news/2010/100510/full/news.2010.232.html?s=news_rss (accessed July 9, 2010).

Box, G.E.P., Hunter, J.S., and Hunter, W.G., 1978. *Statistics for experimenters.* New York: Wiley.

DeAngelis, C.D., and Fontanarosa, P.B., 2008. Impugning the integrity of medical science: The adverse effects of industry influence. *Journal of the American Medical Association* 299(15): 1833–1835.

Dutton, J.M., and Thomas, A., 1984. Treating progress functions as a managerial opportunity. *Academy of Management Review* 9(2): 235–247.

Golomb, B.A., and Evans, M.A., 2008. Statin adverse effects: A review of the literature and evidence for a mitochondrial mechanism. *American Journal of Cardiovascular Drugs* 8(6): 373–418.

Gomes, B., 2008. *Search experiments, large and small.* Official Google blog http://googleblog.blogspot.com/2008/08/search-experiments-large-and-small.html (accessed July 9, 2010).

Harris, G., 2010. Research ties diabetes drug to heart woes. *The New York Times,* February 19.

Hopkins, M.S., 2010. The four ways IT is revolutionizing innovation – Interview with Erik Brynjolfsson. *MIT Sloan Management Review* 51(3): 51–56.

Hunter, W.G., and Kittrell, S.R., 1966. Evolutionary operations: A review. *Technometrics* 8(3): 389–397.

Juran, J.M., and Gryna, F.M., 1988. *Juran's quality control handbook.* 4th ed. New York: McGraw-Hill.

Kean, S., 2010. Besting Johnny Appleseed. *Science* 328(5976): 301–303.

Kipping, D.M., Fossey, S.J., and Campanella, G., 2009. On the detectability of habitable exomoons with Kepler-class photometry. *Monthly Notices of the Royal Astronomical Society* 400(1): 398–405.

Lapré, M.A. 2011. Inside the learning curve: Opening the black box of the learning curve. In *Learning curves: Theory, models, and applications.* ed. M.Y. Jaber. Chapter 2. Boca Raton: Taylor & Francis.

Lapré, M.A., Mukherjee, A.S., and Van Wassenhove, L.N., 2000. Behind the learning curve: Linking learning activities to waste reduction. *Management Science* 46(5): 597–611.

Lapré, M.A., and Van Wassenhove, L.N., 2001. Creating and transferring knowledge for productivity improvement in factories. *Management Science* 47(10): 1311–1325.

Lee, H., Whang, S., Ahsan, K., Gordon, E., Faragalla, A., Jain, A., Mohsin, A., Guangyu, S., and Shi, G., 2003. *Harrah's Entertainment Inc.: Real-time CRM in a service supply chain.* Stanford Graduate School of Business Case Study GS-50.

Leonard-Barton, D., 1988. Implementation as mutual adaptation of technology and organization. *Research Policy* 17(5): 251–267.

Leonard-Barton, D., 1995. *Wellsprings of knowledge: Building and sustaining the sources of innovation.* Cambridge: Harvard Business School Press.

Little, J.D.C., 1966. A model of adaptive control of promotional spending. *Operations Research* 14(6): 1075–1097.

Logothetis, N.K., 2008. What we can do and what we cannot do with fMRI. *Nature* 453:869–878.

March, J.G., and Simon, H.A., 1958. *Organizations.* New York: Wiley.

Mellors, J.W., 1998. Viral-load tests provide valuable answers. *Scientific American* 279:90–93.

Murnane, R., and Nelson, R.R., 1984. Production and innovation when techniques are tacit: The case of education. *Journal of Economic Behavior and Organization* 5(3–4): 353–373.

Nelson, R.R., 2008. Bounded rationality, cognitive maps, and trial and error learning. *Journal of Economic Behavior and Organization* 67(1): 78–87.

Newell, A., and Simon, H., 1972. *Human problem solving.* Englewood Cliffs: Prentice-Hall.

Nissen, S.E., and Wolski, K., 2007. Effect of Rosiglitazone on the risk of myocardial infarction and death from cardiovascular causes. *The New England Journal of Medicine* 356(24): 2457–2471.

Parker-Pope, T. 2008. Great drug, but does it prolong life? *The New York Times*, January 29.

Pearl, J., 2001. *Causality: Models, reasoning, and inference.* Cambridge: Cambridge University Press.

Shankland, S., 2008. We're all guinea pigs in Google's search experiment. *CNET News*, May 29. http://news.cnet.com/8301-10784_3-9954972-7.html (accessed July 9, 2010).

Terwiesch, C., and Bohn, R.E., 2001. Learning and process improvement during production ramp-up. *International Journal of Production Economics* 70(1): 1–19.

Terwiesch, C., and Ulrich, K.T. 2009. *Innovation tournaments.* Boston: Harvard Business School Press.

Thomke, S.H., 1998. Managing experimentation in the design of new products. *Management Science* 44(6): 743–762.

Ulrich, K.T., and Eppinger, S.D., 2008. *Product design and development.* New York: McGraw Hill.

Varian, H.R., 2006. The economics of internet search. *Rivista di Politica Economica* 96(6): 9–23.

Weber, C., 2004. Yield learning and the sources of profitability in semiconductor manufacturing and process development. *IEEE Transactions on Semiconductor Manufacturing* 17(4): 590–596.

12 Linking Quality to Learning – A Review

*Mehmood Khan, Mohamad Y. Jaber,
and Margaret Plaza*

CONTENTS

INTRODUCTION

Managers have long been looking for ways to improve the productivity of their companies. It has become imperative for an enterprise to look for tools that reduce costs and improve productivity at the same time. The world has seen American and Japanese companies struggling to achieve this goal in the final quarter of the last century. It was an eye for enhanced quality that gave Japanese manufacturers an edge. On the other hand, researchers also believe that an organization that learns faster will have a competitive advantage in the future (Kapp 1999).

In the most simplistic model of learning, in which learning is a by-product of doing, cumulative output is functionally related to average cost reduction. However, firms often consciously focus on learning in order to trigger technological advancement and quality improvements beyond simply reducing average cost (Cohen and Levinthal 1989). Malreba (1992) identified six key types of learning in organizations, which can be closely interrelated and play a dominant role in product life cycle: (1) learning by doing, (2) learning by using, (3) learning from advancements in science and technology, (4) learning from inter-industry spillovers, (5) learning by interacting, and (6) learning by searching. Types 1, 2, and 3 are internal to the firm, while the other three types are external. In the case of a simple learning curve, only the first two types are linked together. However, for more complex improvements of processes and products, other types must also be taken care of. For example, since learning by searching is aimed at the improvement/generation of new products or processes, this type is dominant during research and development (R&D). Learning by searching is often coupled with learning by doing and targets improvements of quality, reliability, performance, and compatibility (Malreba 1992).

This chapter will shed light on the relationship between quality and complex learning processes, which may incorporate some, or all, of those learning types. This linkage is crucial in that it has become vital in increasing the productivity of a company in the past few decades.

WHAT IS QUALITY?

There are many ways to define quality (Garvin 1987). A common definition in the industry is: "meeting or exceeding customer expectations" (Sontrop and MacKenzie 1995). Many companies refine their processes/products to meet customer expectations based on their surveys. The customers can be internal or external. For example, the internal customers for a fuel tank would be an assembly line or the paint shop, while its external customers would be a car dealer or the purchaser.

The definition of quality emphasized in this chapter is "conformance to specifications." Specifications are target values and tolerances such as the length of a trunk lid can be 150 ± 1 cm. That is, a conforming length falls in the interval from 149 to 151 cm. Targets and tolerances are set by the design and manufacturing engineers in a plant. The other characteristics of interest can be design configuration, such as the weight, thickness, reliability, ease of fitness, and so forth.

Statistical quality control or statistical process control (SPC) tries to understand and reduce the variations in manufacturing processes (Yang et al. 2009).

The measures of variation can be accuracy, precision, bias, stability, and so on. The role of quality control is to reduce the variability in a characteristic around its target. Quality control involves the employment of established procedures to assess, attain, uphold, and enhance the quality of a product or service by reducing the variability around a target. Therefore, the most useful definition for the technical application of quality control is conformance to specifications.

SPC is the component of a continuous quality improvement system that consists of a set of statistical techniques to analyze a process or its output, and to take appropriate actions to achieve and maintain a state of statistical control. SPC is not a short-term fix. The most successful organizations today have learned that SPC only works when the operating philosophy is that everyone in the organization is responsible for, and committed to, quality. SPC on the methods by which results are generated—on improvements to the processes that create products and services of the least variability. The traditional tools that SPC uses to improve on variability are: (i) flow charts, (ii) cause-and-effect diagrams, (iii) pareto charts, (iv) histograms, (v) scatter plots, and (vi) control charts.

With this notion of quality or quality control in mind, industry today refers to the fraction of defective items in their production, or in the supplier lots, as the "quality" of the lot produced or received. This chapter will explore if this quality improves by virtue of learning or, in other words, by experience and repetition. Let us have a formal understanding of the learning process.

LEARNING BEHAVIOR

Learning in an organization has been receiving more and more attention. Steven (1999) presented examples from electronics, construction, and aerospace industries to conclude that learning curves will gain more interest in high-technology systems. Wright (1936) was the first to model the learning relationship in a quantitative form. This complex behavior has been given different names over the last century, such as start-up curves (Baloff 1970), progress functions (Glover 1965), and improvement curves (Steedman 1970). However, researchers have agreed that the power-form learning curve is the most widely used form to depict the learning phenomenon (Yelle 1979).

It is very hard to define this complex behavior, but practitioners and researchers mostly believe that it can be defined as the trend of improvement in performance, achieved by virtue of practice. The Wright (1936) learning curve states that the time to produce every successive unit in repetition keeps on decreasing until plateauing occurs. Plateauing is a state where a system, or a worker, ceases to improve on their performance. The reason for this could be that the worker ceases to learn, or it could be the unwillingness of the organization to invest any more capital. The mathematical form of Wright's model is given by:

$$T_x = T_1 x^{-b}, \tag{12.1}$$

where x is the tally of the unit being produced, T_x and T_1 are the times to produce the xth and the first unit, respectively, and b is the learning exponent. The learning

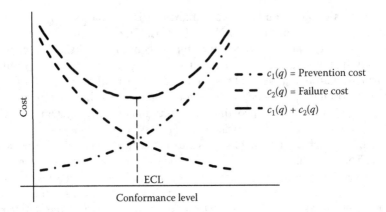

FIGURE 12.1 Quality-related costs.

exponent in this expression is often referred to as an index called the "learning rate" (*LR*). Learning occurs each time the production quantity is doubled, such as:

$$LR = \frac{T_{2x}}{T_{1x}} = \frac{T_1(2x)^{-b}}{T_1(x)^{-b}} = 2^{-b},$$
(12.2)

Thus, following the above learning curve, time to produce x units, $t(x)$, is given by:

$$t(x) = \sum_{i=1}^{x} T_1 i^{-b} \approx \int_{i=0}^{x} T_1 i^{-b} di = \frac{T_1}{1-b} x^{-b}.$$
(12.3)

Although Figure 12.1 and the above expression represent the improvement in the time to process a unit, learning can be shown in the cost, productivity, and other similar measures of a production system. Next, we categorize the linkage between quality and learning by exploring the literature that finds a (i) mathematical and/or (ii) empirical relationship.

LITERATURE ON THE QUALITY-LEARNING RELATIONSHIP

MATHEMATICAL RELATIONSHIP

Wright (1936) model

Wright (1936) was probably the first to come up with a relationship that signifies the importance of experience or learning in a production facility. He studied the variation in production cost with production quantity. This issue was of increasing interest and importance because of a program sponsored by the Bureau of Air Commerce for the development of a small airplane. Wright had started working on the variation of cost with quantity in 1922. A curve depicting such variation was worked up empirically from the two or three points that previous production experience of the same model in differing quantities had made possible. Through the succeeding years, this original curve, which at first showed the variation in labor only, was used for estimation purposes and was corrected as more data became available. This curve was found to take an exponential form.

Wright showed that the factors that make the reduction in cost possible with the increase in production quantity are labor, material, and overheads. The labor cost factor (F) and the production quantity (N) followed a logarithmic relationship:

$$X = \frac{\log F}{\log N}. \tag{12.4}$$

where log is a logarithm of base 10. A plot of this relationship resulted in a value of 0.322 for X; that is, it was an 80% curve. This represents a factor by which the average labor cost in any quantity shall be multiplied in order to determine the average labor cost for a quantity of twice that number of airplanes. On the other hand, material cost also decreases with the increase in quantity because waste is cut down. Wright showed these variations in price with an actual example and compared the savings in cost between the production of cars and airplanes.

It should be noticed that the traditional way to obtain data for a learning cycle is often erroneous, as the individual data are composed of some variance. To counter this, Zangwill and Kantor (1998) proposed to measure the individual improvements directly and to use the learning cycle repeatedly. This would make the management observe the techniques that produce greater improvement and thus, managers would learn how to improve processes faster. They came up with a differential equation that was composed of three forms of learning: power form, exponential form, and the finite form. Zangwill and Kantor (2000) extended on their earlier work and emphasized that traditional learning curves cannot identify which techniques are making improvements and which are not, on a period-by-period basis. Their approach helps to boost the rate of improvement in every cycle and makes learning a dynamic process.

The same analogy can be applied to emphasize the relationship of the quality of a product or service (Anderson 2001) with experience (learning), which is the objective of this chapter.

Fine (1986)

Levy (1965) is believed to be the first to capture the linkage between quality and learning. Fine (1986) advocated that learning bridges quality improvements to productivity increases. He found this to support the observation of Deming (1982) that quality and productivity go together as productivity increases result from quality improvement efforts.

Fine (1986) modeled a quality-learning relationship. He was of the view that product quality favorably influences the rate of cost reduction when costs are affected by quality-based learning. Thus, costs decline more rapidly with the experience of producing higher-quality products. He presented two models for quality-based learning.

The first model assumes that quality-based experience affects the direct manufacturing costs. Whereas the second model assumes that the quality-based experience affects quality control costs. For the second model, Fine (1986) found that the optimal quality level increases over time.

According to Fine (1986), a key feature of the second quality-based model is that it resolves the inconsistency between the cost tradeoff analysis of Juran (1978) to find the optimal quality level and the "zero defects is the optimal quality level" as per Crosby (1979) and Deming (1982).

Fine (1986) claimed that firms that choose to produce high-quality products will learn faster than firms producing lower quality products. He discussed price (cost per unit) and quality relationship of Lundvall and Juran (1974). Figure 12.1 shows that a firm may choose quality (fraction of defectives) levels greater than the economic conformance level (ECL, Juran 1978), but no rational was provided to why a firm would ever choose a quality level smaller than the ECL. Prevention and failure costs—$c_1(q)$ and $c_2(q)$, respectively—are also shown in Figure 12.1.

Fine (1986) found that the prescription of an optimal conformance level with a strictly positive proportion of defects in Figure 12.1 is in direct opposition to the literature (Crosby 1979; Deming 1982) that recommends zero defects as the optimal conformance level.

Fine's quality-based learning theory added a dynamic learning curve effect to the static ECL model so that the modified model is consistent with the slogan that "higher quality costs less."

In the quality-based learning formulation adopted by Fine (1986), two types of learning were modeled: Induced and autonomous. Induced learning is the result of managerial and technical efforts to improve the efficiency of the production system, whereas autonomous learning is due to repetition or learning by doing.

Fine (1986) also reviewed two quality-based learning models—Lundvall and Juran (1974) and Spence (1981). The first model assumed that learning reduces the direct costs of manufacturing output without affecting quality-related costs. The second model assumed that learning improves quality without affecting the direct manufacturing costs. He then presented two quality-based learning models to derive optimal quality, pricing, and production policies.

Fine's (1986) first quality-based learning model was for manufacturing activities. He related the conformance level (q) and experience (z) as:

$$c(q,z) = c_1(q) + c_2(q) + c_3(z), \qquad (12.5)$$

where $c_3(z)$ is positive, decreasing, and convex, and becomes a constant, c_3, when z goes to infinity. In Equation 12.5, cumulative experience affects direct manufacturing costs through the function $c_3(z)$. This suggests that as experience is gained from producing larger volumes at higher quality levels, manufacturing costs are reduced.

In his second model, Fine (1986) assumed that quality-based learning benefits lead to a reduction in the appraisal and prevention expenditures required to attain any given quality level. That is, the learning benefits accumulate in the appraisal and prevention activities. This idea was modeled by assuming a cost function of the form:

$$c(q,z) = a(z)c_1(q) + c_2(q) + c_3. \tag{12.6}$$

The function $a(z)$ represents the learning in appraisal and prevention activities, while the constant a represents a limit to the improvement that is possible, where $a(z)$ is decreasing and convex for $z \in (0, \infty)$. He also assumed that $a(0) = 1$ and $\lim_{x \to \infty} a(x) = a \geq 0$. Note that c_3 is a constant in this formulation.

Figure 12.2 illustrates the effect of increasing experience on the cost function described above. The solid curve represents the quality-related costs, failure costs, and the appraisal and prevention costs for the case where $z = 0$. The dashed curves represent the quality-related costs for $z > 0$; that is, after some learning has taken place. Note that $a(0)c_1(q) + c_2(q)$ has its minimum at q^*—as in the ECL model. For any z, $a(z)c_1(q) + c_2(q)$ will have a unique minimum at $q^*(z)$, where $q^*(z) = q^*$ and $q^*(z)$ increases in z. That is, if the experience level at time t is $z(t)$, then the optimal quality is $q^*(z)$, which rises with $z(t)$.

Fine (1986) concluded that (i) the optimal pricing policy under a quality-based learning curve is qualitatively similar to the optimal pricing policy under a volume-based learning curve; (ii) optimal quality levels decrease (or increase) over time if learning reduces direct manufacturing (appraisal and prevention) costs; and (iii) the optimal quality level under a quality-based learning curve exceeds the optimal quality level in the corresponding static, no-learning case.

Tapiero (1987)

Tapiero (1987) discussed the practice in manufacturing in which quality control is integrated into the process of production in altering both the product design and the

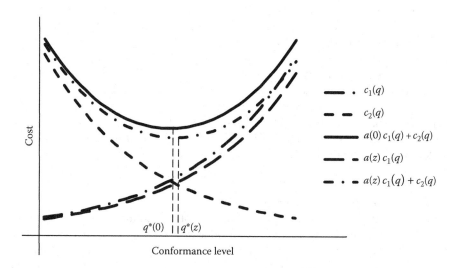

FIGURE 12.2 Quality-related costs.

manufacturing techniques in order to prevent defective units. He developed a stochastic dynamic programming problem for determining the optimal parameters of a given quality control policy with learning. He defined the knowledge in a production facility with quality control as a learning function given by:

$$x_{t+1} = f(x_t, N_t, l_t); \quad x_0 > 0, \tag{12.7}$$

where x_t is the variable for knowledge at time t, N_t is the sales volume, and l_t is the variable denoting the quality control process. This learning function had the following properties:

$\partial f/\partial x < 0$ reflecting the effect of forgetting in the learning process
$\partial f/\partial l > 0$ reflecting the positive effects of quality control on the learning process
$\partial f/\partial N > 0$ reflecting the positive effects of production volumes on the learning process

With this, Tapiero developed a profit function comprising of the costs, sales, failures, and reworks to be solved through stochastic dynamic programming.

The approach and the results obtained were interpreted and extended by Tapiero (1987) in several ways. In particular, the analytical solution for the risk-neutral, curtailed sampling case has shown that:

a. When the costs of sampling are smaller than the expected failure costs, it is optimal to have full sampling, regardless of experience and learning.
b. When the costs of sampling are larger than the expected failure costs, two situations arise. First, the optimal sample size is bang-bang, a function of experience accumulated through quality control learning. "The larger the experience, the smaller the amount of quality control."
c. When the inspection costs are not "that much greater" than the expected failure costs, experience and high quality control can move together. This was also pointed out by Fine (1986).

Thus, he concluded that the optimal quality control policy that a manufacturer may use is not only a function of the costs of inspection and the costs of products' failures, but also of the manufacturer's ability to use the inspected samples in improving (through "experience") the production technology. This observation is in line with the Japanese practice of "full inspection" and of learning "as much as possible" in order to finally obtain a zero-defects production technology.

Fine (1988)
To help explain how one firm can have both higher quality and lower costs than its competitors, Fine (1988) explored the role of inspection policies in quality-based learning, a concept which was introduced by Fine (1986) and extended by Tapiero (1987). He discussed an imperfect production process. Fine (1988) noted that the process may produce defective items for several reasons: poorly designed or poorly constructed materials and/or components, substandard workmanship,

faulty or poorly maintained equipment, or ineffective process controls. The model that Fine (1988) developed permits the production process to be improved through "quality-based learning" by the operators. It has tight quality control standards for the time in the "out-of-control" state, enforced by intensive inspection procedures, thus resulting in faster learning. Fine (1988) found that this leads to lower long-run failure costs than the short-run inspection costs. This work of Fine (1988) is different from that of Fine (1986) and Tapiero (1987) as cumulative output was not used as a measure of learning. Fine (1988) concluded that his model cautions managers responsible for quality policies that they may be choosing suboptimal inspection policies if potential learning benefits from inspection were ignored.

Fine (1988) imagined a work area, its inputs, and its outputs as a station. This station could be either in an "in-control" or an "out-of-control" state, denoted by I and Ω, respectively. Time was indexed by the production of output; each unit produced would advance the clock by one. Before production in each time period, the firm inspects the station to verify that it is not in an out-of-control state. The system is restored if it is so. He denoted X_t, as the state of the tth station prior to the tth decision whether or not to inspect the station. X^t would be the state of the station following any inspection and/or repair at time t but before production of the tth unit. Thus, $X_t = \Omega$ ($X^t = \Omega$) would mean that the state before (or after) the inspection decision at time t is out of control. This way, he defined the state of the station through a time-homogeneous Markov chain as:

$$\begin{array}{c} I \\ \Omega \end{array} \begin{pmatrix} I & \Omega \\ 1-h(n) & h(n) \\ 0 & 1 \end{pmatrix}; \quad 0 \le h(n) \le 1 \; \forall n \ge 0. \tag{12.8}$$

The symbol n serves as an index of cumulative learning and $h(n)$ is the probability that the station will go from the in-control to the out-of-control state. The probability of qth unit being defective in the in-control and out-of-control states is q_1 and q_2, respectively. He assumed that $q_1 < q_2$.

Learning was captured in this model by taking $h(n)$ as a decreasing function of n and incrementing n by one each time the station is inspected and it is found to be out of control. Inspecting the station and finding it out-of-control constitutes a learning event that allows the workers to improve the process so that the probability that the station goes out of control in the future, is reduced. He assumed that $h(n)$ takes the form $h(n) = \gamma^n h(0)$, where $0 \le \gamma \le 1$ and $0 \le h(0) \le 1$. This may be interpreted as a station having a sequence of "layers" of quality problems. Each layer must be discovered and fixed before the next layer of problems can be uncovered.

Before producing the tth unit, Fine (1988) noted that the firm must decide whether or not to expend c (> 0) dollars to inspect the station and learn whether it is, in fact, truly out of control. He denoted this action by a_t; $a_t = 0$ for "do not inspect" and $a_t = 1$ for "inspect." Once the unit of time t has been produced, its quality (denoted by Y_t) is classified as either defective ($Y_t = D$) or non-defective ($Y_t = N$). Fine (1988) also assumed that a defective unit can be easily detected by visual inspection. He denoted the probability that item t is defective given that the station is in control at time t is by q_1—that is, $P\{Y_t = D \mid X_t = I\} = q_1$. Similarly, $P\{Y_t = D \mid X_t = \Omega\} = q_2$, and he assumed that $1 \geq q_2 > q_1 \geq 0$. He assumed the repair/replace cost of each defective item as $d\ (> 0)$. He further assumed that $A_t = 1$ if, and only if, the inspection action was taken ($a_t = 1$), and the station was found to be out of control ($X_t = \Omega$). That is, "learning took place" at time t. Therefore, the history of the process immediately preceding the observation of the output at time t would be:

$$H_t = \{X_0, a_1, A_1, Y_1, a_2, A_2, Y_2, \ldots, a_t, A_t\}. \tag{12.9}$$

Fine (1988) denoted the probability that the system is out of control as:

$$P_t = P\{X^t = \Omega \mid H_t\}. \tag{12.10}$$

He noted that if the "inspect" action is chosen ($a_t = 1$), then the station will be in control for certain, either because it was found to be in control or because it was adjusted into control. Whereas, if the "do not inspect" action is chosen ($a_t = 0$), then no additional information was assumed by Fine (1988) to be available to update the probability that the state is out-of-control. If $P_t = p$, then after the production of the tth unit and the observation of Y_t (the quality of that unit), he assumed that the probability that the station was out of control at the time of production, is updated by using Bayes' theorem:

$$P_t = P\{X^t = \Omega \mid H_t, Y_t = D\} = \frac{q_2 p}{q_2 p + q_2(1-p)}; \quad \forall Y_t = D,$$

$$P_t = P\{X^t = \Omega \mid H_t, Y_t = N\} = \frac{(1-q_2)p}{(1-q_2)p + (1-q_1)(1-p)}; \quad \forall Y_t = N. \tag{12.11}$$

Fine (1988) used transition matrix (12.8) to compute the post-production probability that the station will be out of control for the next unit. That is,

$$P_t = P\{X_{t+1} = \Omega \mid H_t, Y_t = D\} = \frac{q_2 p + h(n)q_1(1-p)}{q_2 p + q_2(1-p)}; \quad \forall Y_t = D,$$

$$P_t = P\{X_{t+1} = \Omega \mid H_t, Y_t = N\} = \frac{(1-q_2)p + h(n)(1-q_1)(1-p)}{(1-q_2)p + (1-q_1)(1-p)}; \quad \forall Y_t = N. \tag{12.12}$$

He reviewed the "no learning" case before examining the effects of learning opportunities on the optimal inspection policy.

Chand (1989)

Chand (1989) discussed the benefits of small lot sizes in terms of reduced setup costs and improved process quality due to worker learning. He showed that the lot sizes do not have to be equal in the optimal solution, even if the demand rate is constant.

Chand (1989) adopted the approach of Porteus (1986) in estimating the number of defective units in a lot. Porteus (1986) assumed that for very small probability (ρ) that a process goes out-of-control, the expected number of defective items in a lot of size Q could be approximated as:

$$d(Q) = \frac{\rho Q^2}{2}.$$ (12.13)

Like Porteus (1986), Chand (1989) assumed that the process is in control at the start of a production lot and no corrective steps are taken if the process goes out of control while producing a lot. Both works also assume that before starting a new production lot, the process is in control. Chand's (1989) expected cost with Nq setups is written as:

$$TC(Nq) = \sum_{n=1}^{N\rho} K(n) + \frac{D}{2N\rho}(H + C\rho D),$$ (12.14)

where K is the cost of each setup, H and C are the costs of the non-defective and defective units, respectively, D is the demand rate per period and the objective is to find an optimal lot size.

If there is learning in the process quality with each setup, then:

$$\rho(1) \geq \rho(2) \geq \rho(3) \geq \ldots,$$ (12.15)

where the minimum and the maximum probability values are $\rho_\infty = \lim_{n \to \infty} \rho(n)$ and $\rho(1)$, respectively. The lot size would be Q_n in the *nth* production. The objective function in Equation 12.14, for this case would be:

$$TC(N) = \sum_{n=1}^{N} K(n) + \sum_{n=1}^{N} \frac{Q_n^2}{2}\left(\frac{H}{D} + C\rho(n)\right),$$ (12.16)

such that $Q_1 + Q_2 + Q_3 + \cdots + Q_N = D$. Improvement in the product quality favors a reduced setup frequency. The learning curve of quality was shown as:

$$\rho(n-1) - \rho(n) \geq \rho(n) - \rho(n+1).$$ (12.17)

Urban (1998)

Urban (1998) assumed the defect rate of a process to be a function of the run length. Using this assumption, he derived closed-form solutions for the economic production quantity (EPQ). He also formulated models to account for either positive or negative learning effects in production processes.

Urban (1998) studied the learning effect of run length on product quality, and on production costs. He examined an EPQ model where the defect rate is a function of the production lot size. In his model, a constant and deterministic demand of a single item was discussed without any backorders. A reciprocal relationship between the defect rate and the production quantity was taken by Urban (1998) as:

$$w = \alpha + \frac{\beta}{Q} \quad 0 \le \alpha \le 1. \tag{12.18}$$

Urban (1998) found this functional form to be very useful for the following reasons:

1. Using appropriate parameters, this functional form can represent the JIT (just-in-time) philosophy ($\beta < 0$), the disruptive philosophy ($\beta > 0$), or a constant defect rate independent of the lot size ($\beta = 0$).
2. It provides a bound for the defect rate—that is, as the lot size increases, the defect rate approaches a given value, $w_\infty = \lim_{Q\to\infty} w_\infty = \alpha$.
3. It is straightforward to estimate the model parameters in practice, using simple linear regression and generally readily available data on lot sizes and defect rates.
4. A closed-form solution can easily be obtained, which can then be examined to gain important insights into the problem.

Under the JIT philosophy, Equation 12.18 suggests that the defect rate decreases as the lot size decreases. At $Q_i = -\beta/\alpha$ the defect rate equals zero. On the other hand, under the disruptive philosophy (where $\beta > 0$), the defect rate increases as the lot size decreases. As $Q_i \to \beta/(1 - \alpha)$, the defect rate approaches one, meaning that all the units are defective. Urban (1998) analyzed four possible scenarios where he only considered the holding costs, setup costs, and the costs associated with the defective units (i.e., scrap, shortage, or rework cost). The first scenario assumes that all defective items are scrapped. That is, $D/(1 - w)$ are produced to satisfy a demand of D units. The total cost for this scenario is:

$$\xi_1 = \left(1 - \frac{D}{P}\right)\frac{ch\{(1-\alpha)Q-\beta\}}{2} + \frac{DK}{\{(1-\alpha)Q-\beta\}} + \frac{(\alpha Q+\beta)cD}{\{(1-\alpha)Q-\beta\}}, \tag{12.19}$$

where:
P = rate of production
c = unit production cost
h = unit holding cost
K = setup cost

The second scenario assumes that the defective units result in shortages at a cost of s dollars per unit. The total cost for this scenario is:

$$\xi_2 = \left(1 - \frac{D}{P}\right)\frac{Qch}{2} + \frac{DK}{Q} + sD\left(\alpha + \frac{\beta}{Q}\right). \tag{12.20}$$

The third scenario assumes that the defective units reach the customer and cost r dollars per unit in compensation. The total cost for this scenario is:

$$\xi_3 = \left(1 - \frac{D}{P}\right)\frac{Qch}{2} + \frac{DK}{Q} + rD\left(\alpha + \frac{\beta}{Q}\right). \tag{12.21}$$

The fourth and last scenario assumes that the defective units are reworked before they reach the customers. The total cost for this scenario is:

$$\xi_4 = \left(1 - \frac{D}{P}\right)\frac{Q(c+w)h}{2} + \frac{DK}{Q} + wD\left(\alpha + \frac{\beta}{Q}\right). \tag{12.22}$$

Jaber and Guiffrida (2004)

Jaber and Guiffrida (2004) presented a quality learning curve (QLC), which is a modification of Wright's learning curve (WLC), for imperfect processes where defectives can be reworked. They incorporated process quality into the learning curve by assuming no improvement in the reworks. They then modeled the same situation by relaxing this assumption. Their assumption was the same as that of Porteus (1986) in that the process remains in control at the beginning of production and generates no defects.

The cumulative time to produce x units is (Equation 12.3)

$$Y(x) = \frac{y_1}{1-b}x^{1-b}, \tag{12.23}$$

where the value of y_1, the time to produce the first unit, is estimated after a subject has had a training period. The total time to produce x units, where there is no learning in the rework process, was given by:

$$Y(x) = \frac{y_1}{1-b}x^{1-b} + r\rho\frac{x^2}{2}, \tag{12.24}$$

with the marginal time

$$t(x) = y_1 x^{-b} + r\rho x, \tag{12.25}$$

where b is the learning exponent in production, ρ is as defined in Porteus (1989) and Chand (1989), and r is the time to rework a unit. Jaber and Guiffrida's (2004) total time to produce x units, where there is learning in the rework process, is:

$$Y(x) = \frac{y_1}{1-b}x^{1-b} + \frac{r_1}{1-\varepsilon}\left(\frac{\rho}{2}\right)x^{2-2\varepsilon}, \tag{12.26}$$

where r_1 is the time to rework the first defective unit and ε is the learning exponent of the rework learning curve.

In the course of learning, there is a point where there is no more improvement in performance. This phenomenon is known as "plateauing." The possible reasons for plateauing may be (i) labor ceases to learn, (ii) management becomes unwilling to invest in learning efforts, and (iii) management becomes skeptical that learning improvement can continue. There is no strong empirical evidence to either support or contradict these hypotheses.

The composite learning model of Jaber and Guiffrida (2004) resulted in the following findings:

a. For learning in reworks such that $0 \leq \varepsilon < 0.5$, a composite learning curve was found to be convex with a local minimum x^* that represents the cumulative production in a given cycle.
b. For $\varepsilon = 0.5$, the composite learning curve would plateau at a value of $2r_1\sqrt{\rho/2}$, as the cumulative production approaches infinity.
c. When $0.5 < \varepsilon < 1$, learning was found to behave in a similar manner to that of Wright (1936). That is, as the cumulative production approaches infinity, the time to produce a unit would approach zero.
d. In a case where there is learning in production only, if the cumulative production exceeds x^*, then the time to produce each additional unit beyond x^* will start increasing. That is, it results in a convex learning curve.

Jaber and Guiffrida (2004) noted that their work has the following limitations:

a. It cannot be applied to cases where defects are discarded.
b. The rate of generating defects is constant.
c. The process can go out of control with a given probability each time an item is produced (Porteus 1986).
d. There is only one stage of production and rework considered.

Jaber and Guiffrida (2008)

In this paper, Jaber and Guiffrida (2008) assumed that an imperfect process can be interrupted in order to restore quality. In this way they addressed the shortcomings in WLC model—that is, (i) the learning phenomenon continues indefinitely, and (ii) all units produced are of acceptable quality. This model also addressed the third limitation in the work of Jaber and Guiffrida (2004), as stated above.

They assumed that (i) a lot of size x is divided into n equal sub-lots corresponding to $(n - 1)$ interruptions, and (ii) the restoration time is a constant percentage $(0 < \alpha < 1)$ of the production time. That is,

$$S(x) = \alpha Y(x), \qquad (12.27)$$

and the expected number of defectives in the lot of x was computed as:

$$d(x,n) = \sum_{i=1}^{n} \rho \frac{(x/n)^2}{2} = \rho \frac{x^2}{2n}. \qquad (12.28)$$

The total restoration or interruption time was given by:

$$S(x,n) = \frac{\alpha y_1}{1-b}\left(\frac{n-1}{n}\right)^{1-b} x^{1-b}.$$ (12.29)

The equations for the total time in Jaber and Guiffrida (2004), to produce x units were modified for the cases with and without learning in reworks, as:

$$T(x,n) = \frac{y_1}{1-b}x^{1-b} + \frac{\alpha y_1}{1-b}\left(\frac{n-1}{n}\right)^{1-b} x^{1-b} + r\rho\frac{x^2}{2n},$$ (12.30)

and

$$T(x,n) = \frac{y_1}{1-b}x^{1-b} + \frac{\alpha y_1}{1-b}\left(\frac{n-1}{n}\right)^{1-b} x^{1-b} + \frac{rn}{1-\varepsilon}\left(\frac{\rho}{2n}\right)^{1-\varepsilon} x^{2-2\varepsilon},$$ (12.31)

respectively. The results of Jaber and Guiffrida (2008) indicated that introducing interruptions into the learning process to restore the quality of the production process improves the system's performance. They found this possible when the percentage of the total restoration time is smaller than the production time. Otherwise, they recommended $n = 1$ (Jaber and Guiffrida 2004). One important outcome of Jaber and Guiffrida (2008) was that restoring the production process breaks the plateau barrier, thereby providing opportunities for improving the performance.

Jaber and Khan (2010)
In this paper, Jaber and Khan relaxed the first and the fourth assumption of the work of Jaber and Guiffrida (2004) as pointed out above. That is, they considered the scrap after production and the rework of a lot in a serial production line. They also elaborated on the impact of splitting a lot into smaller ones on the performance of the process. Jaber and (Khan 2010) defned the overall performance as the sum of the average processing time and the process yield. They assumed the time to produce and rework at every stage in a series was assumed to follow learning.

Accordingly, Jaber and Khan (2010) wrote the total time to process x_i units, at stage i in a serial line as:

$$T_i(x_i) = Y_i(x_i) + R_i(x_i) = \frac{y_{1i}}{1-b_i}x_i^{1-b_i} + \frac{r_{1i}}{1-\varepsilon_i}\left(\frac{\rho_i}{2}\right)^{1-\varepsilon_i} x_i^{2-2\varepsilon_i}.$$ (12.32)

They noted that x_i is the number of non-defective units that enter stage i. Their average processing time for N stages with n sub-lots would be:

$$P(x) = P(x_{N+1}) = \frac{T_{nN}(x_{N+1})}{nx_{N+1}}.$$ (12.33)

Two performance measures were used by Jaber and Khan (2010). One for processing time and the other for process yield (or quality). The first performance measure was given as:

$$Z_1 = 1 - \frac{P(x)}{P_0}, \tag{12.34}$$

where P_0 is the processing time with no learning in production and reworks and with no lot splitting. The second performance measure was given as:

$$Z_2 = \pi_p = 1 - \frac{\sum_{i=1}^{N} s_i(x_i)}{x_1}, \tag{12.35}$$

where $s_i(x_i)$ is the number of scrap items at stage i, when a lot size x_1 enters the first stage. Jaber and Khan (2010) concluded the following:

a. The optimal performance improves as learning in production becomes faster, and deteriorates when learning in reworks becomes faster.
b. The time spent in production or reworks was also found to affect the performance.
c. The system's performance deteriorates as the number of stages in the serial production increases.

EMPIRICAL RELATIONSHIP

Li and Rajagopalan (1997)
In this paper, Li and Rajagopalan (1997) collected about three years of data on quality levels, production, and labor hours from two manufacturing firms. The aim of their study was to answer three questions related to the impact of quality on learning. These questions were—namely: (1) How well does the cumulative output of defective versus good units explain learning curve effects? (2) Do defective units explain learning curve effects better than good units? (3) How should cumulative experience be represented in the learning curve model when the quality level may have an impact on learning effects? The data were taken from two plants making tire tread and medical instruments (kits and fixtures), respectively. The study of Li and Rajagopalan (1997) resulted in the following findings: (1) the learning rate slows down as quality improves; (2) over time, as defects are less frequent, the opportunities for learning are also fewer; and (3) managers pay less attention to process improvement at higher levels of experience.

Li and Rajagopalan (1997) used defect levels as a substitute for the effort devoted to process improvement. In a latter study, Li and Rajagopalan (1998) showed that the optimal investment in a process improvement effort is proportional to the defect levels. This is contrary to the traditional learning curve model (Wright 1936) where cumulative production volume, which contains good and defective units, was used as a proxy for knowledge or experience. They proposed a regression model complementary to the one in Fine (1986).

Li and Rajagopalan (1997) showed that if the defect level in a period is very high, then it immediately gets attention and considerable effort is directed at identifying the source of the defectives. They found that if defect levels continue to be high for a few consecutive periods, then increased attention is paid and additional resources are devoted to investigating the cause of the defects. In their opinion, these efforts lead to a better understanding of the process variables and interactions, which is useful in helping to avoid such defects in the future. They also found that in the later stages, where defects are less frequent, the opportunities for learning from the analysis of these defects are also fewer. Li and Rajagopalan (1997) concluded that defective and good units do not explain learning curves equally. They further elaborated that defective units are statistically more significant than good units in explaining learning curve effects.

Foster and Adam (1996)

In this paper, Foster and Adam (1996) included the speed of quality improvement in Fine's (1986) quality-based learning curve model. Their model demonstrated that, under different circumstances, rapid quality improvement effects are either beneficial or unfavorable to improvement in quality-related costs.

They also demonstrated that sustained and permanent rapid quality improvement can lead to higher levels of learning. However, they found that under certain conditions the rapid speed of quality improvement can also impede organizational learning. In their opinion this suggests that when management imposes higher goals for the reduced number of defects while systems are not in place to achieve those goals, costs will increase. Foster and Adam (1996) developed two hypotheses from this analysis and tested them in an automotive parts manufacturing company with five similar plants.

They found that fast quality improvements result in reducing the rate at which improvement in quality-related costs occur. The opposite to this behaviour was found to be true. They referred to this behavior as "organizational learning." A type of learning that Foster and Adam (1996) found in many organizations. They also observed that with the passage of time, (i) inspection-related costs are reduced, (ii) the need for the acceptance sampling of raw materials is reduced, (iii) prevention-related costs decline, and (iv) prevention activities become more focused and specific.

Foster and Adam (1996) cautioned that some quality-related efforts may be ineffective. Foster and Adam (1997) recommended that companies adopt slower and steadier rates of quality improvements. Their findings were supported empirically.

Forker (1997)

In this paper, the results of a survey of 348 aerospace component manufacturers were examined in order to investigate the factors that affect supplier quality performance. Forker discussed the process view of quality to depict the inconsistencies between practice and performance in a supplier firm. Foker (1997) noted that by linking quality management with process it would be possible to address issues of effectiveness and efficiency in firms. He linked the quality performance of a supplier with a variety of dimensions, such as features, reliability, conformance, durability, serviceability, and aesthetics.

Foker (1997) emphasized in his study the importance of human learning for the following reasons: (i) the learning curve affected the transaction cost of different supplies, and (ii) suppliers' attitudes toward learning, and thus their efficiency, impacted on the quality magnitude.

Forker (1997) showed that as processes become more streamlined and capable, firms should invest their resources in product design and in training all employees in quality improvement concepts and techniques.

Badiru (1995)

In this paper, it was claimed that quality is a hidden factor in learning curve analysis. Badiru (1995) considered quality to be a function of performance, which in turn is a function of the production rate. He found that forgetting affects the product quality in the sense that it can impair the proficiency of a worker in performing certain tasks. Badiru (1995) further explained that the loss in worker performance due to forgetting is reflected in the product quality through poor workmanship. He also noted that forgetting can take several different forms:

1. Intermittent forgetting (i.e., in scheduled production breaks)
2. Random forgetting (e.g., machine breakdown)
3. Natural forgetting (i.e., effect of ageing)

Badiru (1995) emphasized that there are numerous factors that can influence how fast, how far, and how well a worker or an organization learns within a given time span. The multivariate learning curve he suggested was given by:

$$C_x = K \prod_{i=1}^{n} c_i x_i^{b_i},$$ (12.36)

where
C_x = cumulative average cost per unit
K = cost for the first unit
n = number of variables in the model
x_i = specific value of the ith variable
b_i = learning exponent for the ith variable
c_i = coefficient for the ith variable

Badiru (1995) tested the learning model in Equation 12.37 on the four-year record of a troublesome production line. The production line he investigated was a new addition to an electronics plant and was thus subject to significant learning. Badiru (1995) observed that the company used to stop the production line temporarily if quality problems would arise. He also hypothesized that the quality problems could be overcome if the downtime (forgetting) could be reduced so that workers could have a more consistent operation. The four variables of interest in his study were: (i) the production level (X_1), (ii) number of workers (X_2), (iii) the number of hours of production downtime (X_3), and (iv) the dependent variable was the average production cost per unit C_x. Badiru (1995) showed in his analysis of variance of the regression model that the fit is highly significant, with 95% variability in the average cost per unit.

Badiru (1995) noticed that the average cost per unit would be underestimated if the effect of downtime hours is not considered. He concluded that the multivariate model provides a more accurate picture of the process performance.

Mukherjee et al. (1998)

Mukherjee et al. (1998) studied why some quality improvement projects are more effective than others. They explored this by studying 62 quality improvement projects undertaken in one factory over the course of a decade, and identified three learning constructs that characterize the learning process—namely, scope, conceptual learning, and operational learning. The purpose of their study was to establish a link between the pursuit of knowledge and the pursuit of quality.

Mukherjee et al. (1998) followed the approach in Kim (1993) to distinguish between two types of effort—conceptual learning and operational learning. They defined conceptual learning as, trying to understand why events occur (i.e., the acquisition of know-why). Operational learning, in their view, consists of implementing changes and observing the results of these changes. They further added that operational learning is the process of developing a skill of how to deal with experienced events (i.e., the acquisition of know-how).

Mukherjee et al. (1998) recommended that in order to establish links between learning and quality, field researchers should try to control potentially confounding factors such as variations in product and resource markets, general management policies, corporate culture, production technology, and geographical location. They also recommended that field researchers must have access to detailed data about the systems that are used to improve quality. Mukherjee et al. (1998) further recommended that the plant take the following total quality management (TQM) measures to enhance quality:

- Training managers in problem-solving concepts and ensuring process capability
- Investing heavily in the training of plant personnel
- Creating a functional TQM organization, consisting of a plant-level TQM-steering committee and departmental TQM teams
- Introducing SPC for the control of a few key parameters and attributes
- Adding accuracy and precision indices to the existing quality index
- Training foremen in creating, establishing, and monitoring standard operating procedures (SOPs)
- Installing a new information system, which economically provides standardized daily, weekly, and monthly production and quality data
- Emphasizing the behavioral (instead of the technical) component of TQM

Mukherjee et al. (1998) concluded the following: (i) management plays a role in addressing 80%–85% of quality problems, (ii) in dynamic production environments a cross-functional project team is in a better position to create technological knowledge, and (iii) that operational and conceptual learning have different kinds of potential in a plant.

Lapré et al. (2000)

In this paper, the learning curve of TQM in a factory was explored. The link between learning and quality was extended from a cross-sectional, project-level analysis to a longitudinal, factory-level analysis.

Lapré et al. (2000) cautioned that the power form of learning has many fundamental shortcomings. Using an exponential model, Lapré et al. (2000) focused on waste, which is a key driver of both quality and productivity. They modeled the improvement in the waste as:

$$\frac{dW(z)}{dz} = \mu[W(z) - P], \tag{12.37}$$

where $W(z)$, P, z, and μ are the current waste rate, desired waste rate, proven capacity, and the learning rate, respectively. Lapré et al. (2000) took the learning rate as a function of autonomous and induced learning. They also assumed that $y_{1t}, y_{2t}, \ldots, y_{nt}$ are the managerial factors that affect the learning rate (e.g., the cumulative number of quality improvement projects), and modeled the learning rate at time t as:

$$\mu_t = \beta_0 + \sum_{i=1}^{n} \beta_i y_{it}. \tag{12.38}$$

Lapré et al. (2000) combined Equations 12.37 and 12.38 to get:

$$\ln W(z_t) = a + \left(\beta_0 + \sum_{i=1}^{n} \beta_i y_{it} \right) z_t, \tag{12.39}$$

where a is a constant, while β_0 and $\sum_{i=1}^{n} \beta_i y_{it}$ measure the autonomous and induced part of the learning rate, respectively. The parameters of Equation 12.39 were determined by analyzing a number of projects in a factory. Lapré et al. (2000) coded these projects on questions that dealt with their learning process and their performance by giving a response on a five-point Likert scale (Bucher 1991). This allowed them to provide a systemic explanation, based on the dimensions of the learning process they used, on why induced learning that yields both know-why and know-how enhances the learning rate, while induced learning that only yields know-why disrupts the learning process.

Jaber and Bonney (2003)

Jaber and Bonney (2003) used the data in Badiru (1995) to show that the electronics production line follows two hypotheses:

1. The time required to rework a defective item reduces as the production increases and the rework times conform to a learning relationship.
2. Quality deteriorates as forgetting increases, due to interruptions in the production process.

To validate the first hypothesis, Jaber and Bonney (2003) analyzed the effect of the cumulative production level, X, on the average time to rework a unit, Y, to fit the model

$$Y = \beta_1 X^{-\beta_2}.$$
(12.40)

The p values for the intercept β_1 and the slope β_2 indicated that the regression fit is significant. The analysis of variance showed that 84% of the variability in the average rework time per unit is explained by cumulative production as an independent variable.

Similarly, to validate the second hypothesis, Jaber and Bonney (2003) analyzed the impact of forgetting due to production downtime, D, on the average rework time for a unit, Y, and cumulative production level, X. The fitted regression model of Jaber and Bonney (2003) was of the form

$$Y = \beta_1 X^{\beta_2} D^{\beta_3}.$$
(12.41)

Again, the p values for the intercept β_1 and the slopes β_2, β_3 indicated that the regression fit is significant. Their analysis of variance showed that 88% of the variability in the average rework time per unit is explained by cumulative and production downtime as independent variables.

Hyland et al. (2003)

In this paper, Hyland et al. (2003) reported the research on continuous improvement (CI) and learning in the logistics of a supply chain. This research is based on a model of continuous innovation in the product development process and a methodology for mapping learning behaviors.

Hyland et al. (2003) took learning to be crucial to innovation and improvement. To build innovative capabilities, Hyland et al. (2003) suggested that organizations need to develop and encourage learning behaviors. Hyland et al. (2003) believed that capabilities could only be developed over time by the progressive consolidation of behaviors, or by strategic actions aimed at the stock of existing resources.

Hyland et al. (2003) identified four key capabilities that are central to learning and CI in a supply chain. These capabilities are: (1) the management of knowledge; (2) the management of information; (3) the ability to accommodate and manage technologies and the associated issues; and (4) the ability to manage collaborative operations.

Jaber et al. (2008)

Salameh and Jaber (2000) came up with a new line of research concerning the defective items in an inventory model. They recommended the screening and disposal of the defective items in the basic economic order quantity model. This model has recently been widely extended to address the issues of shortages/backorders, quality, fuzziness of demand, and supply chains. However, Salameh and Jaber assumed the fraction of defectives to be following a known probability density function. Jaber et al. (2008) noticed that this fraction in an automotive industry reduces according to a learning curve, over the number of shipments.

They tried to fit several learning curve models to the collected data and found that the S-shaped logistic learning curve fitted the data well:

$$w_n = \frac{a}{g + e^{bn}},$$

(12.42)

where a and g are the model parameters, b is the learning exponent, and w_n is the fraction of imperfect items in the nth shipment.

Jaber et al. (2008) developed two models similar to that of Salameh and Jaber (2000). One for an infinite planning horizon, and one for a finite planning horizon. They found that in the infinite planning model, the number of defective units, the shipment size, and the total cost reduces with an increase in learning. For the finite learning model, they found that an increase in learning recommends larger lot sizes less frequently.

FUTURE DIRECTIONS OF RESEARCH

In this chapter, a number of scenarios have been presented. These scenarios relate the quality of a product to a number of parameters in a production facility—that is, the production quantity, production run, number of supplies, number of setups, and so on. Though we found a number of papers expressing the relationship between quality and learning in one way or another, a formal correlation between the two still seems to be missing. This association can be beneficial not only for inventory control, but also for the enhancement of coordination in a supply chain.

There is a dire need to clearly understand the difference between the enhancements in quality of a product through induced and motor learning. The literature in the field of inventory and supply chain management has been dealing with these issues on distinct grounds, but the impact of joint learning on quality has never been examined. This can provide a rich field of research to be investigated. The issues of investment in different types of learning and sharing of their benefits in a supply chain are also a vast arena to be studied.

There has also been constant debate on the problem of how much to produce/order in manufacturing setups. The literature provides a substantial background in this area. As discussed in the chapter, the quantity and run length in production are directly linked to the quality of a product. Combined with experience (learning), this subject should be of particular interest to engineers and managers in the field of production management.

A number of researchers have been investigating errors in screening, though not many have studied such errors in production environments. The relationship between quality and screening and investment (in either, or both) is a topic that needs further research. This research should be of special importance for practitioners in the field of supply chain management as it calls for the individual and mutual benefits of all the stakeholders.

Quality should also be considered at the strategic level, where various investment decisions about new product development, process improvements, or system implementations are made. Those decisions are based on several assumptions, including the performance of a development project that is expected to deliver a product at the required quality level. In contrast to industrial projects, which are usually repeated many times, learning in development projects causes increased productivity during the course of each activity and is referred to as "activity-specific learning" (Ash and Daniels 1999). Since development projects are hugely affected by learning (Robey et al. 2002; Plaza 2008), they are often delivered with significant schedule delays (Vendevoorde and Vanhoucke 2006). In an attempt to recover the schedule and reduce additional expenses, the testing of a prototype is reduced. If a prototype with a larger number of defects that would normally be acceptable is released, this impacts on the quality of the product down the road.

Therefore, it becomes critical to incorporate learning curves into a standard project management methodology, as this would allow for the assessment of the impact of learning curves on the quality of a new product, a new process, or a system. In contrast to industrial projects, which are usually repeated many times, learning on development projects causes increased productivity during the course of each activity and is, in turn, referred to as "activity-specific learning" (Ash and Daniels 1999). Learning influences the effectiveness of teams and increases future performance efficiency (Edmondson 2003); the strength of its impact is sufficient to warrant (Sarin and McDermott 2003) the employment of careful management (Garvin 1993).

ACKNOWLEDGMENTS

The authors thank the Natural Sciences and Engineering Research Council (NSERC) of Canada for supporting their research. They also thank Professor Michael A. Lapré of the Owen Graduate School of Management at Vanderbilt University for his valuable comments and suggestions.

REFERENCES

Anderson, E.G., 2001. Managing the impact of high market growth and learning on knowledge worker productivity and service quality. *European Journal of Operational Research* 134(3): 508–524.

Ash, R., and Daniels, D.E.S., 1999. The effect of learning, forgetting, and relearning on decision rule performance in multiproject scheduling. *Decision Sciences* 30(1): 47–82.

Badiru, A.B., 1995. Multivariate analysis of the effect of learning and forgetting on product quality. *International Journal of Production Research* 33(3): 777–794.

Baloff, N., 1970. Startup management. *IEEE Transactions on Engineering Management* EM-17(4): 17132–141.

Bucher, L., 1991. Consider a likert scale. *Journal for Nurses in Staff Development* 7(5): 234–238.

Chand, S., 1989. Lot sizes and setup frequency with learning in setups and process quality. *European Journal of Operational Research* 42(2): 190–202.

Cohen, W.M., and Levinthal, D.A., 1989. Innovation and learning: The two faces of R & D. *The Economic Journal* 99(397): 569–596.

Crosby, P.B., 1979. *Quality is free.* New York: McGraw-Hill.

Deming, W.E., 1982. Quality, productivity, and competitive position. M.I.T. Center for Advanced Engineering Study, USA.

Edmondson, A.C., 2003. Framing for learning: Lessons in successful technology implementation. *California Management Review* 45(2): 34–54.

Fine, C.H., 1988. A quality control model with learning effects. *Operations Research* 36(3): 437–444.

Fine, C.H., 1986. Quality improvement and learning in productive systems. *Management Science* 32(10): 1301–1315.

Forker, F.B., 1997. Factors affecting supplier quality performance. *Journal of Operations Management* 15(4): 243–269.

Foster, S.J., and Adam, E.E., 1996. Examining the impact of speed of quality improvement on quality-related costs. *Decision Sciences* 27(4): 623–646.

Garvin, D.A., 1993. Building a learning organization. *Harvard Business Review* 71(4): 78–91.

Garvin, D.A., 1987. Competing on the eight dimensions of quality. *Harvard Business Review* 65(6): 101–109.

Glover, J.H., 1965. Manufacturing progress functions I: An alternative model and its comparison with existing functions. *International Journal of Production Research* 4(4): 279–300.

Hyland, P.W., Soosay, C., and Sloan, T.R., 2003. Continuous improvement and learning in the supply chain. *International Journal of Physical Distribution and Logistics Management* 33(4): 316–335.

Jaber, M.Y., and Khan, M., 2010. Managing yield by lot splitting in a serial production line with learning, rework and scrap. *International Journal of Production Economics* 124(1): 32–39.

Jaber, M.Y., Goyal, S.K., and Imran, M., 2008. Economic production quantity model for items with imperfect quality subject to learning effects. *International Journal of Production Economics* 115(1): 143–150.

Jaber, M.Y., and Guiffrida, A.L., 2008. Learning curves for imperfect production processes with reworks and process restoration interruptions. *European Journal of Operational Research* 189(1): 93–104.

Jaber, M.Y., and Guiffrida, A.L., 2004. Learning curves for processes generating defects requiring reworks. *European Journal of Operational Research* 159(3): 663–672.

Jaber, M.Y., and Bonney, M., 2003. Lot sizing with learning and forgetting in set-ups and in product quality. *International Journal of Production Economics* 83(1): 95–111.

Juran, J.M., 1978. Japanese and western quality - A contrast. *Quality Progress* (December): 10–18.

Kapp, K.M., 1999. Transforming your manufacturing organization into a learning organization. *Hospital Material Management Quarterly* 20(4): 46–55.

Kim, D.H., 1993. The link between individual and organizational learning. *Sloan Management Review* 35(1): 37–50.

Lapré, A.M., Mukherjee, A.S., and Wassenhove, L.N.V., 2000. Behind the learning curve: Linking learning activities to waste reduction. *Management Science* 46(5): 597–611.

Levy, F., 1965. Adaptation in the production process. *Management Science* 11(6): 136–154.

Li, G., Rajagopalan, S., 1998. Process improvement, quality and learning effects. *Management Science* 44(11): 1517–1532.

Li, G., and Rajagopalan, S., 1997. The impact of quality on learning. *Journal of Operations Management* 15(3): 181–191.

Lundvall, D.M., and Juran, J.M., 1974. Quality costs. In *Quality control handbook*, ed. J.M. Juran. Third Edition. San Francisco: McGraw-Hill, pp. 1–22.

Malreba, F., 1992. Learning by firms and incremental technical change. *The Economic Journal* 102(413): 845–859.

Mukherjee, A.S., Lapré, M.A., Wassenhove, L.N.V., 1998. Knowledge driven quality improvement. *Management Science* 44(11): 36–49.

Plaza, M., 2008. Team performance and information systems implementation: Application of the progress curve to the earned value method in an information system project. *Information Systems Frontiers* 10(3): 347–359.

Porteus, E.L., 1986. Optimal lot sizing, process quality improvement and setup cost reduction. *Operations Research* 34(1): 137–144.

Robey, D., Ross, J.W., and Boudreau, M.C., 2002. Learning to implement enterprise systems: An exploratory study of the dialectics of changes. *Journal of Strategic Information Systems* 19(1): 17–46.

Salameh, M.K., and Jaber, M.Y., 2000. Economic production quantity model for items with imperfect quality. *International Journal of Production Economics* 64(1): 59–64.

Sarin, S., and McDermott, C., 2003. The effect of team leader characteristics on learning, knowledge application, and performance of cross-functional new product development teams. *Decision Sciences* 34(4): 707–739.

Sontrop, J.W., and MacKenzie, K., 1995. Introduction to technical statistics and quality control. Addison Wesley: Canada

Spence, A.M., 1981. The learning curve and competition. *Bell Journal of Economics* 12(1): 49–70.

Steedman, I., 1970. Some improvement curve theory. *International Journal of Production Research* 8(3): 189–206.

Steven, G.J., 1999. The learning curve: From aircraft to spacecraft? *Management Accounting* 77(5): 64–65.

Tapiero, C.S., 1987. Production learning and quality control. *IIE Transactions* 19(4): 362–370.

Urban, T.L., 1998. Analysis of production systems when run length influences product quality. *International Journal of Production Research* 36(11): 3085–3094.

Vandevoorde, S., and Vanhoucke, M., 2006. A comparison of different project duration forecasting methods using earned value metrics. *Journal of Project Management* 24(4): 289–302.

Wright, T.P., 1936. Factors affecting the cost of airplanes. *Journal of the Aeronautical Sciences* 3(2): 122–128.

Yang, L., Wang, Y., and Pai, S., 2009. On-line SPC with consideration of learning curve. *Computers and Industrial Engineering* 57(3): 1089–1095.

Yelle, L.E., 1979. The learning curve: Historical review and comprehensive survey. *Decision Sciences* 10(2): 302–328.

Zangwill, W.I., and Kantor, P.K., 2000. The learning curve: A new perspective. *International Transactions in Operational Research* 7(6): 595–607.

Zangwill, W.I., and Kantor, P.B., 1998. Toward a theory of continuous improvement and the learning curve. *Management Science* 44(7): 910–920.

13 Latent Growth Models for Operations Management Research: A Methodological Primer

Hemant V. Kher and Jean-Philippe Laurenceau

CONTENTS

INTRODUCTION

With an increased emphasis on empirical research in the field of operations management (OM), researchers are increasingly turning to the use of structural equation modeling (SEM) as a preferred method of data analysis. Shah and Goldstein (2006) reviewed nearly 100 papers in OM over two decades (1984–2003) that used SEM methodology. SEM allows researchers to study hypothesized relationships between unobservable constructs (i.e., latent variables) that are typically measured using two or more observable measures (i.e., manifest or indicator variables). The purpose of this chapter is to illustrate the use of a special type of SEM called a "latent variable growth curve model" (abbreviated as LCM by Bollen and Curran 2006; and LGM by Duncan et al. 2006), for studying longitudinal changes (i.e., changes over time) in observed or latent variables of interest to researchers in the field of OM.

Traditional approaches to analyzing longitudinal data include techniques such as repeated measures analysis of variance (RANOVA), analysis of covariance (ANCOVA), multivariate analysis of variance (MANOVA), multivariate analysis of covariance (MANCOVA), and autoregressive models (for a review, see Hancock and Lawrence 2006). However, these methods have been used to describe changes at the *group level* of analysis rather than changes at the *individual level* of analysis. These traditional approaches are limited when the variation in the constructs describing change at the intra-individual level and inter-individual differences in intra-individual change are under focus. The following example illustrates the difference between traditional group level techniques and latent growth model (LGM) for analyzing longitudinal change.

Suppose a call center is interested in measuring the improvements in productivity for its customer service representatives. Productivity improvements may be documented using objective measures such as the number of calls answered, accuracy in providing requested information, courtesy in handling calls, and so forth (Castilla 2005). If we were only interested in describing the changes in productivity at the group level, we could follow the two-step approach suggested in Uzumeri and Nembhard (1998). In Step 1, we would fit a common mathematical function (using a curve-fitting technique) to longitudinal productivity data for every individual in the sample. Then, in Step 2, we would use the resulting set of best-fit parameter estimates to describe the changes in productivity for the entire group. The approach adopted by Uzumeri and Nembhard may have been appropriate in their study if their intention was to fit a single functional form for growth to the "population of learners."

However, if we also want to identify whether groups of individuals in the sample have growth patterns that are different relative to those fit to the entire sample and identify reasons for these differences, then the above-mentioned case-by-case approach is limited. Bollen and Curran (2006, 33–34) note five technical limitations of the case-by-case approach to using the ordinary least squares (OLS) regression procedure for estimating growth trajectories over time for sample members. First, overall tests of model fit are not readily available. Second, OLS imposes a restrictive structure on error variances. Third, the estimation of variances for random intercepts and slopes is difficult, and their significance tests are complicated. Fourth, inclusion of certain covariates is possible, but with the assumption that their measurement is error free, which is not always possible. Finally, the inclusion of time-varying covariates is impossible with the OLS for a case-by-case approach. Latent variable curve growth models can replicate the OLS regression results for the case-by-case approach while overcoming the above-listed limitations. The strength of this technique is that it allows researchers to *quantify statistically* the variations in sample elements on intra-individual (e.g., changes in each individual's productivity over time) as well as inter-individual levels (i.e., changes in productivity across different individuals in the sample; Nesselroade and Baltes 1979).

Latent growth models (LGMs) allow researchers to answer three important questions about the observed measure or the latent construct under focus (Bollen and Curran 2006). The first question concerns the nature of the trajectory at the group level. This question is answered by fitting an *unconditional* LGM, and of interest in this analysis is the functional form of longitudinal changes in the manifest variable or the latent construct (e.g., linear, quadratic, cubic, exponential, etc.), along with the associated parameter estimates (e.g., intercept and slope for linear LGM).

The second question allows researchers to determine whether a distinct trajectory of longitudinal changes in the observed measure or latent construct is needed for each case in the sample. As an example, statistically significant variability in the intercept and slope for a linear LGM would suggest that the starting point and the rate of change in productivity for each individual may be different from that fitted to the entire sample.

Finally, the third question pertains to identifying the predictors of change in trajectories for the different cases. For example, if the answer to the first question suggests that a linear model fits changes in individual productivity over time, and the answer to the second question indicates that different individuals in the sample may have different linear trajectories, then the answer to the third question consists of identifying the reasons underlying the different trajectories for the different individuals in the sample. Identifying these variations may be important to understanding whether some individuals start out with a high (or low) level of productivity and show a high (or low) rate of productivity changes over time. For example, do individuals that received extensive training prior to starting their jobs have a higher starting productivity? Is the rate of improvement in the productivity for these individuals higher than for others that did not receive such training? As noted above, the unconditional LGM answers the first two questions, while a *conditional* LGM answers the third question.

Stated somewhat differently, Willett and Sayer (1994, 365) note that "logically, individual change must be described before inter-individual differences in change can be examined, and inter-individual differences in change must be present before one can ask whether any heterogeneity is related to predictors of change." Here, the individual change (i.e., Level 1) model posits that the form of change is the same for all individuals in the sample, while the inter-individual change (i.e., Level 2) model posits that different individuals may have different starting points and rates of change.

LGMs have grown significantly in popularity across the behavioral and social sciences since the 1990s. Several textbooks (we have cited two—Bollen and Curran 2006; and Duncan et al. 2006) and hundreds of papers, both technically oriented and demonstrating applications, have been printed on the topic. LGMS have been used extensively in the fields of psychology, sociology, health care, and education. Some examples of focal variables where change (growth or decline) has been studied in these fields include: (1) the behavioral aspects that cause increases in the consumption of tobacco, alcohol, and drugs (Duncan and Duncan 1994); (2) the relationship between crime rates and weather patterns (Hipp et al. 2004); (3) the behavioral aspects that cause changes in the level of physical activity among teenagers/adolescents (Li et al. 2007); and (4) the attitudes of middle- and high-school students towards science courses (George 2000).

They are also starting to appear in the management literature where change is being assessed in focal variables such as: (1) the adjustment of new employees to the work environment (Lance et al. 2000); (2) the manner in which new employees seek information and build relationships (Chan and Schmitt 2000); (3) employee commitment and the intention to quit (Bentein et al. 2005); and (4) individual productivity (Ployhart and Hakel 1998). To the best of our knowledge, researchers in OM have not yet used LGMs for longitudinal analysis.

The purpose of this chapter is to advocate for the use of LGMs for OM research and to provide a primer. Towards this end, we provide an overview of LGMs, and discuss the issues related to data requirements, model identification, estimation methods, sample size requirements, and model fit assessment statistics. We then illustrate the application of LGMs using simulated longitudinal data. We conclude by noting the advantages to using LGMs over other more traditional longitudinal approaches, and highlight areas in OM where researchers can use this technique effectively.

OVERVIEW OF LATENT GROWTH MODELS

Although a documented interest in modeling group and individual-level growth models exists from the early twentieth century, work in the area of LGMs is more recent. According to Bollen and Curran (2006), Baker (1954) presented the first known factor analytic model on repeated observations. They also note that Tucker (1958) and Rao (1958, 13) provided the approach to "parameterizing these factors to allow for the estimation of specific functional forms of growth." Finally, Meredith and Tisak (1990) are credited for placing trajectory modeling in the context of confirmatory factor analysis.

UNCONDITIONAL AND CONDITIONAL LINEAR LATENT GROWTH MODELS

The unconditional, linear LGM is shown in Figure 13.1. Following the path diagram convention used to represent SEMs, circles or ellipses represent the latent construct, while rectangular boxes represent observed (manifest) variables. Thus, in Figure 13.1, the rectangular boxes represent observed productivity for the years 2001 through 2005. Loadings of these measured variables on the intercept and slope construct are as shown. All variables have a constant loading (i.e., 1.0) on the intercept construct, while the loadings from observed productivity going from year 2001 to 2005 start at 0.0, and increase in steps of 1.0, representing linear growth. Finally, E1 through E5 represent the errors associated with observed productivity for the years 2001 through 2005, respectively.

The double-headed arrow indicates the covariance between the intercept and slope constructs. Statistically significant variances for the intercept and slope construct suggest significant variability of the individuals in the sample with regards to initial productivity (i.e., in year 2001), and the rate at which they improve, respectively. A statistically significant covariance between the slope

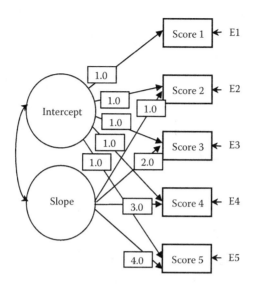

FIGURE 13.1 Unconditional linear LGM.

and intercept construct establishes the relationship between them; e.g., a negative covariance indicates that individuals with a high initial productivity improve at a slower rate.

A conditional, linear LGM is depicted in Figure 13.2a. Whether or not the individual received training prior to starting their job is included as the *time invariant* predictor in this model, and is included as a 0/1 dummy variable. As with standard regression analysis, it is possible to use continuous as well as categorical predictors with LGMs. In addition to time invariant predictors, one can also use *time variant* predictors with LGMs, as denoted in Figure 13.2b, where changes in individual productivity are hypothesized to covary with the extent of feedback (E-2001 through E-2005) provided by supervisors.

Data Requirements, Trajectory Modeling, and Invariance

The basic requirement for LGMs is that data on the same observed measure (or the same latent construct measured using the same indicators) should be collected over time. Researchers can use LGMs flexibly. Where appropriate, using an exploratory approach, it may be possible to assume that a single functional form governs growth, but without an a priori specification of this functional form. Thus, the researcher may model different functional forms to identify the one that provides the best fit. As an example, Ployhart and Hakel (1998) tested the no-growth, linear, quadratic, and cubic (S-curve) trajectories and identified the latter as the best-fitting curve for describing the growth in sales commissions for a sample of sales persons. On the other hand, the researcher can hypothesize and confirm a specific functional form for the growth in individuals and the group. Moreover, if the functional form of change is expected to be different during one set of time points versus another, non-linear forms of change can also be approximated using piecewise growth modeling Flora (2008).

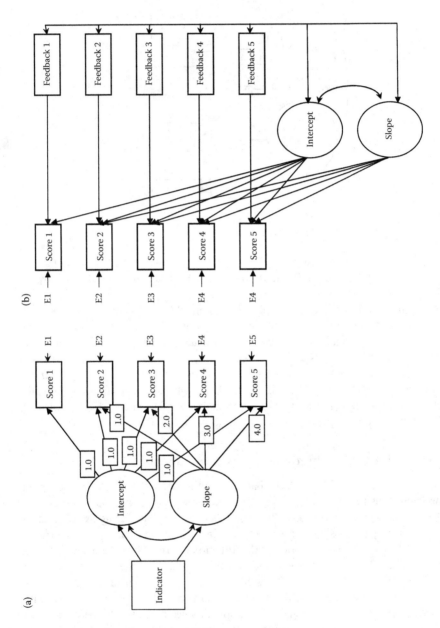

FIGURE 13.2   (a) Conditional linear LGM with a time invariant covariate, and (b) time varying covariate.

MODEL IDENTIFICATION

Shah and Goldstein (2006) note significant concerns with just-identified and under-identified models in research utilizing SEM in OM, and recommend using over-identified models (degrees of freedom [DF] > 0) whenever possible. With LGMs (as with SEMs), the degrees of freedom are a function of the number of data waves (or measured variables in SEMs), parameters to be estimated, and the parameter constraints placed on the model (e.g., assuming equal error variances versus allowing for a free estimation of the error terms). The expression $(1/2) \times \{p(p + 1)\} - q$, where p is the number of data waves and q is the number of parameters, is useful in calculating the degrees of freedom for LGMs. Because LGMs incorporate both means and covariance structure analyses, the degrees of freedom given by the above-stated expression are increased by p, since there are as many observed means as there are data waves. The degrees of freedom calculation allows researchers to determine the minimum number of data waves that they should work with.

As an example, to fit a linear LGM, Bollen and Curran (2006) recommend using at least three waves of data (they note that this is a necessary, although not a sufficient, condition to fit LGM). Figure 13.1 is useful in understanding the reason behind this. The LGM shown in Figure 13.1 is based on five waves of data ($p = 5$). These five waves provide a total of 15 variances and covariances ($(1/2) \times \{5 \times (5 + 1)\} = 15$), plus five means, resulting in 20 observed pieces of information. Using these, we must estimate five error variances (one for each data wave), two factor variances, one covariance between the two factors, and two factor means, which represent 10 estimated pieces ($q = 10$). Thus, our model will have 10 degrees of freedom. If all error variances were constrained to be equal, then our model would have 14 degrees of freedom (20 observed pieces—six estimated pieces). With only two waves of data, we would have a total of five observed pieces of information, which are insufficient to estimate six parameters, even for the restricted model with equal error variances.

As the number of parameters to be estimated increases, so should the number of data waves. Thus, for an LGM describing a quadratic trajectory, the parameters to estimate would include the mean and variance for the intercept, linear, and quadratic slope terms plus three covariances (one each between the intercept and linear slope, the intercept and quadratic slope, and the linear and quadratic slope), and error variances. With only three data waves, we would have nine observed pieces of information (six variances and covariances plus three means), which are not sufficient to estimate 12 parameters. Thus, for the quadratic LGM, at least four waves of data are recommended. Using a similar approach, we can see that for a cubic LGM (which involves constructs for the intercept, and slopes for the linear, quadratic, and cubic components), at least five waves of data are recommended. More repeated measures, however, will increase the reliability estimation of the parameters of change (Singer and Willett 2003).

ESTIMATION, SAMPLE SIZE, AND MEASURES OF MODEL FIT

Maximum likelihood is the commonly used approach to estimating LGMs. This approach assumes that the repeated measures come from a continuous, or a reasonably continuous, distribution. If repeated data are collected on ordinal measures, such as a dichotomous variable for a yes/no scale, or where subjects rate variables of interest on categorical scales (e.g., rating a car's reliability on a five-point scale ranging from one = poor to five = excellent), data have to be preprocessed before fitting the model. The first step involves calculating polychoric correlations and standard deviations, followed by generating an asymptotic covariance for the variables. Then, maximum likelihood (ML) or another estimator—such as weighted least squares (WLS) or the diagonally weighted least squares (DWLS)—can be used to fit LGMs. We illustrate this approach in an example in the next section. Duncan et al. (2006) identify the approaches to fitting LGMs for categorical variables currently available in different SEM software packages.

Muthén and Muthén (2002) provide guidelines on the influence of sample size on the statistical power of confirmatory factor analysis (CFA) and LGMs. For the latter, their investigation considered unconditional models, as well as conditional models, with a single, dichotomous time invariant predictor (covariate). Other conditions that varied in their study included the absence and presence of missing data, and the size of the population regression slope coefficient (low = 0.1, high = 0.2). The results of their simulation experiments (see Muthen and Muthen 2002, Table 2, 607) suggest that for unconditional models, samples with as few as 40 observations can provide a statistical power of 0.80 or better. The presence of a covariate increases the sample size to about 150. Sample size requirements increase further in the presence of missing data and smaller values of the population regression coefficient.

Hamilton et al. (2003) investigate the relationship between sample size (varied from $n = 25, 50, 100, 200, 500$, and 1000), model convergence rates, and the propensity to generate improper solutions under conditions created by varying the number of data waves (time points—modeled at four, five, and six), variance of the intercept (high vs. low), and the variance of the slope parameter (high vs. low). Their results indicate that the convergence rate increases and the chances of generating improper solutions decrease with increased sample size. In the majority of cases, a sample size of 100 was sufficient to achieve a 100% convergence rate combined with a reduction in the number of instances where improper solutions were generated. An increased number of data waves and lower levels of intercept and slope variance also reduced the sample size requirements for proper convergence and reduced the instances of improper solutions.

Because LGMs are a special case of SEM, the indices used to assess model fit are the same as those highlighted in the review of SEM research by Shah and Goldstein (2006). Specifically, measures such as the chi-squared test statistic, goodness of fit index (GFI), adjusted goodness of fit index (AGFI), normed fit index (NFI), non-normed fit index (NNFI), comparative fit index (CFI), root mean squared error of approximation (RMSEA), and the root mean squared residual (RMR or SRMR), among others, are routinely reported in the context of assessing model fit. In addition, Bollen and Curran (2006) recommend using the Tucker–Lewis Index (TLI), incremental fit index (IFI), Akaike information criterion (AIC), and the Bayesian information criterion (BIC). It is interesting to note that different SEM software packages tend to report different fit measures. Duncan et al. (2006) provide a useful list of fit indices reported by current versions of Amos, EQS, LISREL, and Mplus. Default and independent model chi-squared statistics, AIC, CFI, NFI, and RMSEA appear to be reported by all four programs.

ILLUSTRATION OF LATENT GROWTH MODEL APPLICATION

We use simulated data to illustrate LGM-based analysis. The simulated data could represent a number of different scenarios relevant for OM researchers. As an example, the data sets could represent changes in the user acceptance of software designed to aid operational decisions in a manufacturing or service setting. Alternatively, the data could also represent the changes in productivity in a manufacturing setting (e.g., the number of units produced by employees) or a service setting (e.g., the number of calls answered per period for call center employees, or the amount of sales recorded by salespersons per period, etc.). All data sets are generated using the Monte Carlo data simulation facilities in Mplus (Muthén and Muthén 1998–2004). The syntax and/or data are available to interested readers on request.

By using simulated data sets we illustrate how OM researchers can identify: (1) the form of growth that occurs at the group level (e.g., linear growth, non-linear growth, etc.); (2) whether two different groups with the same form of growth (e.g., linear) have similar growth parameters (e.g., intercepts, slopes, etc.); (3) predictors of growth (e.g., does training influence growth parameters); (4) whether an intervention in the learning process (e.g., introducing new technology or methods) has an influence on growth parameters (e.g., the growth rate); and (5) whether the growth in one domain affects the growth in another (i.e., parallel growth model).

ILLUSTRATION OF THE UNCONDITIONAL LATENT GROWTH MODEL

We start by illustrating the unconditional LGM using a simulated data set that contains observations for four time points on 100 subjects. Data modeling follows the linear system of equations (e.g., see Hancock and Lawrence 2006, 175). In this data set, the intercept is generated from a normally distributed population with a mean of 30, while the growth rate is generated from a normal distribution with a mean of 10. We also decided to generate uncorrelated intercept and slope means. In our second example, we consider another simulated data for linear growth. When data sets from the first two examples are merged, the intercept and slope means of the resulting merged data turn out to be correlated, and we provide an interpretation of this correlation in that illustration.

Researchers using the LGM technique usually plot and examine raw individual and summarized data in order to decide which functional form best fits the data. We show the correlation and average values for the time points in the first data set in Table 13.1. A researcher examining these results, especially the averages for different times, would likely conclude that a linear model might fit these data. On the other hand, if the researcher decides to adopt an exploratory approach, different models can be fit to the data in order to identify the best-fitting form. Following Chan and Schmitt (2000), we fit three different LGMs—the no-growth model (strict stability model suggested by Stoolmiller 1994), the free-form model, and the linear model. The no-growth model contains only the intercept, with a loading of 1.0 to each measured time point. A good fit for this model would support the hypothesis that there is no growth in observed values

TABLE 13.1

Correlations and Means for a Linear Growth Data Set Generated With Intercept Mean = 30 and Slope Mean = 10

	Time-0	Time-1	Time-2	Time-3
Time-0	1.00			
Time-1	0.52	1.00		
Time-2	0.41	0.73	1.00	
Time-3	0.40	0.73	0.88	1.00
	29.62	39.99	50.10	60.69

over time. In the free-form model, the loadings from all measured time points to the intercept are fixed at 1.0. The loadings for the first two time points to the slope are fixed at 0.0 and 1.0, respectively. The loadings from the subsequent time points to the slope are set free and are estimated by the SEM program, thus allowing for non-linear shapes of change. The loadings for the linear LGM were explained earlier (see Figure 13.1).

When selecting the best-fitting model we compare alternate nested models using the likelihood ratio test (also known as the chi-square difference test). For example, the no-growth model is nested in the linear model, and the linear model is nested within the free-form model. The chi-square difference test is implemented by calculating the difference in model chi-square and degrees of freedom for a pair of models. If the chi-square difference (which is itself chi-square distributed) is significant, then, one of the two models has a significantly better fit. The chi-square test results in Table 13.2 show that both the linear and free-form models provide the best fit compared to the no-growth model. The differences in model fit statistics are not significant for the linear and free-form models. As such, we could adopt the linear LGM as a final choice, given that fewer parameters are estimated and thus it is more parsimonious relative to the free-form LGM (this same rationale is used by Chan and Schmitt 2000, in choosing the linear LGM over other forms).

Table 13.2 also shows model fit measures that include RMSEA, SRMR AGFI, CFI, and NNFI. For the first two measures, values close to 0.0 indicate a good fit, while values greater than 0.10 are taken to imply a poor fit. For AGFI, CFI, and NNFI, values over 0.90 imply a good fit. Fit statistics associated with linear LGM (as indicated by RMSEA, CFI, NNFI and SRMR values of 0.0, 1.0, 1.0, and 0.034, respectively) appear to be quite good and justify the choice of the linear form over other forms. Some recommended guidelines for model fit can be found in Hu and Bentler (1999). The estimated means for the intercept and slope are 29.60 and 10.35, respectively. The variances for both constructs (intercept and slope) are statistically significant, while the covariance between the intercept and slope constructs is, as expected, not significant (because we had generated uncorrelated intercept and slope values).

LATENT GROWTH MODELS FOR TWO GROUPS: MULTI-GROUP ANALYSIS

We also simulated another sample of data based on the linear growth for 100 cases with four observations per case. For this set of data, the intercept was generated from a

TABLE 13.2
Unconditional Latent Growth Models for Simulated Linear Growth Data Set

	Chi-squared	DF	p-value	Model Comparison	Change in Chi-squared	Change in DF	p-value	AGFI	NNFI	CFI	SRMR	RMSEA
1 No-growth model	*544.20*	8	.00	–	–	–	–	–5.11	–2.13	0.00	0.550	0.820
2 Linear	3.75	5	.59	1 vs. 2	*540.45*	3	.00	0.97	1.00	1.00	0.034	0.000
3 Free form	2.88	3	.41	2 vs. 3	0.87	2	.65	0.95	1.00	1.00	0.037	0.000

$n = 100$.

Numbers in **bold italics** are significant at 5%.

Data set generated such that intercept mean = 30 and slope mean = 10.

Log likelihood test does not show a statistically significant improvement between the linear and free-form models.

Linear model is chosen over free-form model as it offers a more parsimonious fit.

Mean and variance for the intercept are 29.60 and 1.78, respectively; both are statistically significant ($p < .01$).

Mean and variance for the slope are 10.35 and 2.38, respectively; both are statistically significant ($p < .01$).

Correlation between the intercept and slope is 0.23; this is not statistically significant ($p > .10$).

normal distribution with a mean of 50, while the slope was generated from a normal distribution with a mean of 5. These data could represent a group that is performing a task similar to that in the first illustration, but in a different environment (e.g., this group may contain employees that were trained before starting their job; or this group may work in an environment that has different technology, management, etc.).

What would the researcher find if the two data sets were combined and a linear LGM was fit to the data? The first conclusion would be that a linear LGM fits the combined data set quite well. The estimated mean and variance for the intercept and slope constructs for the combined data are statistically significant (the estimated mean is 39.60 for the intercept and 7.84 for the slope). Note that the mean values for the combined data for the intercept and slope are halfway between those generated for the two sets independently. There is a also statistically significant negative covariance between the intercept and slope constructs, suggesting the existence of groups that have low and high intercepts coupled with high and low growth rates, respectively (specifically, as we know, the group with an intercept of 30 has a slope of 10, while the group with an intercept of 50 has a slope of 5).

When the combined data contains two distinct groups that can be distinguished by a grouping variable, we can use the multi-group approach for analyzing the data. Here, in Step 1, we test whether the same form of growth (linear in our case) governs both sets of data. Furthermore, we examine whether the means and variances of the intercept and the slope, as well as the error variances for the different time points, are the same across the two groups. This is the "constrained" model, since the growth form and growth constructs are constrained to be equal across the two groups. This model is then compared to other less constrained models in order to identify which parameters vary across the groups. Fit statistics for the analysis performed on the two simulated linear growth data sets is shown in Table 13.3.

As might be expected, the constrained model in Step 1 demonstrates a poor fit because the data sets were created such that intercept and slope means and variances across the two samples are different. As seen in Table 13.3, model fit statistics improve significantly in Step 2 as we release constraints such that construct means and the associated error variances are estimated independently for the two groups. There is further statistically significant improvement in fit statistics in Step 3 when the error variances for the different time points are estimated independently for the two groups. The aforementioned chi-squared difference test is used to determine the superiority of the model in Step 2 relative to that in Step 1, and the model in Step 3 relative to the one in Step 2. A conclusion of this analysis is that a linear form of growth exists for individuals in both data sets; however, the growth construct means and the error variances differ across groups.

CONDITIONAL LATENT GROWTH MODEL

As an alternative to the multi-group approach, one can also use an indicator variable to denote group membership (we set the indicator variable to 0 for the group with mean intercept = 30 and mean slope = 10, and to 1 for the second group), and use the conditional LGM as shown in Figure 13.2a. By using this model we can show that the same form of linear growth applies to the two groups and that the groups differ with respect to the intercept and slope constructs. The path coefficients from

TABLE 13.3
Multi-Group Analysis for Two Linear Growth Data Sets

	Chi-squared	DF	p-value	Change in Chi-square	Change in DF	p-value	NNFI	CFI	RMSEA
Model Step 1	*215.91*	19	.00	–	–		0.040	0.000	0.320
Model Step 2	21.24	15	.13	*194.67*	4	.00	0.990	0.980	0.065
Model Step 3	7.09	11	.79	*14.15*	4	.01	1.000	1.000	0.000

Model Step 1: Constrained model assumes that growth constructs and error variances are identical across groups.

Model Step 2: Model allows growth constructs to be estimated separately, but error variances are assumed identical across groups.

Model Step 3: Model allows growth constructs and error variances to be estimated separately for each group.

Data set 1 generated such that intercept mean = 30 and slope mean = 10.

Data set 2 generated such that intercept mean = 50 and slope mean = 5.

For each data set, $n = 100$.

Log likelihood test is used to determine if difference between model chi-square is significant at 5%.

Numbers in **bold italics** are significant at 5%.

the indicator variable to the intercept and slope constructs are statistically significant, suggesting that growth constructs vary across groups. The path coefficient from the indicator variable to the intercept is 19.98 ($p < .01$), which is the numerical difference between the estimated intercept mean for Group 2 and Group 1. Similarly, the path coefficient from the indicator variable to the slope is −5.01 ($p < .01$), which is the numerical difference between the slope means for Group 2 and Group 1. Other measures show that the model fits the data well (chi-square = 8.94 with 6 degrees of freedom, $p > .10$, RMSEA = 0.05, SRMR = 0.024, and AGFI, NNFI, and CFI are greater than 0.95).

PIECEWISE LINEAR LATENT GROWTH MODEL

One area of interest in OM research involves the understanding and modeling of the effect of interventions in the learning process (e.g., Bailey 1989; Dar-El et al. 1995; Shafer et al. 2001). Let us assume that the effect of an intervention can be expected to impact on the performance immediately following the intervention in different ways. If the intervention was designed to improve performance, and is implemented successfully, then the rate of learning following the intervention may increase. On the other hand, there may be a drop in performance following a disruptive interruption.

A piecewise LGM can be useful for capturing the effect of an intervention that follows a baseline period Flora (2008). We illustrate both the positive and negative effects of intervention on subsequent performance using two different data sets. In the first set of simulated data, we generated five repeated observations for 100 cases where the linear growth for the first three time points has a mean intercept of 30, and a mean slope of 5. An intervention occurs between the third and fourth time points, which causes the average rate of growth at the group level to double (i.e., the new slope for periods four and five has a mean of 10). In the second data set, the mean intercept and mean slope for the first three periods are the same (30 and 5, respectively); however, in this case the intervention between the third and fourth periods has a negative effect, causing the post-intervention growth rate to decline (modeled using a mean slope of 3 for the last two periods).

The piecewise linear model uses three latent constructs—intercept, the slope for the first three periods (Slope 1), and the slope for the last two periods (Slope 2). There are two possible ways of fitting the piecewise model beyond this point. In both approaches, the coefficient of the intercept is fixed at 1 at all time points (as we did before with all linear LGMs). With the first approach, the coefficients for Slope 1 for the periods one to five would be 0, 1, 2, 2, and 2, respectively, while the coefficients for Slope 2 for the same periods would be 0, 0, 0, 1, and 2, respectively. Thus, effectively, the growth attained at period three is "frozen in" and serves as the intercept for the second phase of the growth (Hancock and Lawrence 2006). With the second approach, the only difference is that the coefficients for Slope 1 for the five periods are 0, 1, 2, 3, and 4, respectively, while the coefficients for Slope 2 are the same as in the first approach. The value of Slope 2 with the second approach is interpreted as the "added growth" (Hancock and Lawrence 2006).

As shown in Table 13.4a, a researcher attempting to fit a linear or free-form model to these data would find the fit statistics to be very poor (i.e., large chi-square, RMSEA, and SRMR values). Fit statistics improve significantly when the data are modeled using the two above-mentioned piecewise growth modeling approaches. Table 13.4b shows that the mean values for the intercept and Slope 1, as well as the model fit statistics, are identical for the two approaches. However, the mean value of Slope 2 differs based on the approach used, although its numerical value remains the same. As an example, consider the data set where the intercept and slope for growth during the first three periods have means of 30 and 5, respectively, while the slope has a mean of 10 for the last two periods. For both approaches the intercept and Slope 1 means are 30.02 ($p < .01$) and 5.16 ($p < .01$), respectively. With the first modeling approach, the Slope 2 mean is estimated at 10.30 ($p < .01$), and with the "additional growth" approach, the Slope 2 mean is estimated at 5.14 ($p < .01$).

Multivariate Latent Growth Models: Cross-Domain Analysis of Change

As a final illustration of LGMs, we present a model for the cross-domain analysis of change (Singer and Willett 2003; Willett and Sayer 1996). Also referred to as the "parallel" or "dual growth" model, this multivariate approach allows researchers to test whether growth in one domain (e.g., performing a particular type of task) is associated with growth in another domain (e.g., performing another task that shares traits with the first task). The richness of this analysis comes from the way in which the model is specified (the parallel growth model is shown in Figure 13.3). Specifically, for two separate growth processes (linear in our example), the researcher can test if the intercept and slope associated with one process impact on the intercept and slope associated with the other process.

We constructed a data set to illustrate the dual growth for a sample of 100 cases. For the first process, linear growth was modeled with a mean intercept of 30 and a mean slope of 10. For the second process, the intercept mean was generated with loadings of 1 each from the intercept and slope for Task 1, and the slope was generated with loadings of 0.1 and 0.3 from the intercept and slope of Task 1, respectively. Thus, the expected values of the mean intercept and mean slope for the second growth process would be 40 ($1 \times 30 + 1 \times 10$) and 6 ($0.1 \times 30 + 0.3 \times 10$), respectively.

Statistics for this model suggest an adequate fit with the p-value for the model chi-square being greater than 10%, RMSEA and SRMR less than 0.06, while AGFI, NNFI, and CFI are equal to or greater than 0.90. Path coefficients (shown in Figure 13.3) from the intercept and slope of the first growth process to the intercept and slope of the second growth process are close to the expected values and are statistically significant. Using the means and standard errors, we can show that the loadings of the intercept and slope from the first growth process on the intercept of the second growth process are not different; similarly, for the slope of the second growth process, the corresponding loadings are not different from 0.1 and 0.3, respectively. These results show that the intercept and slope constructs associated with the first linear growth process have a statistically significant impact on the intercept and slope constructs associated with the second linear growth process (which may suggest a transfer of knowledge across processes).

TABLE 13.4a

Evaluation of Fit Statistics for Data with Intervention in the Learning Process

	Chi-Squared	DF	p-value	Delta-Chi-Squared	Delta-DF	p-value	AGFI	NNFI	CFI	SRMR	RMSEA
Linear growth model	285.2	10	.00	–	–	–	−0.18	0.43	0.43	0.16	0.53
Free-form growth model	*17.05*	7	.02	*268.15*	3	.00	0.86	0.97	0.98	0.083	0.12
Piecewise growth model	6.34	6	.39	*10.71*	1	.00	0.95	1.00	1.00	0.046	0.024

$n = 100$.

Numbers in **bold italics** are significant at 5%.

Free-form is compared to the linear growth model.

Piecewise model is compared to the free-form model.

Log likelihood test is used to determine if difference between model chi-square is significant at 5%.

TABLE 13.4b

Mean Values for the Intercept and Slopes for the Piecewise Growth Modeling Approaches

	Intercept	Slope-1	Slope-2
Data-1 Approach-1	30.02	5.16	10.3
Data-1 Approach-2	30.02	5.16	5.14
Data-2 Approach-1	30.01	5.16	3.21
Data-2 Approach-2	30.01	5.16	−1.95

For both data, $n = 100$.

Expected intercept and Slope 1 means are 30 and 5, respectively, for both data.

For Data 1, expected Slope 2 mean is 10.

For Data 2, expected Slope 2 mean is 3.

With Approach 1, performance at the 3rd time point serves as the intercept for time points 4 and 5.

With Approach 2, the slope for time points 4 and 5 is interpreted as "additional growth" relative to the slope for time points 1, 2, and 3.

SUMMARY OF LATENT GROWTH MODEL ILLUSTRATIONS

We illustrated the application of LGMs to simulated data sets. Our examples showed how LGMs can be used to identify the functional form for growth occurring for the entire sample using unconditional models, as well as the differences in the growth parameters for distinct groups within the overall sample using the multi-group and conditional models. We showed how the effect of intervention on the growth process can be captured using LGMs. Finally, our illustration also showed how a researcher can test whether growth in one domain affects growth in another domain using the dual or parallel growth models (i.e., cross-domain analysis).

OTHER LONGITUDINAL DATA-ANALYSIS METHODS

LGMs are not the only way to study longitudinal changes in the measures and latent constructs of interest. According to Meredith and Tisak (1990), repeated measures ANOVA (RANOVA) and mixed-linear models (MLM) are special cases of LGMs. Duncan et al. (2006) perform a comparison of three analytic approaches for longitudinal data: RANOVA in SPSS, mixed-linear models in HLM (hierarchical linear models, version 6, 2004), and LGM (using EQS version 6, 2005). They note that with a set of common constraints, the three approaches produce identical results for an unconditional model. The three approaches also provide comparable results when estimating conditional growth models that include predictors of growth. In HLM, the conditional growth model is referred to as the "intercept and slope as outcomes" model. However, the authors note that when modeling complex models of growth that simultaneously include predictors and sequelae (i.e., distal outcomes) of growth, the current versions of the RANOVA and HLM cannot be used. Moreover, RANOVA and HLM cannot handle latent variables as predictors of intercepts and slopes. Latent growth modeling, on the other hand, is flexible and allows for the testing of such models.

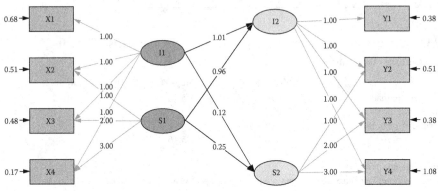

Chi-square = 32.59, df = 25, *p*-value = 0.14154, RMSEA = 0.055

Structural equations

I2 = 1.01*I1 + 0.96*S1, Error var. = 1.04, R² = 0.89
      (0.019)   (0.057)   (0.23)
      52.48   16.97   4.52

S2 = 1.12*I1 + 0.25*S1, Error var. = 0.87, R² = 0.23
      (0.015)   (0.045)   (0.14)
      7.69   5.65   6.23

FIGURE 13.3 Path diagram for cross-domain analysis and related structural equations from LISREL.

Chan (1998), and Hancock and Lawrence (2006) note an important limitation of the RANOVA procedure for handling growth data in a sample that includes individuals with variable growth rate parameters. The limitation involves the sphericity assumption inherent in RANOVA which requires that variances and covariances in the sample across different time units should be equal. Thus, the very existence of intra-individual differences violates the statistical assumption of the RANOVA procedure, and hence it is not recommended when one wishes to statistically quantify the individual differences in change over time.

A major difference between HLM and LGM is the treatment of measurement error terms. While HLM imposes constraints by assuming that measurement error terms are independent, LGMs allow the researcher to model measurement errors quite flexibly. For example, the researcher may fix the error terms to specific values, let the software estimate them by treating them as unknowns, allow for auto-correlated error terms, and so forth. The key differences between LGM and HLM, with the advantages and disadvantages of both approaches over each other, are documented by Chan (1998), while Curran (2003) provides a more technical and detailed comparison.

LIMITATIONS AND POSSIBLE REMEDIES FOR LATENT GROWTH MODELS

Despite the many advantages, LGM suffers from many of the limitations that are inherent to the SEM approach. In addition, as with longitudinal methods, a major limitation of the LGM approach is the extended time frame necessary to follow participants when acquiring longitudinal data. The cohort sequential design allows for compressing the time frame over which data can be collected. Duncan et al. (1996) show that by using a cohort sequential design, longitudinal data for a single cohort that would normally require six years can be collected over a three-year period using multiple cohorts. In their example, comparisons are made between the annual tracking of a single cohort of individuals from the starting age of 12 until they reach the age of 17. With a single cohort, this approach would require six years of tracking (observations taken when the cohort is 12, 13, 14, 15, 16, and 17). However, with four cohorts (with the starting ages of 12, 13, 14, and 15), data can be collected effectively over a three-year period, with the assumption that the cohorts are comparable. Their study compares the growth trajectories for the single cohort and multi-cohort samples and finds the differences to be statistically insignificant.

Longitudinal studies also suffer from issues related to missing data and attrition. For example, if a researcher designs data collection to occur over four time points, it may not always be possible to obtain data from every sample unit at every occasion. There is also the concern that some subjects may participate in providing data initially but may drop out before the study has been completed. Advances in LGM allow for procedures that can account for some of these problems. Bollen and Curran (2006) advise researchers against using the two commonly adopted procedures of dealing with missing data in practice—namely, list-wise deletion and pair-wise deletion. Instead, they recommend using either the direct maximum likelihood or the multiple imputation method in dealing with missing data (for details see Schafer and Graham 2002). Enders (2006) provides an excellent additional discussion on the handling of missing data within the context of SEM, while Duncan and Duncan (1994) demonstrate fitting LGMs in the presence of missing data.

Lastly, the temporal design (Collins and Graham 2002) of a longitudinal study is an important but often under-considered factor in methodological designs. Temporal design refers to the timing and spacing of repeated assessments in a longitudinal design and has consequences for a study's findings. Let us say that the true course of change for a given outcome is cubic or S-shaped. However, if the assessments are taken only at two points in time, the functional form of change can only be linear (i.e., a straight line), suggesting a smooth, continuous change in an outcome over time, which is a misleading conclusion.

LATENT GROWTH MODEL ANALYSIS IN OM RESEARCH

To begin with, as an SEM-based methodology, LGMs allow researchers to study change in both observed variables and latent variables. To the best of our knowledge, where OM researchers have focused on change, they have always done so by focusing on changes in observable measures (e.g., Uzumeri and Nembhard [1998] fit a learning curve model to a population of learners where the observed measure is the amount of work done in standard hours; Pisano et al. [2001] study the improvements in the time required to perform cardiac surgery as a measure of organizational learning in order to understand if/why learning rates vary across organizations, i.e., hospitals). LGM methodology allows OM researchers the opportunity to study longitudinal changes in observed variables as well as in multidimensional constructs such as product or service quality, customer service, employee satisfaction, and so forth.

Multilevel LGMs (MLGMs) allow for the possibility of modeling change as a combination of its latent components. For example, researchers have viewed individual learning in industrial settings as a combination of motor and cognitive skill-related learning (e.g., Dar-El et al. 1995). Using instruments to measure cognitive and motor learning for relevant tasks over time, it may be possible to fit LGMs for each learning component individually at Level 1. Overall learning can then be viewed as a combination of learning in both of these components. Such a model would have a common learning intercept and a common learning slope at Level 2. A conceptual diagram of this learning model is shown in Figure 13.4. Duncan et al. (2006) refer to such models as the "factor of curves" model (in our example, the growth in motor and cognitive learning refer to the "curves" at Level 1, and the common intercept and growth constructs are the "factor" at Level 2).

In modeling performance changes as a function of experience (Wright, 1936) Yelle (1979) notes that non-linear forms like the S-curve and exponential functions have received extensive attention in the OM literature on learning. Ployhart and Hakel (1998) provide an application of the cubic (S-curve shaped) LGM to capture the growth in individual productivity, measured in sales commission (dollars), in a sample of 303 sales persons over a two-year (eight quarters) period. Blozis (2004,

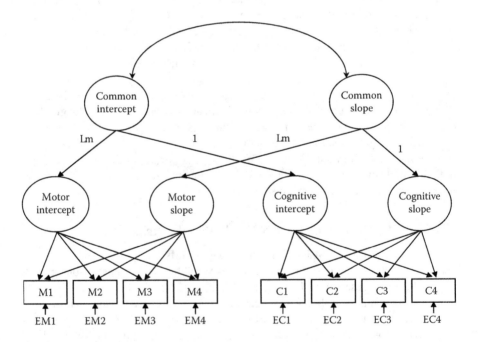

FIGURE 13.4 Multilevel latent growth model (MLGM) depicting a factor of curves. (Adapted from Duncan et al., *An introduction to latent variable growth curve modeling: Concepts, issues and applications.* Mahwah: Lawrence Erlbaum, 2006, p. 70.)

343) provides an example of fitting data on verbal and quantitative skills acquisition for 204 individuals using a structured LGM with a "negatively accelerated exponential function." Thus, OM researchers have several choices with regard to modeling the form of longitudinal change using LGMs.

In addition to modeling individual-level learning effects, LGMs also allow for the modeling of interruptions (discontinuities) in learning. In recent years, several studies have focused on understanding the impact of interruptions in the learning process and how they can lead to forgetting effects (e.g., Bailey [1989] studied these effects in a laboratory setting; on the other hand, Shafer et al. [2001] used organizational data to study these effects). As demonstrated in our illustrations of the piecewise growth models, OM researchers can use LGMs for modeling the effects of interventions that affect the rate of learning (e.g., the introduction of new technology or methods during the training period may help individuals learn at a faster rate).

Researchers in OM and management have also been interested in understanding the learning and growth that occurs at the organizational level (i.e., within and across organizations). Weinzimmer et al. (1998) provide a review study on organizational growth within management. Their summarized results show that while researchers have collected repeated measures on financial and other publicly available data for samples of companies, data analysis has typically been based on a single measure, and performed using an OLS approach. The LGM methodology can allow the use of one or more measures (e.g., multiple measures can be aggregated to form a composite) in order to assess growth. The benefits of LGM over OLS were noted in the introduction to this chapter.

With organizations as the unit of analysis, another area where growth models can help is in analyzing the long-term effects of specific program adoptions on company performance. Consider, for example, the issue of the successful management of environmental responsibilities. If we were to focus on the effects of companies' adopting programs aimed at improving their environmental performance, it would then make sense to investigate the long-term impacts of such actions on performance levels (e.g., financial performance) simply because it often takes a long time to adopt such programs. Such a study would be based on tracking performance over a longer time frame (rather than analyzing cross-sectional data). Companies are now being encouraged to make public their efforts toward environmental management, which allows researchers access to such data. For example, data on organizations reporting their environmental performance is now maintained by the U.S. Environmental Protection Agency's National Environmental Performance Track program, which currently lists 450 reporting organizations. Similarly, there are other strategic programs that companies adopt which also warrant longitudinal analysis. Examples include examining the long-term effects of investing in enterprise resource planning (ERP) systems, adopting a just-in-time (JIT) manufacturing system, adopting quality improvement programs such as Six Sigma, the ISO 9000 series, and so on.

CONCLUSIONS

Motivated by the recent popularity of SEM in OM research, our chapter illustrated the application of LGMs to simulated data sets. Using an exploratory approach, our illustrations show that the researcher can evaluate the efficacy of modeling different forms of growth in order to identify the one that fits best. We demon-strated how conditional LGMs can be employed to identify groups of cases within the overall sample which have trajectories that differ from the one fitted for the entire sample. Finally, we also illustrated cross-domain analysis whereby LGMs are used to determine whether the growth in one measured variable (or latent construct) is influenced by the growth in another measured variable (or latent con-struct) over the same time frame. The primary benefits of using an LGM-based longitudinal analysis are listed in Table 13.5.

In promoting the use of LGMs, we also noted the similarities and differences with other longitudinal analysis approaches. Existent research shows that there are certain types of advantages that only the LGM approach is able to provide, at least for now (future developments in other areas may offset these advantages). In particular, the flexibility of LGMs in allowing certain variables to be included simultaneously as dependent and independent variables in the same model, and the ability to allow for cross-domain analysis both provide LGMs with advantages over other approaches. We also noted the limitations inherent in LGMs, and the remedies recommended by researchers to overcome them.

TABLE 13.5

Benefits of Using Latent Growth Models

Latent Growth Models:

1. *Allow for the identification of a single form of growth for the entire group.* For example, we can attempt to fit a hypothesized functional form for the growth, or use an exploratory approach to determine the best model for the group. The SEM framework allows us to compare model fits to identify best functional form.

2. *Provide an estimate of variance in the constructs for the form of growth across individuals.* For example, with a linear growth model, a significant variance in the intercept suggests that individuals' vary considerably with regards initial performance. Similarly, significant variance in the slope would suggest that individuals in the group vary with respect to their rate of growth. Growth parameter variances can then be explained by covariates.

3. *Provide an estimate of covariance among constructs that define the form of growth.* For example, with a linear growth model, a significant positive covariance between the intercept and slope would suggest that high initial performance is associated with higher learning rates.

4. *Allow for testing the effects of intervention.* For example, did the intervention have a positive or negative effect on the post-intervention rate of growth?

5. *Allow for testing whether the same form of growth exists across multiple groups.* For example, if two different groups are learning the same task in similar environments, then, is the form of growth (e.g., linear, non-linear, etc.) the same for these two groups?

6. *Allow for testing whether the growth constructs vary across groups where growth follows the same form.* For example, we can test whether the intercepts and slopes are similar for two groups experiencing linear growth.

TABLE 13.5 (Continued)
Benefits of Using Latent Growth Models

7. *Allow for the inclusion of time invariant and time variant predictors of growth.* Suppose one group was briefed about the task before learning commenced and the other group was not. A dummy variable denoting the presence/absence of training can be included as a time invariant predictor of growth. Suppose also that individuals in a group were provided feedback on their performance for that period, and the amount of feedback provided declined as individuals became more proficient at the task over time. In this case we may include the amount of feedback for each period as the time variant predictor of growth.

8. *Allow for the inclusion of outcomes of growth.* For example, individual performance appraisal after learning may be statistically linked to the growth constructs – in this case the performance appraisal is said to be a distal outcome of growth. This type of modeling cannot be conducted in HLM.

9. *Allow for testing whether growth in one domain affects growth in another domain.* For example, individuals that are learning the task well over time also happen to show increases in job satisfaction over the same time periods; the cross-domain model allows for a statistical test of linkages among the parameters that characterize growth across these two domains.

We highlighted some research areas within OM where LGM applications appear to be most promising. These include studying learning at the individual and organizational levels, examining the antecedents of growth, and examining the longitudinal effects of strategic programs that companies employ to better their performance. We believe that there are many other areas where LGMs will help in advancing our knowledge about growth processes. We conclude with a quote from Hancock and Lawrence (2006, 160) that "LGM innovations are both exciting and too numerous to address." With an increased emphasis on SEM, and growing expertise of OM researchers in SEM in general, LGMs offer the opportunity to analyze the longitudinal changes in measures and latent constructs that are relevant to OM.

ACKNOWLEDGMENTS

The first author would like to thank the Alfred Lerner School of Business, University of Delaware, for supporting this work through its summer research grant.

REFERENCES

Bailey, C., 1989. Forgetting and the learning curve: A laboratory study. *Management Science* 35(3): 340–352.

Baker, G.A., 1954. Factor analysis of relative growth. *Growth* 18(3): 137–143.

Bentein, K., Vandenberg, R., Vandenberghe, C., and Stinglhamber, F., 2005. The role of change in the relationship between commitment and turnover: A latent growth modeling approach. *Journal of Applied Psychology* 90(3): 468–482.

Blozis, S.A., 2004. Structured latent curve models for the study of change in multivariate repeated measures. *Psychological Methods* 9(3): 334–353.

Bollen, K.A., and Curran, P.J., 2006. *Latent curve models: A structural equation perspective.* Hoboken: John Wiley.

Castilla, E.J., 2005. Social networks and employee performance in a call center. *American Journal of Sociology* 111(5): 1243–1283.

Chan, D. (1998). The conceptualization and analysis of change over time: An integrative approach incorporating longitudinal mean and covariance structures analysis (LMACS) and multiple indicator latent growth modeling (MLGM). *Organizational Research Methods* 1(4): 421–483.

Chan, D., and Schmitt, N., 2000. Inter-individual differences in intra-individual changes in proactivity during organizational entry: A latent growth modeling approach to understanding newcomer adaptation. *Journal of Applied Psychology* 85(2): 190–210.

Collins, L.M., and Graham, J.W., 2002. The effect of the timing and spacing of observations in longitudinal studies of tobacco and other drug use: Temporal design considerations. *Drug and Alcohol Dependence* 68(1): S85–S96.

Curran, P.J., 2003. Have multilevel models been structural equation models all along? *Multivariate Behavioral Research* 38(4): 529–569.

Dar-El, E.M., Ayas K., and Gilad, I., 1995. A dual-phase model for the individual learning process in industrial tasks. *IIE Transactions* 27, 265–271.

Duncan, T.E., and Duncan, S.C., 1994. Modeling incomplete longitudinal substance use data using latent variable growth curve methodology. *Multivariate Behavioral Research* 29(4): 313–338.

Duncan, S.C., Duncan, T.E., and Hops, H., 1996. Analysis of longitudinal data within accelerated longitudinal designs. *Psychological Methods* 1(3): 236–248.

Duncan, T.E., Duncan, S.C., and Strycker, L.A., 2006. *An introduction to latent variable growth curve modeling: Concepts, issues and applications.* Mahwah: Lawrence Erlbaum.

Enders, C.K., 2006. Analyzing structural equation models with missing data. In *A second course in structural equation modeling*, eds. G.R. Hancock and R.O. Mueller, 313–344. Greenwich: Information Age.

Flora, D.B. (2008). Specifying piecewise latent trajectory models for longitudinal data. *Structural Equation Modeling*, 15(3): 513–533.

George, R., 2000. Measuring change in students' attitudes toward science over time: An application of latent variable growth modeling. *Journal of Science Education and Technology* 9(3): 213–225.

Hamilton, J., Gagne, P.E., and Hancock, G.R., 2003. The effect of sample size on latent growth models. Paper presented at the annual meeting of American Educational Research Association, April 21–25, Chicago, IL.

Hancock, G.R., and Lawrence, F.R., 2006. Using latent growth models to evaluate longitudinal change. *Structural equation modeling: A second course*, 171–196. Greenwich: Information Age.

Hipp, J.R., Bauer, D.J., Curran, P.J., and Bollen, K.A., 2004. Crimes of opportunity or crimes of emotion? Testing two explanations of seasonal change in crime. *Social Forces* 82(4): 1333–1372.

Hu, L., and Bentler, P.M., 1999. Cutoff criteria for fit indexes in covariance structure analysis: Conventional criteria versus new alternatives. *Structural Equation Modeling* 6(1): 1–55.

Lance, C.E., Vandenberg, R.J., and Self, R.M., 2000. Latent growth models of individual change: The case of newcomer adjustment. *Organizational Behavior and Human Decision Processes* 83(1): 107–140.

Li, C., Goran, M.I., Kaur, H., Nollen, N., and Ahluwalia, J.S., 2007. Developmental trajectories of overweight during childhood: Role of early life factors. *Obesity* 15(3): 760–771.

Meredith, W., and Tisak, J., 1990. Latent curve analysis. *Psychometrika* 55(1): 107–122.

Muthén, L.K., and Muthén, B.O., 2002. How to use a Monte Carlo study to decide on sample size and determine power. *Structural Equation Modeling* 9(4): 599–620.

Muthén, L.K., and Muthén, B.O., 1998–2004. *Mplus user's guide*. Third Edition. Los Angeles: Muthén and Muthén.

Nesselroade, J.R., and Baltes, P.B. (eds.), 1979. *Longitudinal research in the study of behavior and development*. New York: Academic Press.

Pisano, G.P., Bohmer, R.M.J., and Edmondson, A.C., 2001. Organizational differences in rates of learning: Evidence from the adoption of minimally invasive cardiac surgery. *Management Science* 47(6): 752–768.

Ployhart, R.E., and Hakel, M.D., 1998. The substantive nature of performance variability: Predicting inter-individual differences in intra-individual performance. *Personnel Psychology* 51(4): 859–901.

Rao, C.R., 1958. Some statistical methods for comparison of growth curves. *Biometrics* 14(1): 1–17.

Schafer, J.L., and Graham, J.W., 2002. Missing data: Our view of the state of the art. *Psychological Methods* 7(2): 147–77.

Shafer, S.M., Nembhard, D.A., and Uzumeri, M.V., 2001. The effects of worker learning, forgetting, and heterogeneity on assembly line productivity. *Management Science* 47(12): 1639–1653.

Shah, R., and Goldstein, S.M., 2006. Use of structural equation modeling in operations management research: Looking back and forward. *Journal of Operations Management* 24(2): 148–169.

Singer, J.D., and Willett, J.B., 2003. *Applied longitudinal data analysis: Modeling change and event occurrence*. Oxford: Oxford University Press.

Stoolmiller, M., 1994. Antisocial behavior, delinquent peer association and unsupervised wandering for boys: Growth and change from childhood to early adolescence. *Multivariate Behavioral Research* 29(3): 263–288.

Tucker, L.R., 1958. Determination of parameters of a functional relation by factor analysis. *Psychometrika* 23(1): 19–23.

Uzumeri, M., and Nembhard, D., 1998. A population of learners: A new way to measure organizational learning. *Journal of Operations Management* 16(5): 515–528.

Weinzimmer, L.G., Nystrom, P.C., and Freeman, S.J., 1998. Measuring organizational growth: Issues, consequences and guidelines. *Journal of Management* 24(2): 235–262.

Willett, J.B., and Sayer, A.G., 1994. Using covariance structure analysis to detect correlates and predictors of change. *Psychological Bulletin* 116(2): 363–381.

Willett, J.B., and Sayer, A.G., 1996. Cross-domain analyses of change over time: Combining growth modeling with covariance structure modeling. In *Advanced structural equation modeling: Issues and techniques*, eds. G.A. Marcoulides and R.E. Schumacker, 125–157. Mahwah: Lawrence Erlbaum.

Wright, T., 1936. Factors affecting the cost of airplanes. *Journal of Aeronautical Science* 3(2): 122–128.

Yelle, L.E., 1979. The learning curve: Historical review and comprehensive survey. *Decision Science* 10(2): 302–328.

Part II

Applications

14 The Lot Sizing Problem and the Learning Curve: A Review

Mohamad Y. Jaber and Maurice Bonney

CONTENTS

INTRODUCTION

Traditional models for determining the economic manufacture/order quantity assume a constant production rate. One result of this assumption is that the number of units produced in a given period is constant. In practice, the constant production rate assumption is not valid whenever the operator begins the production of a new product, changes to a new machine, restarts after a break, or implements a new production technique. In such situations, learning cannot be ignored.

Learning suggests that the performance of a person or an organization engaged in a repetitive task improves with time. This improvement results in a decrease in the manufacturing cost of the product, but if the savings due to learning are significant, the effect on production time—and hence inventory—can also be significant. Factors contributing to this improved performance include more effective use of tools and machines, increased familiarity with operational tasks and the work environment, and enhanced management efficiency.

Today, manufacturing companies are adopting an "inventory is waste" philosophy using just-in-time (JIT) production, which usually combines the elements of total quality control and demand pull to achieve high productivity. JIT turns the economic manufacture/order quantity (EMQ/EOQ) formula around. Instead of accepting set-up times as fixed, companies work to reduce the set-up time and reduce lot sizes. The JIT concept of continuous improvement applies primarily to repetitive manufacturing processes where the learning phenomenon is present. This chapter surveys work that deals with the effect of learning on the lot size problem. It also explores the possibility of incorporating some of the ideas adopted by JIT (reduction in set-up times, zero defectives, total preventive maintenance, etc.) to such models with the intention of narrowing the gap between the "inventory is waste" and the "just-in-case" philosophies.

THE LEARNING PHENOMENON

Early investigations of learning focused on the behavior of individual subjects. These investigations revealed that the amount of time required in order to perform a task declined as experience with the task increased (Thorndike 1898; Thurstone 1919; Graham and Gagné 1940). The first attempt made to formulate relations between learning variables in quantitative form (by Wright in 1936) resulted in the theory of the "learning curve." The learning curve can describe group as well as individual performance, and the groups can comprise direct and indirect labor. Technological progress is a kind of learning. The industrial learning curve thus embraces more than the increasing skill of an individual by repetition of a simple operation. Instead, it describes a more complex organism—the collective efforts of many people, some in line and others in staff positions, but all aiming to accomplish a common task progressively better. This may be why the learning phenomenon has been called "progress functions" (Glover 1965), "start-up curves" (Baloff 1970), and "improvement curves" (Steedman 1970). The "aircraft learning curve" originated by Wright is known to some as "Wright's model of progress." Wright's power function formulation can be represented as:

$$T_Q = T_1 Q^{-b}, \qquad (14.1)$$

where T_Q is the time to produce the Qth unit of production, T_1 is the time to produce the first unit, and b is the learning exponent. Note that the cumulative time to produce Q units is determined from (1) as $\sum_{i=1}^{Q} T_1 i^{-b} \cong T_1 \int_0^Q x^{-b} dx = T_1 Q^{1-b} / (1-b)$. Figure 14.1 illustrates the Wright learning curve. It is worth noting that a few researchers (e.g., Glover 1966, 1967) adopted graphical solutions for negotiating the prices for military contracts based on Wright's learning curve.

Hackett (1983) compares a number of models, including that of Wright, and concludes that the time constant model is a good choice for general use since it nicely fills a wide range of observed data. Although the time constant model may be a better one with regard to its predictive ability, it is more complicated to use and is less commonly encountered than the Wright learning curve (1936). Jordan (1958),

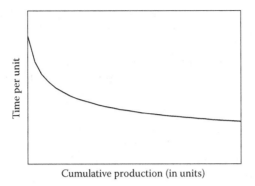

FIGURE 14.1   Wright's learning curve.

and Carlson and Rowe (1976) state that the learning curve is non-linear and that, in practice, the learning function is an S-shaped curve. The task "life cycle" can be described as consisting of three phases, as shown in Figure 14.2. The incipient phase, Phase 1, is the phase during which the worker is getting acquainted with the set-up, the tooling, instructions, blueprints, the workplace arrangement, and the conditions of the process. In this phase, improvement is slow. Phase 2—the learning phase—is where most of the improvement takes place (e.g., reduction in errors, changes in the workplace, and/or changes in the distance moved). The third and final phase (Phase 3) represents maturity, or the levelling of the curve. Looking at the graph in Figure 14.2, it will be seen that the learning rate during the incipient or start-up phase is slower than during the learning phase. At the level of the firm, the learning curve is considered as an aggregate model since it includes learning from all sources of the organization. Learning curves, besides describing changes in the labor performance, also describe changes in materials input, process or product technologies, or managerial technologies, from the level of the process to the level of the firm. Hirsch (1952) in his study found that approximately 87% of the changes in direct labor requirements were associated with changes in technical knowledge, which can be considered as a form of organizational learning. Hirschmann (1964) observed the learning curve in the petroleum-refining industry where direct labor

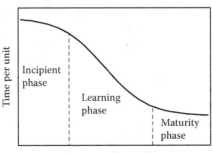

FIGURE 14.2   The S-shaped learning curve.

is non-existent. He suggested that learning is due to organizational learning as a result of the continuous investment in new technology in order to reduce the cost per unit of output.

When does the learning process cease? Crossman (1959) claimed that the process continues even after 10 million repetitions. Hirschmann (1964) made the point that skepticism on the part of management that improvement can continue may in itself be a barrier to its continuance. Conway and Schultz (1959) noted that two products that had plateaued in one firm continued down the learning curve when transferred to another firm. Baloff (1970) concluded that plateauing is much more likely to occur in machine-intensive industries than it is in labor-intensive industries. One possible explanation for this is that plateauing could be strongly associated with labor ceasing to learn or machines having a fixed cycle that limits the reduction in time. Corlett and Morcombe (1970) related plateauing either to the necessity to consolidate what has already been learnt before making further progress, or to forgetting.

Although there is almost unanimous agreement by scientists and practitioners that the form of the learning curve presented by Wright (1936) is an effective representation of learning, a full understanding of the behavior and factors affecting the forgetting process has not yet been developed. Many researchers in the field have attempted to model the forgetting process mathematically. Carlson and Rowe (1976) described the forgetting or interruption portion of the learning cycle by a negative decay function comparable to the decay observed in the electrical losses in condensers. Hoffman (1968) and Adler and Nanda (1974a, 1974b) presented two refined mathematical techniques for incorporating the effects of production breaks into planning and control models. Globerson et al. (1989) indicated that the degree of forgetting is a function of the break length and the level of experience gained prior to the break. Jaber and Bonney (1996a) developed a mathematical model for the learning-forgetting relationship, referred to as the "learn-forget curve model" (LFCM). The LFCM was tested and shown to be consistent with the model fitted to the experimental data of Globerson et al. (1989) with less than 1% deviation (Jaber and Bonney 1997a). The forgetting slope was assumed to be dependent on four factors: (1) the length of the interruption period, (2) the equivalent accumulated output by the point of interruption, (3) the time over which it is assumed that total forgetting will occur, and (4) the learning exponent. Unlike the "forgetting factor" model of Towill (1985), it is believed that the LFCM model resolved the behavioral nature of the forgetting factor as a function of the break time. The LFCM has been shown to be a potential model to capture the learning-forgetting process. For further background information, readers may refer to Jaber and Bonney (1997a), Jaber et al. (2003), and Jaber and Sikström (2004a, 2004b).

LOT SIZING WITH LEARNING (AND FORGETTING) IN PRODUCTION

The use of the learning curve has been receiving increasing attention due to its application to areas other than traditional learning. In general, learning is an important consideration whenever an operator begins the manufacture of a new product, changes to a new machine, or restarts production after some delay. Learning implies that the time (cost) needed to produce a product will reduce as the individual (or group) becomes more proficient with the new product or process. For previous reviews of learning curves, see Yelle (1979) and Jaber (2006a). Optimal lot size formulae are typically developed and used on the assumption that manufacturing unit costs are constant. With learning, this approximation is true only when the production rate has stabilized. This section categorizes the publications on learning curve effects related to the lot sizing problem into two groups. The first group includes those authors who studied the effects of learning on the economic manufacturing quantity (EMQ) with the assumption of instantaneous replenishment. The second group includes those authors who studied the effects of learning on the EMQ with a finite replenishment rate.

Effects of Learning on the Economic Manufacturing Quantity with Instantaneous Replenishment

Keachie and Fontana (1966) provided an extension of the theory of lot sizes to include "learning" related to three different cases that can occur. These cases are: (1) there is a total transfer of learning from period to period, (2) there is no transmission of learning, and (3) there is a partial transmission of learning. The total inventory cost function, C_T, developed by Keachie and Fontana (1966) is represented as:

$$C_T = S + h\frac{Q^2}{2D} + c_1\frac{T_1 Q^{1-b}}{1-b}, \tag{14.2}$$

where S is the set-up cost per lot, h is the holding cost per unit per unit of time, c_1 is the labor cost per unit of time, D is the demand rate, and T_1 and b are as defined in Equation 14.1. They assumed constant demand, no shortages occur, and their set-up costs are independent of the number of pieces in the lot. Spradlin and Pierce (1967) describe lot sizing production runs whose production rate can be described by a learning curve. They considered the infinite planning horizon case. Their solution included the possibility of variable lot sizes, as well as the possibility that all units are produced in one lot. They assumed that the production rate is high enough, relative to the demand rate, that for the purpose of calculating holding costs, one can assume that a lot is not started before the preceding lot is depleted. They assumed a constant regression of learning (forgetting rate), which is related to the rate of learning at the point at which the interruption occurs. This is accomplished by taking the regression equal to the amount learned during the production of the last M units. At each interruption the rate of learning reverts back to the rate M units.

Steedman (1970) extended the work of Keachie and Fontana by investigating the properties of the solution of Equation 14.2 and demonstrating that $Q^* > Q_0 = \sqrt{2SD/h}$, where Q_0 is the economic order quantity (EOQ) (Harris 1990). The improvement phenomenon makes the optimum lot size greater than that given by the traditional square-root formula. Steedman noted that the optimum lot size decreases as learning becomes faster.

Carlson (1975) provided a method to determine "lost time" based on the classical learning curve relationship, which he added to the cost function. Carlson altered the EOQ formula accordingly as:

$$Q^* = \sqrt{\frac{2D(S+LT)}{U(Q)I}},$$

(14.3)

where LT is lost time cost, $U(Q)$ is the average unit labor cost when producing a lot of Q units, and I is the carrying cost rate per time period. While the approach of Carlson (1975) considers the role of $U(Q)$ in the carrying cost, it ignores its impact on the production cost.

Wortham and Mayyasi (1972) proposed the use of the classical square-root formula for demonstrating the impact of learning on the EOQ and system inventory cost, using a decreasing average value of the holding cost. Muth and Spremann (1983) extended the works of Keachie and Fontana (1966), Steedman (1970), and Wortham and Mayyasi (1972). These extensions are: (1) production costs consist of two portions—one is learning dependent, the other is linear; (2) it is shown that the ratio Q^*/Q_0 is a function of two parameters, the progress rate and the cost ratio; (3) an explicit numerical solution procedure is given for the transcendental equation defining Q^*/Q_0 and specific values of the solution are tabulated; and (4) a simple approximation for Q^* is given in the classical square-root format whose form is:

$$Q^* \approx \sqrt{\frac{2D}{h}\left[S + c_1 \frac{T_1 Q_0^{1-b}}{2}\right]}.$$

(14.4)

Kopcso and Nemitz (1983) explored the effect of learning in production on the lot sizing problem. They developed two models to describe two demand situations: dependent (Model 1) and independent (Model 2). They found that by excluding the material costs (for an assembly operation, the cost of all components), the optimal lot size was seen to vary linearly with demand and inversely with the carrying cost rate. Kopcso and Nemitz also found that when material costs were included, a smaller optimal lot size was derived. Their cost function, with labor cost, is given as:

$$C_T = c_1 \frac{T_1 Q^{-b}}{1-b}\left(\frac{Q}{2}I + D\right),$$

(14.5)

where I is the inventory carrying cost rate per time period (i.e., $h = c_1 I$) and $Q = Q_L = 2bD/I(1-b)$ is the optimal lot size, where Q_L is linearly related to D and to $1/I$. This is contrary to the EOQ model where the order quantity is proportional to their square root. Kopcso and Nemitz (1983) modified Equation 14.5 to include the material cost, c_2, as:

$$C_T = \left(c_2 + c_1 \frac{T_1 Q^{-b}}{1-b} \right) \times \left(\frac{Q}{2} I + D \right).$$ (14.6)

No closed-form expression was found for $Q = Q_M$ (including material cost) that optimizes Equation 14.6, which could be determined using numerical search techniques. However, they provided a relationship between Equations 14.5 and 14.6 where $(Q_L - Q_M)/Q_M = c_2/(c_1 T_1 Q_M^{-b})$. Kopcso and Nemitz (1983) ignored the set-up cost by referring to Freeland and Colley (1982), who proposed an improved variation on part period balancing, combining lots until the incremental (not total) carrying cost equals the savings resulting from eliminating one set-up.

EFFECTS OF LEARNING ON THE ECONOMIC MANUFACTURING QUANTITY WITH FINITE DELIVERY TIME

Adler and Nanda (1974a, 1974b), analyzed the effect of learning on production lot sizing models for both the single- and multiple-product cases. A general equation is developed for the average production time per unit, for producing the annual demand in batches with some percentage of learning not retained between lots. Two models are developed —the first restricted to equal lot sizes; $Q_1 = Q_2 = Q_3 = \cdots = Q_n = Q$, and the second case restricted to equal production intervals; $t_1 = t_2 = t_3 = \cdots = t_n = t$. Adler and Nanda defined the average production rate, $\bar{P}_i$, as the reciprocal of the average time per unit for the ith lot of Q units. They developed a technique for estimating the drop in labor productivity as a decrease in the average production rate, as a result of production breaks. Adler and Nanda concluded that, not considering learning, $b = 0$, $n^* = \sqrt{hQ(1 - D/\bar{P}_i)/2S}$, and when maximum learning is considered, $b = 1$, $n_0^* = \sqrt{hQ/2S}$, resulting in less than the optimal number of lots, $n^* < n_0^*$. In the case of equal production intervals, when $b = 0$, then $n^* = \sqrt{hQ(1 - D/\bar{P}_i)/2S}$, whereas when $b = 1$, the cost function yielded an infeasible solution.

Sule (1978) developed a method for determining the optimal order quantity for the finite production model, which incorporates both learning and forgetting effects. Sule based his decision on a forget relationship assumed by Carlson (1975). The forgetting rate was arbitrarily selected and the production quantity was assumed to be equal for all lots. Sule suggested a linear relationship between the production time and the quantity produced for known learning and forgetting rates. Axsäter and Elmaghraby (1981) argued that Sule's model is based on some assumptions that were either not demonstrated by the author or were of questionable validity. They showed that Sule's cost function is based on an approximation of the inventory accumulation during the production interval, in that he assumed a linear accumulation, while in reality it is a convex function of time. This consideration increases the invalidity of Sule's optimizing procedure. Sule (1981) responded to the argument of Axsäter and Elmaghraby by reemphasizing that, under steady-state conditions, the time required to produce Q units (i.e., t) is given by a linear function of Q. He agreed with Axsäter and Elmaghraby that the expression he used for inventory accumulation in his basic model is a linear approximation. However, he argued that the substitution of the exact equation (a convex function) does not add any significant accuracy to the determination of the economic manufactured quantity and restated that the linear relationship $t = \alpha + \beta Q$ is a very good estimate of the production time. It is both easy to use and is well within the acceptable norms of accuracy.

Fisk and Ballou (1982) presented models for solving the manufacturing lot sizing problem under two different learning situations. The first situation assumes that learning conforms to the power function formulation introduced by Wright. The second recognized approach for modeling learning is the bounded power function formulation proposed by de Jong (1957). Fisk and Ballou developed a dynamic programming approach to solve for the optimal production lot size. When regression in learning is operative, the recursive relation does not apply. Regression in learning refers to the loss of labor efficiency that can occur during the elapsed time between successive runs of a product. A different dynamic recursion and cost relationship is required. Using their models, Fisk and Ballou found that, relative to the classical production order quantity models, significant cost savings occur whenever the production rate is high relative to the demand rate. They suggested that an application of their models to an infinite horizon problem would yield satisfactory results since, as the cumulative number of units produced becomes large, production lot sizes tend to stabilize. Fisk and Ballou (1982) assumed forgetting to be a constant percentage; that is, $u_i = \sum_{j=1}^{i-1} fQ_j$, where $0 \leq f \leq 1$, $i \geq 2$ and $u_1 = 0$.

Smunt and Morton (1985), unlike Fisk and Ballou (1982), relaxed the equal lot size and fixed carrying cost assumptions. Smunt and Morton's relaxation of the carrying cost assumption implies that holding costs can approach zero as the total number of units requested, Q, increases, resulting in lot sizes that increase without bound as n gets large. Smunt and Morton also showed that the effect of forgetting was significant in the determination of optimal lot sizes, thus indicating that operations managers should consider longer, not shorter, production runs. They suggested the use of robotics and flexible manufacturing systems to overcome the need for larger lot sizes, which also reduces the forgetting effects.

Klastorin and Moinzadeh (1989) considered the production of a single product with static demand and no shortages. They treated the problems defined by Fisk and Ballou (1982) by allowing lot size quantities to be real rather than integer numbers. Klastorin and Moinzadeh show that: (1) lot sizes monotonically decrease over time, and (2) that lot sizes approach the traditional EOQ in the limit as the total number of units ordered (N) approaches infinity. Two algorithms are presented for finding the lot sizes under a transitory learning effect. The first algorithm finds a solution for the case of no learning regression between production cycles; that is, production efficiency is the same at the end of cycle $j - 1$ and at the beginning of cycle j. The second algorithm finds a solution for the case of a constant amount of forgetting occurring between order periods.

INTERMEDIATE RESEARCH (1990–2000)

Elmaghraby (1990) reviewed and critiqued two previously proposed models, corrected some minor errors in them, and expanded one of them to accommodate a finite horizon. These models are those of Spradlin and Pierce (1967) and Sule (1978), respectively. He also proposed a different forgetting model from that of Carlson and Rowe (1976), which he suggested is more consistent with the learning-forgetting relationship. He then applied this to the determination of the optimal number and size of the lots in the finite and infinite horizon cases. Elmaghraby (1990) used a forgetting function, variable regression invariant forgetting model (VRIF), $\hat{T}_1 x^f$, that assumes a forgetting exponent $(f_i = f)$ and the intercept $(\hat{T}_{1i} = \hat{T}_1)$ do not vary between cycles $(i = 1, 2, ...)$. Jaber and Bonney (1996a) showed that using the LFCM, which assumes f_i and $\hat{T}_{1i}$ to vary from cycle to cycle, produces better results, as using the VRIF overestimates the total cost. The limitations of the VRIF are discussed in Jaber and Bonney (1997a) and Jaber et al. (2003).

Salameh et al. (1993) incorporated Wright's learning curve equation into the total inventory system cost resulting in a mathematical model that required a numerical search technique to determine the economic manufactured quantity (EMQ). In their model, the cycles were treated independently from one another. This assumption simplified the mathematical formulation. They assumed a new learning curve in each production cycle with T_1 reducing while b remained constant. The EMQ was shown to decrease in successive lots as labor productivity increases because of learning. Their unit time cost function was given as:

$$c_T = \frac{SD}{Q_i} + c_2 D + c_1 \frac{T_{1i} D Q_i^{-b}}{1-b} + h \frac{Q_i}{2} - h \frac{T_{1i} D Q_i^{1-b}}{(1-b)(2-b)}, \qquad (14.7)$$

where Q_i is the lot size for cycle i and $T_{1i} = T_1 \left(\sum_{j=1}^{i-1} Q_j + 1 \right)^{-b}$, where $T_{11} = T_1$. Jaber and Bonney (1996a) extended the work of Salameh et al. (1993) by incorporating the learn-forget model (LFCM) into the total inventory system cost. The effect of both learning and forgetting on the optimum production quantity and minimum total inventory cost was demonstrated by a numerical example. They showed that forgetting had an adverse effect on the productivity of labor-intensive manufacturing and on the total labor cost as a result of the drop in labor productivity. The difference between the two is in computing T_{1i}, where it was given as $T_{1i} = T_1(u_i+1)^{-b}$ and $0 \le u_i \le \sum_{j=1}^{i-1} Q_j$ (the partial transfer of learning) representing the equivalent number of units of experience remembered from $i-1$ cycles at the beginning of cycle i, with $u_i = 0$ and $u_i = \sum_{j=1}^{i-1} Q_j$ representing no or full transfer of learning, respectively. The mathematics of the LFCM is provided in Jaber and Bonney (1996a). So far, shortages have not been discussed. Jaber and Salameh (1995) extended the work of Salameh et al. (1993) by assuming shortages to be allowed and backordered, with the assumption of full transmission of learning between successive lots. Jaber and Bonney (1997b) assumed a case where it is sometimes possible that the improvement in the production rate is slow due to the complexity of the task performed and it is possible for the demand rate to exceed the production rate for a portion of the production cycle. This situation was referred to as intra-cycle backorders. It has been shown that the presence of intra-cycle backorders results in longer cycle runs that mean additional labor and inventory costs. These costs tend to be more critical when there is a partial transmission of learning, since intra-cycle shortages can appear in consecutive production runs. This shows the need for the proper estimation of Wright's (1936) power learning curve parameters.

A common theoretical drawback of Wright's model is that the results obtained are not meaningful as the cumulative production approaches infinity. A correction of the zero asymptote was made by adding a "factor of compressibility" as a third parameter to Wright's model (see de Jong 1957). This forces the learning curve to plateau at a value above zero. Although plateauing is usually observed in learning data, there is no consensus among researchers as to what causes the learning curve to plateau. Further discussion on this point is found in Jaber and Guiffrida (2004, 2008).

Fisk and Ballou (1982) studied the manufacturing lot size problem under both the unbounded (Wright 1936) and the bounded (plateau) power function of de Jong's (1957) learning situations. Jaber and Bonney (1996b) presented an efficient approximation for the closed algebraic form of the total holding cost expression developed by Fisk and Ballou (1982). This approximation simplified the solution of the lot size problem with bounded learning. This simplicity is characterized by relying on one numerical search rather than two. The estimated total cost recorded a maximum deviation of 0.5% from the estimate of Fisk and Ballou, which is almost negligible. It was also shown that, as the factor of incompressibility increases, the effect of learning decreases. This could be due to the decrease in the proportion of labor time out of the total time required to produce each consecutive unit.

Shiue (1991) developed a model for determining the batch production quantity where a modified learning curve is used to fit the manufacture of products in the growth phase of its life cycle. The total cost function that was developed was dependent on two decision variables—the storage capacity and the batch size quantity—and was based on the values of set-up cost, holding cost, shortage cost, and production cost.

Zhou and Lau (1998) claimed that the approximation of Jaber and Bonney (1996) contained an error, which they had corrected in a previous paper (in Chinese). They presented a similar model to that of Jaber and Bonney (1996b) where shortages are allowed and backordered. The claim by Zhou and Lau (1998) resulted in a response by Jaber and Bonney (2001a) and cross responses (Jaber and Bonney 2001b; Zhou and Lau 2001). In response, Jaber and Bonney (2001a) compared the models of those of Fisk and Ballou (1982), Jaber and Bonney (1996b), and Zhou and Lau (1998). Their paper rebuts the claims of Zhou and Lau that the approximation of Jaber and Bonney contained mathematical errors that rendered the developed model invalid, and they demonstrated that the results produced by Jaber and Bonney (1996b) are satisfactory. The cost function of Jaber and Bonney (1996b) is given as:

$$C_T = \frac{SD}{Q_i} + c_2 D + c_1 \left(T_{11}MD + \frac{(1-M)DT_{1i}}{1-b} Q_i^{-b} \right)$$

$$+ h \left(\frac{Q_i}{2} - \frac{T_{11}MDQ_i}{2} - \frac{1-M}{2(1-b)} DT_{1i}Q_i^{1-b} \right) \qquad (14.8)$$

$$- h \frac{(1-M)bD}{2(2-b)} \left(T_{11}MQ_i + \frac{1-M}{1-b} T_{1i}Q_i^{1-b} \right),$$

where $T_{11} = T_1$, $0 < M < 1$ is the factor of compressibility in de Jong's learning curve, and T_{1i} is as defined earlier.

Chiu (1997) investigated lot sizing with learning and forgetting for the assumption of discrete time-varying demand. For the purpose of his study, he chose three lot sizing models from the literature, which are the EOQ (Harris 1990), the least unit cost (Gorham 1968), and the maximum part-period gain (Karni 1981) and then used the extended dynamic optimal lot sizing model (Wagner and Whitin 1958) to generate optimal solutions. Chiu's numerical results suggested using the maximum part-period gain heuristic due to its simplicity and good performance. Chiu (1997) assumed the forgetting rate to be either a given percentage or an exponential function of the break length. These assumptions were not justified in the paper. Chiu and Chen (1997) investigated the work of Chiu (1997) for the effect of the time-value of money. Their computational results indicated that the learning coefficient, the forgetting rate, and the real discount rate all have significant effects on the determination of lot sizes and relevant costs. Eroglu and Ozdemir (2005) argued that the use of the Wagner-Whitin problem is not generating an optimal solution for this problem, and developed a new recurrence relationship for this purpose.

Jaber and Bonney (1998) described three mathematical models that incorporate the effects of learning and forgetting (LFCM) in the calculation of the economic manufactured quantity (EMQ). The first model assumes an infinite planning horizon, while the other two assume a finite planning horizon; one investigates the equal lot size while the other model investigates the unequal lot size. The cost function for the first model is presented as:

$$C_T = \frac{SD}{Q_i} + c_2 D + c_1 \frac{T_1 D}{1-b} \left[\frac{(Q_i + u_i)^{1-b} - u_i^{1-b}}{Q_i} \right] + h \frac{T_1 D u_i^{1-b}}{1-b}$$

$$+ h \frac{Q_i}{2} - h \frac{T_1 D}{(1-b)(2-b)} \left[\frac{(Q_i + u_i)^{2-b} - u_i^{2-b}}{Q_i} \right]. \tag{14.9}$$

Their results showed that under the partial transmission of learning, the optimal policy was to carry less inventory in later lots, and to extend the cycles' lengths within the planned production horizon so that the total cost per unit time and the total cost is at a minimum.

RECENT RESEARCH (2001–2010)

The interest of some researchers in the lot sizing problem with learning (and forgetting) effects has continued into the new millennium. The researchers have extended the range of problems considered using different assumptions to gain additional insights.

Jaber and Bonney (2001c) examined whether, when learning is considered, it is reasonable to ignore the effect of the continuous time discounting of costs by investigating the effect of learning and time discounting, both on the economic manufacturing quantity and the minimum total inventory cost. Their results indicated that, although discounting and learning affect the optimal batch size—suggesting that one should make items in smaller batches more frequently—changing the batch size does not greatly affect the total discounted cost. They concluded that from a decision-making point of view there is considerable flexibility in the choice of lot size.

Jaber and Abboud (2001) considered a situation where on completion of the production run, the facility may not be available for a random amount of time due to several reasons, or that the facility is leased by different manufacturers and the demand for the facility is random. As a result of machine unavailability, stock-out situations might arise. They extended the work of Abboud et al. (2000), who excluded the production cost component from the cost function, by assuming that the production capacity is continuously improving over time because of learning. By so doing, this would be the first work that adds a stochastic component to the lot sizing problem with learning and forgetting. Abboud et al. (2000) assumed machine unavailability time is a uniformly distributed random variable. The results of Jaber and Abboud (2001) indicated that, as the production rate improves, the production policy was to produce smaller lots more frequently, resulting in shorter cycles. This enhanced service levels and machine availability. Forgetting had an adverse effect on service levels and machine availability, since with forgetting the production-inventory policy was to produce larger lots less frequently. Job complexity, as a measure of forgetting level, had an effect on machine availability. Results showed that machine availability increased as job complexity decreased.

Ben-Daya and Hariga (2003) developed a continuous review inventory model where lead time is considered as a controllable variable. They assumed that the lead time is decomposed into all its components: set-up time, processing time, and non-productive time. These components reflect the set-up cost reduction, lot size lead time interaction, and lead time crashing, respectively. The learning effect in the production process was also included in the processing time component of the lead time. They argued that the finite investment approach for lead time and set-up cost reduction and their joint optimization, in addition to the lot size lead time interaction, introduce a realistic direction in lead time management and control. They found that the learning effect reflected by the parameter, b, had no significant effect on the expected total cost. This awkward result had to do with their choice of input parameters where the labor cost is overshadowed by the other costs, making learning insignificant.

Balkhi (2003) studied the case of the full transmission of learning for the production lot size problem with an infinite planning horizon for the following assumptions: (1) items deteriorate while they are produced or stored; (2) both demand and deterioration rates are (known) functions of time; (3) shortages are allowed, but are partially backordered; and (4) the production rate is defined as the number of units produced per unit time. Although the paper presented interesting extensions, it lacked solid conclusions and meaningful managerial insights.

Alamri and Balkhi (2007) studied the effects of learning and forgetting on the production lot size problems for the case of infinite planning horizon and deteriorating items. They presented a generalization of the LFCM (GLFCM) that allows variable total forgetting breaks. Their forgetting slope was given as $f_i = b \log(u_i + Q_i)/\log(1 + 1/T_{1i})$, which is different from that of the LFCM (Jaber and Bonney 1996a). Their forgetting model was found to be in conformance with six of the seven characteristics of forgetting that should be considered when developing forgetting curves

(Jaber et al. 2003). Their numerical results suggested an optimal policy of producing small lots. This suggestion was found to have two properties: (1) it decreased the cycle length; and (2) it increased the experience gained, causing a further decrease in the time required to produce the unit for each consecutive cycle. Although the LFCM was tested against empirical data (Jaber and Sikström 2004b), the GLFCM was not.

All the works that studied the effect of learning on the lot size problem commonly assumed an invariant learning slope throughout the production-planning horizon. Jaber and Bonney (2007) argued that when learning rates are dependent on the number of units produced in a production cycle, then the assumption of invariant learning rates might produce erroneous lot size policies. The paper investigated the effect of lot size-dependent learning and forgetting rates on the lot size problem by incorporating the dual-phase learning-forgetting model (DPLFM; Jaber and Kher 2002). Jaber and Kher (2002) developed the DPLFM by combining the dual-phase learning model (DPLM, Dar-El et al. 1995) with the learn-forget curve model (LFCM; Jaber and Bonney 1996b). The dual-phase learning model (DPLM) proposed by Dar-El et al. (1995) is a modification of Wright's learning curve, which aggregates two curves, one cognitive and one motor. Jaber and Bonney (2007) did so by extending the work of Salameh et al. (1993), which used invariant learning and forgetting rates. Their results indicated that ignoring the cognitive and motor structure of a task can result in lot size policies with a high percentage of errors in costs. This finding suggests that earlier work investigating the lot size problem in conjunction with learning and forgetting in production, may be unreliable, and therefore should be revisited and possibly revised.

All the models that investigated the effect of learning in production on the lot sizing problem have limitations. Jaber and Guiffrida (2007) addressed two of these limitations. The first limitation is that the models found in the literature do not address the problem of when the learning exponent b approaches or exceeds the value 1, where these models are mathematically invalid for the special cases of $b = 1$ and 2, and not investigated for the cases $1 < b < 2$ and $b > 2$. The second limitation is that the models found in the literature assume that the holding cost per unit is fixed even though the unit production cost is decreasing because of learning. Jaber and Guiffrida (2007) addressed the first limitation; the mathematics of the economic order quantity model with learning and forgetting was reworked to address the learning exponent approaching and exceeding the value of $b = 1$. The numerical results suggested that Wright's learning curve may not be the appropriate curve to capture learning in processes characterized by excessively small initial processing time and/ or very large learning exponent values, since the production time becomes insignificant. This implies adding a flattening (plateauing) factor to the learning curve in order to attain a minimum value of production time. They addressed the second limitation by allowing for a holding cost function that decreases as a result of learning effects. The numerical results also indicated that assuming a fixed holding cost underestimates the lot size quantity and slightly overestimates the total cost.

Teyarachakul et al. (2008) analyzed the steady-state characteristics of a batch production time for a constant-demand lot sizing problem with learning and forgetting in production time. They report a new type of convergence, the alternating convergence, in which the batch production time alternates between two different values. This is different from the literature that reports that batch production time converges to a unique value; for example, Sule (1978), Axsäter and Elmaghraby (1981), and Elmaghraby (1990) used forgetting models in which the amount of forgetting is unbounded. They assumed forgetting to follow the model of Globerson and Levin (1987), which does not allow the amount of forgetting to exceed the amount of learning. Teyarachakul et al. (2008) also developed several mathematical properties of the model to validate convergence. In a follow-up paper, Teyarachakul et al. (2011) allowed delayed forgetting where forgetting starts slowly and then becomes faster with time. Their computational results show that it may be better to produce in smaller batches in the presence of learning and forgetting.

Chen et al. (2008) investigated how the effect of learning (with no forgetting) on the unit production time affected the lot sizing problem when the production system is imperfect in the presence of shortages. They assumed that the process may go out of control and start producing defective items (Rosenblatt and Lee 1986), which are reworked at a cost. They also considered that the process generates defective items when in control too, but at a lower rate than in the previous situation.

The most recent paper along this line of research is that of Jaber et al. (2009), who introduced the concept of entropy cost to estimate the hidden costs of inventory systems, which are discussed in Jaber (2009). Jaber et al. (2004) used the laws of thermodynamics to model commodity (heat) flow (or demand) from the production (thermodynamic) system to the market (surrounding), where the price is analogous to temperature. The demand rate is of the form $D(t) = D = -K(P-P_0)$, $\forall t > 0$, an assumption that is consistent with that of the EMQ/EOQ of constant demand. K (which is analogous to a thermal capacity) represents the change in the flux for a change in the price of a commodity and is measured in additional units of demand per year per change in unit price—e.g., units/year/\$. $P(t) = P$ is the unit price at time t, and $P_0(t) = P_0$ is the market equilibrium price at time t, where $P(t) < P_0(t)$. They suggested adding a third component, representing the entropy cost, to the order quantity cost function. They noted that when $P < P_0$, the direction of the commodity flow is from the system to the surroundings, and the entropy generation rate must satisfy $S(t) = K(P/P_0+P_0/P-2)$. The entropy cost per cycle is computed as $E(T) = \int_0^T D(t)dt \big/ \int_0^T S(t)dt = -P_0P\big/(P-P_0)$, where $T = Q/D$ is the cycle time. Jaber et al. (2009) added the term $E(T)/T$ to Equation 14.9. They used $E(T)/Q$ as a cost measure of controlling the flow of one unit of commodity from the system to the market. Results indicate that not accounting for entropy cost may result in more expensive commodity control policies; in particular, for inventory policies that promote producing smaller batches of materials or products more frequently. As production becomes faster, the control cost increases. Forgetting was found to reduce the commodity flow cost (entropy) as it recommends producing larger batches. The results from this paper suggest that a firm that is unable to estimate its cost parameters properly may find ordering in larger lots an appropriate policy to counter entropy effects

LOT SIZING WITH LEARNING IN SET-UPS

The Japanese approach to productivity demands producing in small lots. This can only be achieved if the set-up time is reduced. Instead of accepting set-up times as fixed, they attempted to reduce the set-up time, thereby allowing lot sizes to be reduced. Their success in this area has motivated many researchers to think about the effect of decreasing set-ups. A simple EOQ cost function with learning in set-ups would be given as:

$$C_{T,i} = \frac{S(i)D}{Q_i} + h\frac{Q_i}{2}, \qquad (14.10)$$

where $S(i) = Si^{-a}$ is the set-up learning curve with a being the learning exponent and i the set-up number. The optimal lot size quantity is determined from Equation 14.10 as $Q_i = \sqrt{2S(i)D/h}$, where $Q_1 \geq Q_2 \geq Q_3 \geq \cdots \geq Q_n$ for $i \in [1,n]$ and $Q_{n+1} = Q_{n+2} = \cdots = Q_\infty$ when $S(i) = S_{\min}$. The following is a survey of the work that investigated the effects of learning (and forgetting) in set-ups on the lot sizing problem. Porteus (1985) developed an extension of the EOQ model in which the set-up cost is viewed as a decision variable, rather than as a parameter. He emphasized that lowered set-up costs can occur not only as a result of engineering effort. Porteus considered other benefits of lowered set-up costs (and times) and associated reduced lot sizes with improvements in quality control, flexibility, and effective capacity.

Karwan et al. (1988) proposed a model for joint worker/set-up learning. The computational results suggest that optimal lot sizing policies continue to exhibit decreasing lot sizes over time under the total transmission of learning. Also, as the rate of set-up learning increases, optimal lot sizing policies tend to exhibit an increasing number of lots. Chand (1989) studied the effect of learning in set-ups and process quality on the optimal lot sizes and the set-up frequency. Chand's work differs from that of Karwan et al. (1988) in the following respects: (1) it permits learning in process quality in addition to the learning in set-ups, (2) it allows any form of the learning function, (3) it provides an efficient algorithm for finding the optimal lot sizes as opposed to the computationally demanding dynamic programming algorithm, and (4) it also provides a mathematical analysis and computational results to study the effect of the rate of learning on lot sizes and the set-up frequency. Chand's mathematical and computational results showed that the presence of learning in set-ups, and the effect of changes to the fraction of defectives (when the fraction of defectives for a production lot is proportional to the lot size), increased the optimal set-up frequency. He also showed that this effect can be quite significant for a company with a high production volume, high cost per defective unit, and a large rate of learning. Chand found that learning in process quality has no effect on the optimal set-up frequency since it does not seem to have a definite pattern. The results of Chand's work support the arguments given by various authors in favor of stockless production, zero inventories, or the JIT approach.

Replogle (1988) presented a revised EOQ model that recognizes the effect of learning on set-up costs, and permits the calculation of lot sizes that minimize the total inventory cost over any period. Reduced lot sizes mean more frequent set-ups, thereby moving more rapidly down the set-up learning curve and improving the competitive position of the firm. Cheng (1991) argued that the reduction in lot size and savings in total inventory cost based on Replogle's model seem to be overestimated due to the way in which Replogle defines the learning curve, which is different from the traditional definition. That is, he defined the total set-up cost as Sn^{1-a} rather than $Sn^{1-a}/(1-a)$. Cheng (1994) considered learning in batch production and set-ups in determining the EMQ. The results of the numerical examples solved by Cheng, strongly indicate that the assumption of equal manufacturing lot sizes not only simplifies the process of determining the optimal solutions, but also provides close approximations to the optimal solutions. Cheng (1991) did not refer either to Karwan et al. (1988) or to Chand (1989).

Chand and Sethi (1990) considered the dynamic lot sizing problem of Wagner-Whitin with the difference that the set-up cost in a period depends on the total number of set-ups required thus far and not on the period of set-up. The total cost of n set-ups is a concave non-decreasing function of n, which could arise from the worker learning in set-ups and/or technological improvements in set-ups methods. He showed that the minimum holding cost for a given interval declines at a decreasing rate for an increasing number of set-ups. Pratsini et al. (1994) investigated the effects of set-up time and cost reduction through learning on optimal schedules in the capacitated lot sizing problem with time-varying demand. They illustrated how set-up cost and time reduction, through learning curve effects, can change the optimal production plans in the capacity-constrained setting. The reduction of set-up time, and proportionally the set-up cost, can cause an increase in the prescribed number of set-ups as they become more cost effective, resulting in less inventory. It was also found that the learning effect can be dominated by the capacity effect.

Rachamadugu (1994) considered the problem of determining the optimal lot sizes when set-up costs decrease over time because of learning, and the running or processing costs remain constant. To avoid forecasting the number of set-ups in the remaining planning horizon, they adopted a myopic policy (part period balancing) that sets the holding cost for the lot size in a cycle equal to the set-up cost in the same cycle. According to Rachamadugu (1994), this policy has the following advantages: (i) it is intuitively appealing to practitioners, (ii) it is easy to compute, and (iii) it does not require information on the future set-up costs. His computational results suggested that the performance of the myopic policy is influenced by the learning rate when the ratio of maximum to minimum set-up values is low, and showed that the myopic policy yielded good results even when the product has a short life cycle. In another paper, Rachamadugu and Schriber (1995) addressed the problem of determining optimal lot sizes when reductions in set-up costs persist due to the emphasis on continuous improvement, worker learning, and incremental process improvements. They suggested two heuristic procedures: (i) current set-up cost lot sizing policy (CURS), and (ii) minimum set-up cost lot sizing policy (MINS) (CURS and MINS policies), which can be used when information about set-up cost reduction trends is not available. Their results showed that CURS is better suited for situations in which improvements in set-up costs occur at a slow pace. For other situations, the MINS policy was found to be more appropriate. Their computational results have shown that such a policy can result in lot sizing costs that exceed the optimum by a considerable percentage. This implies that the average cost analysis could be inadequate for non-stationary cost parameter situations, such as when a production process is used for long periods of time while the reductions in set-up costs continue to occur over time due to an emphasis on kaizen (continuous improvement) and worker learning. Along the same line of research, Rachamadugu and Tan (1997) addressed the issue of determining lot sizes in finite horizon environments when learning effects and/or emphasis on continuous improvement result in decreasing set-up costs. They analyzed and evaluated a myopic lot sizing policy for finite horizons that does not require any information about future set-up costs. Their analytical results and computational experiments show that the policy is a good choice for machine-intensive environments. They further showed that the myopic policy yielded good results even when set-up cost changes cannot be completely modeled by the stylized learning curves used in earlier research studies.

LOT SIZING WITH LEARNING IN PRODUCTION

Li and Cheng (1994) is perhaps the first work to investigate the lot sizing problem for learning and forgetting in set-ups and production. They developed an EMQ model that accounts for the effects of learning and forgetting on both set-up and unit variable manufacturing time. Their EMQ model is basically the EOQ model, modified for the gradual delivery or availability of a product, and is justified for a completely integrated flow line and perhaps for a very fast-response JIT system. Li and Cheng (1994) modeled the learning effect in terms of the reduction in direct labor hours as production increases. They found this approach to result in a simpler formulation of the total cost, solvable by dynamic programming. Their computational results indicated that assuming equal lot sizes simplifies the process of determining solutions and provides close approximations to the optimal solutions. Chiu et al. (2003) extended the work of Chiu (1997) by: (1) considering learning and forgetting in set-ups and production, (2) assuming that the forgetting rate in production is a function of the break length and the level of experience gained before the break, and (3) assuming that each production batch is completed as close as possible to the time of delivery in order to reduce the inventory carrying cost. They found that the effect of set-up forgetting increases with the set-up learning effect and with the horizon length, and also that production learning has the greatest effect on the total cost among all the effects of learning and forgetting in set-ups and production. Near-optimal solutions were found for the case(s) when the planning horizon and/or the total demand are/is moderately large. Chiu and Chen (2005) studied the problem of incorporating both learning and forgetting in set-ups and production into the dynamic lot sizing model to obtain an optimal production policy, including the optimal number of production runs and the optimal production quantities during the finite period planning horizon. They considered the rates of: (i) learning in set-ups, (ii) learning in production, (iii) forgetting in set-ups, and (iv) forgetting in production. The results indicated that the average optimal total cost increased with an increase in any of the exponents associated with the four rates. Their results also showed that the optimal number of production runs and the optimal total cost were insensitive to the demand pattern.

LOT SIZING WITH IMPROVEMENT IN QUALITY

Urban (1998) investigated a production lot size model that explicitly incorporates the effect of learning on the relationship—positive or negative—between the run length and the defect rate. Urban intentionally kept the model simple in order to isolate the learning effect of this relationship on the optimal lot size and the resulting production costs, as well as—and perhaps most importantly—to identify important implications of this relationship. He found that reductions in inventory levels can be achieved without corresponding reductions in set-up costs, as long as there is a significant inverse relationship between the run length and product quality.

Jaber and Bonney (2003) investigated the effects that learning and forgetting in set-ups and product quality have on the economic lot sizing problem. Two quality-related hypotheses were empirically investigated using the data from Badiru (1995), which are: (1) the time to rework a defective item reduces if production increases conform to a learning relationship, and (2) quality deteriorates as forgetting increases due to interruptions in the production process. Unlike the work of Chand (1989), Jaber and Bonney (2003) assumed a cost for reworking defective items, referred to as the "product quality cost", which is an aggregate of two components. The first represents a fixed cost e.g., the material cost of repairing a defective item, whereas the second component represents the labor cost of reworking that defective item, taking account of learning and forgetting. Their results indicated that with learning and forgetting in set-ups and process quality, the optimal value of the number of lots is pulled in opposite directions. That is, learning in set-ups encourages smaller lots to be produced more frequently. Conversely, learning in product quality encourages larger lots to be produced less frequently. However, the total cost was shown not to be very sensitive to the increasing values of the learning exponent, which means that it is possible to produce in smaller lots relative to the optimum value without incurring much additional cost.

Jaber (2006b) investigated the lot sizing problem for reduction in set-ups, with reworks and interruptions to restore process quality. In a JIT environment, workers are authorized to stop production if a quality or a production problem arises; for example, the production process going out of control. He assumed that the rate of generating defects benefits from the changes to eliminate the defects, and thus reduces with each quality restoration action. Jaber (2006b) revised the work of Chand (1989) by assuming a realistic learning curve. The results indicate that learning in set-ups and improvement in quality reduces the costs significantly. His results also showed that accounting for the cost of reworking defective units when calculating the unit holding cost may not be unrealistic, given that some researchers suggest using a higher holding cost than the cost of money.

LOT SIZING WITH CONTROLLABLE LEAD TIME

Pan and Lo (2008) investigated the impact of the learning curve effect on set-up costs for the conterminous review inventory model with controllable lead time and a mixture of backorder and partial lost sales. They assumed that the inventory lead time is decomposed into multiple components, each having a different crashing cost for the shortened lead time. They compared their model to that of Moon and Choi (1998), which assumes no learning in set-ups. They found that the expected total inventory cost tends to increase as the learning rate increases, while all the other parameters remain unchanged, and it decreased as the backorder ratio increased while all the other parameters stay fixed. Although this finding is interesting, the authors did not provide any discussion as to why their model behaved in this manner. Their results also showed that the lead time became shorter as the learning rate became faster for a given backorder ratio.

LEARNING CURVES IN SUPPLY CHAINS AND REVERSE LOGISTICS

Supply chain management emerged in the late 1990s and the beginning of this millennium as a source of sustainable competitive advantage for companies (Dell and Fredman 1999). Supply chain management involves functions such as production, purchasing, materials management, warehousing and inventory control, distribution, shipping, and transport logistics. Like supply chains, reverse logistics encompasses the same functions where products flow in the opposite direction (from downstream to upstream). To maintain a sustainable competitiveness in these functions, the managers of these operations could benefit from introducing continuous improvement to foster organizational learning. Historically, learning curve theory has been applied to a diverse set of management decision areas, such as inventory control, production planning, and quality improvement (e.g., Jaber 2006a). Each of these areas exists within individual organizations of the supply chain and, because of the interdependencies among chain members, across the supply chain as a whole. By modeling these learning effects, management may then use established learning models to utilize capacity better, manage inventories, and coordinate production and distribution throughout the chain. Despite its importance, there are only a few quantitative studies that investigate learning in supply chain and reverse logistics contexts. These are surveyed below.

Supply Chain Management

Coordination among players in a supply chain is the key to a successful partnership. Some researchers showed that coordination could be achieved by integrating lot sizing models (e.g., Goyal and Gupta 1989), with the idea of joint optimization for buyer and vendor believed to have been introduced by Goyal (1977). Jaber and Zolfaghari (2008) review the literature for quantitative models for centralized supply chain coordination that emphasize inventory management. Nanda and Nam (1992) were the first to develop a joint manufacturer-retailer (vendor-buyer) inventory (referred to as a two-level supply chain) model for the case of a single buyer. Production costs were assumed to reduce according to a power form learning curve (Wright 1936) with forgetting effects caused by breaks in production. A quantity discount schedule was proposed based on the change of total variable costs of the buyer and manufacturer. To meet the demand of the buyer, the manufacturer considers either a lot-for-lot (LFL) production policy, or a production quantity that is a multiple of the buyer's order quantity (e.g., Jaber and Zolfaghari 2008). Nanda and Nam (1992) assumed a LFL policy, learning in production, no defectives are produced, and like Fisk and Ballou (1982), assumed forgetting to be a constant percentage. They found that the joint total cost is decreased significantly when learning is fast. They also found that the joint total cost savings realized by the manufacturer are in the production and joint lot size inventory holding components when significant learning and learning retention are expected. They extended their work in a subsequent paper (Nanda and Nam 1993) to include multiple retailers.

Kim et al. (2008) is the first study in the literature that examined the benefits of buyer-vendor partnerships over lot-for-lot (i.e., single set-up single delivery [SSSD]) systems and suggests two policies that the supplier can pursue in order to meet customers' needs: (1) single set-up multiple delivery (SSMD), and (2) multiple set-up multiple delivery (MSMD). They found that if the buyer's fixed set-up cost is relatively high, the vendor would prefer to implement SSMD and produce an entire order with one set-up. However, if the vendor can reduce the set-up cost (because of learning) and the vendor's capacity is greater than the threshold level (production rate equals twice the demand rate), it was found to be more beneficial for the vendor to implement the multiple set-ups and multiple deliveries (MSMD) policy, even though he/she pays for more frequent set-up costs because the savings in inventory holding costs are greater than the increased set-up costs.

Jaber et al. (2008) extended the work of Salameh and Jaber (2000) by assuming the percentage defective per lot reduces according to a learning curve, which was empirically validated by data from the automotive industry. Salameh and Jaber (2000) developed an inventory situation where items received are not of perfect quality (defective), and after 100% screening, imperfect quality items are withdrawn from the inventory and sold at a discounted price. Their data showed that the percentage defective per lot reduced for each subsequent shipment following an S-shape learning curve similar to the one described in Jordan (1958). The developed learning curve was incorporated into the model of Salameh and Jaber (2000). Two models were developed. The first model of Jaber et al. (2008), like Salameh and Jaber (2000), assumes an infinite planning horizon, while the second model assumed a finite planning horizon. The results of the first model showed that the number of defective units, the shipment size, and the cost reduce as learning increases following a form similar to the logistic curve. For the case of a finite horizon, results show that as learning becomes faster it is recommended to order in larger lots less frequently. Although the model was discussed in a vendor-buyer context, Jaber et al. (2008) did not investigate a joint-order policy. This remains as an immediate extension of their work.

Jaber et al. (2010) investigated a three-level supply chain (supplier-manufacturer-retailer) where the manufacture undergoes a continuous improvement process. The continuous improvement process is characterized by reducing set-up times, increasing the production capacity, and eliminating rework (Jaber and Bonney 2003). The cases of coordination and no coordination were investigated. Traditionally, with coordination, the manufacturer entices the retailer to order in larger lots than its economic order quantity. In their recent paper, the opposite was shown to be true as the manufacturer entices the retailer to order in smaller quantities than the retailer's economic order quantity. As improvement becomes faster, the retailer is recommended to order in progressively smaller quantities as the manufacturer offers larger discounts and profits. The results also showed that coordination allows the manufacturer to maximize the benefits from implementing continuous improvements. It was also shown that forgetting increases the supply chain cost.

Reverse Logistics

As shorter product life cycles became the norm for manufacturers to sustain their competitive advantage in a dynamic and global marketplace, the forward flow of products quickened, resulting in faster rates of product waste generation and the depletion of natural resources. Consequently, manufacturing and production processes have been viewed as culprits in harming the environment (e.g., Beamon 1999; Bonney 2009). This concern gave rise to the concept of reverse logistics (RL) or the backward flow of products from customers to manufacturers to suppliers (e.g., Gungor and Gupta 1999) for recovery. Recovery may take any of the following forms: repair, refurbishing, remanufacturing, and recycling (e.g., King et al., 2006).

Schrady (1967) is believed to be the first to investigate the EOQ model in production/ procurement and recovery contexts. This line of research was revived in the mid-1990s by the work of Richter (1996a, b). Although the works of Schrady and Richter were investigated for different inventory situations, two have considered the effects of learning.

Maity et al. (2009) developed an integrated production-recycling system over a finite time horizon, where demand is satisfied by production and recycling. Used units are collected continuously from customers, either to be recycled (repaired to an as-good-as-new state) or disposed of (if not repairable). They assumed that the set-up cost reduces over time following a learning curve, and also that the rates of production and disposal are functions of time. Regrettably, no significant results regarding the effect of learning in set-ups on the production-recycling system were presented or discussed.

The paper by Jaber and El Saadany (2011) extended the production, remanufacture, and waste disposal model of Dobos and Richter (2004) by assuming learning to occur in both production and remanufacturing processes. They assumed that improvements due to learning require capital investment. Their results showed that there exists a threshold learning rate beyond which investing in learning may bring savings. That is, unless the learning process proceeds beyond the threshold value, investment in learning may not be worthwhile. It was also shown that faster learning in production lowers the collection rate of used items; however, should there be governmental legislation for firms to increase their collection rates, accelerating the learning process may not be desirable. The results also showed that it may be better to invest in speeding up the learning process in remanufacturing than in production as this reduces the costs of remanufacturing, making it attractive to recover used items. It was generally found that learning reduces the lot size quantity and, subsequently, the time interval over which new items are produced and used ones are remanufactured. Their results suggested that, with learning, a firm can have some degree of flexibility in setting some of its cost parameters that are otherwise difficult to estimate.

SUMMARY AND CONCLUSIONS

This chapter is an updated version of Jaber and Bonney (1999) and includes references that Jaber and Bonney (1999) were unaware of at that time and those that have appeared since 1999 up to 2011. The Jaber and Bonney (1999) paper surveyed work that deals with the effect of learning (and learning and forgetting) on the lot size problem. In addition, this chapter explores the possibility of incorporating some of the ideas adopted by JIT to such models with the intention of narrowing the gap between the "inventory is waste" and the "just-in-case" philosophies.

A common feature among all of the above-surveyed work is the simplicity in modeling the cost function; that is, the papers assume a single-stage production with the set-up and holding costs of a finished product. In reality, production is usually performed in multiple stages, so besides the holding cost of a finished product, modeling should account for the individual costs at each stage, such as set-up, learning in production, holding of work-in-process, reworks, scrap, and so on. One way of addressing this research limitation is to extend the work of Jaber and Khan (2010) to account for the inventory costs of work-in-process and finished items.

Beside inventory management, learning curve theory has been applied to a diverse set of management decision areas such as production planning and quality improvement. Each of these areas exist within the individual organizations of the supply chain, but because of the interdependencies among chain members, they also exist across the supply chain as a whole. By modeling these learning effects, management may then use established learning models to utilize capacity better, manage inventories, and coordinate production and distribution throughout the chain. However, there have been few quantitative studies that investigate learning in a supply chain and reverse logistics contexts (Jaber et al., 2010; Jaber and El Saadany in press). An interesting question that remains to be looked at is: what impact does the quality-learning relationship have on coordination and profitability in supply chains?

Another interesting area of study is to quantify the cost of disorder (entropy) in a production-inventory system by using the second law of thermodynamics (Jaber et al. 2004). This line of research needs further study. For example, is the analogy between business systems and physical systems sufficiently close for valid conclusions to be drawn from the analysis, and are there places where the analogy does not hold? Can learning reduce the entropy of business systems? These research questions will be addressed in future works.

Finally, most of the above models remain nice mathematical exercises generally performed in the absence of actual learning and forgetting data. Many researchers have attempted to acquire such data with little success (e.g., Elmagharaby 1990). To develop models that represent reality faithfully, it is necessary that firms be more open to providing access to their data. Otherwise, and as Elmagharaby (1990) put it, many of the studies related to learning and forgetting will continue to be mainly armchair philosophizing.

REFERENCES

Abboud, N.E., Jaber, M.Y., and Noueihed, N.A., 2000. Economic lot sizing with the consideration of random machine unavailability time. *Computers and Operations Research* 27(4): 335–351.

Adler, G.L., and Nanda, R., 1974a. The effects of learning on optimal lot determination – Single product case. *AIIE Transactions* 6(1): 14–20.

Adler, G.L., and Nanda, R., 1974b. The effects of learning on optimal lot size determination – Multiple product case. *AIIE Transactions* 6(1): 21–27.

Alamri, A.A., and Balkhi, Z.T., 2007. The effects of learning and forgetting on the optimal production lot size for deteriorating items with time–varying demand and deterioration rates. *International Journal of Production Economics* 107(1): 125–138.

Axsäter, S., and Elmaghraby, S., 1981. A note on EMQ under learning and forgetting. *AIIE Transactions* 13(1): 86–90.

Badiru, A.B., 1995. Multivariate analysis of the effect of learning and forgetting on product quality. *International Journal of Production Research* 33(3): 777–794.

Balkhi, Z.T., 2003. The effects of learning on the optimal production lot size for deteriorating and partially backordered items with time-varying demand and deterioration rates. *Applied Mathematical Modeling* 27(10): 763–779.

Baloff, N., 1970. Startup management. *IEEE Transactions on Engineering Management* 17(4):132–141.

Beamon, B.M., 1999. Designing the green supply chain. *Logistics Information Management* 12(4): 332–342.

Ben-Daya, M., and Hariga, M., 2003. Lead-time reduction in a stochastic inventory system with learning consideration. *International Journal of Production Research* 41(3):571–579.

Bonney, M., 2009. Inventory planning to help the environment. In *Inventory management: Non-classical views*, M.Y. Jaber ed. 43–74, Boca Raton: CRC Press (Taylor and Francis Group).

Carlson, J.G.H., 1975. Learning, lost time and economic production (The effect of learning on production lots). *Production and Inventory Management* 16(4): 20–33.

Carlson, J.G., and Rowe, R.G., 1976. How much does forgetting cost? *Industrial Engineering* 8(9): 40–47.

Chand, S., 1989. Lot sizes and set-up frequency with learning and process quality. *European Journal of Operational Research* 42(2): 190–202.

Chand, S., and Sethi, S.P., 1990. A dynamic lot sizing model with learning in set-ups. *Operations Research* 38(4): 644–655.

Chen, C.K., Lo, C.C., and Liao, Y.X., 2008. Optimal lot size with learning consideration on an imperfect production system with allowable shortages. *International Journal of Production Economics* 113(1): 459–469.

Cheng, T.C.E., 1991. An EOQ model with learning effect on set-ups. *Production and Inventory Management* 32(1): 83–84.

Cheng, T.C.E., 1994. An economic manufacturing quantity model with learning effects. *International Journal of Production Economics* 33(1–3): 257–264.

Chiu H.N., 1997. Discrete time-varying demand lot-sizing models with learning and forgetting effects. *Production Planning and Control* 8(5): 484–493.

Chiu, H.N., and Chen, H.M., 1997. The effect of time-value of money on discrete time-varying demand lot-sizing models with learning and forgetting considerations. *Engineering Economist* 42(3): 203–221.

Chiu, H.N., and Chen, H.M., 2005. An optimal algorithm for solving the dynamic lot-sizing model with learning and forgetting in set-ups and production. *International Journal of Production Economics* 95(2): 179–193.

Chiu, H.N., Chen, H.M., and Weng, L.C., 2003. Deterministic time-varying demand lot-sizing models with learning and forgetting in set-ups and production. *Production and Operations Management* 12(1): 120–127.

Conway, R., and Schultz, A., 1959. The manufacturing progress function. *Journal of Industrial Engineering* 10(1): 39–53.

Corlett, N., and Morcombe, V.J., 1970. Straightening out the learning curves. *Personnel Management* 2(6): 14–19.

Crossman, E.R.F.W., 1959. A theory of acquisition of speed skill. *Ergonomics* 2(2): 153–166.

Dar-El, E.M., Ayas, K., and Gilad, I., 1995. A dual-phase model for the individual learning process in industrial tasks. *IIE Transactions* 27(3): 265–271.

de Jong, J.R., 1957. The effect of increased skills on cycle time and its consequences for time standards. *Ergonomics* 1(1): 51–60.

Dell, M., and Fredman, C., 1999. *Direct from Dell: Strategies that revolutionized an industry*. London: Harper Collins.

Dobos, I., and Richter, K., 2004. An extended production/recycling model with stationary demand and return rates. *International Journal of Production Economics* 90(3): 311–323.

Elmaghraby, S.E., 1990. Economic manufacturing quantities under conditions of learning and forgetting (EMQ/LaF). *Production Planning and Control* 1(4): 196–208.

Eroglu, A., and Ozdemir, G., 2005. A note on "The effect of time-value of money on discrete time-varying demand lot-sizing models with learning and forgetting considerations." *Engineering Economist* 50(1): 87–90.

Fisk, J.C., and Ballou, D.P. 1982. Production lot sizing under a learning effect. *AIIE Transactions* 14(4): 257–264.

Freeland, J.R., and Colley, J.L., Jr., 1982. A simple heuristic method for lot sizing in a time-phased reorder system. *Production and Inventory Management* 23(1): 15–22.

Globerson, S., and Levin, N., 1987. Incorporating forgetting into learning curves. *International Journal of Operations and Production Management* 7(4): 80–94.

Globerson, S., Levin, N., and Shtub, A., 1989. The impact of breaks on forgetting when performing a repetitive task. *IIE Transactions* 21(4): 376–381.

Glover, J.H., 1965. Manufacturing progress functions: An alternative model and its comparison with existing functions. *International Journal of Production Research* 4(4): 279–300.

Glover, J.H., 1966. Manufacturing progress functions II: Selection of trainees and control of their progress. *International Journal of Production Research* 5(1): 43–59.

Glover, J.H., 1967. Manufacturing progress functions III: Production control of new products. *International Journal of Production Research* 6(1): 15–24.

Gorham, T., 1968., Dynamic order quantities. *Production and Inventory Management* 9(1): 75–81.

Goyal, S.K., 1977. An integrated inventory model for a single supplier-single customer problem. *International Journal of Production Research* 15(1): 107–111.

Goyal, S.K., and Gupta, Y.P., 1989. Integrated inventory models: The buyer-vendor coordination. *European Journal of Operational Research* 41(3): 261–269.

Graham, C.H., and Gagné, R.M., 1940. The acquisition, extinction, and spontaneous recovery of conditioned operant response. *Journal of Experimental Psychology* 26(3): 251–280.

Gungor, A., and Gupta, S.M., 1999. Issues in environmentally conscious manufacturing and product recovery: A survey. *Computers and Industrial Engineering* 36(4): 811–853.

Hackett, E.A., 1983. Application of a set of learning curve models to repetitive tasks. *The Radio and Electronic Engineer* 53(1): 25–32.

Harris, F.W., 1990. How many parts to make at once? *Operations Research* 38(6): 947–950. [Reprinted from Factory: The Magazine of Management, Vol. 10, no. 2, 1913, pp. 135–36]

Hirsch, W.Z., 1952. Manufacturing progress function. *The Review of Economics and Statistics* 34(2): 143–155.

Hirschmann, W.B., 1964. Profit from the learning curve. *Harvard Business Review* 42(1): 125–139.

Hoffman, T.R., 1968. Effect of prior experience on learning curve parameters. *Journal of Industrial Engineering* 19(8): 412–413.

Jaber, M.Y., 2006a. Learning and forgetting models and their applications. In *Handbook of industrial and systems engineering*, ed. A.B. Badiru, Chapter 30, 1–27, Baco Raton: CRC Press (Taylor and Francis Group).

Jaber, M.Y., 2006b. Lot sizing for an imperfect production process with quality corrective interruptions and improvements, and reduction in set-ups. *Computers and Industrial Engineering* 51(4): 781–790.

Jaber, M.Y., 2009. Modeling hidden costs of inventory systems: A thermodynamics approach. In: *Inventory management: Non-classical views*, ed. M.Y. Jaber, pp. 199–218, Baco Raton: CRC Press (Taylor and Francis Group).

Jaber, M.Y., and Abboud, N.E., 2001. The impact of random machine unavailability on inventory policies in a continuous improvement environment. *Production Planning and Control* 12(8): 754–763.

Jaber, M.Y., and Bonney, M., 1996a. Production breaks and the learning curve: The forgetting phenomenon. *Applied Mathematical Modeling* 20(2): 162–169.

Jaber, M.Y., and Bonney, M., 1996b. Optimal lot sizing under learning considerations: The bounded learning case. *Applied Mathematical Modeling* 20(10): 750–755.

Jaber, M.Y., and Bonney, M.C., 1997a. A comparative study of learning curves with forgetting. *Applied Mathematical Modeling* 21(8): 523–531.

Jaber, M.Y., and Bonney, M., 1997b. The effect of learning and forgetting on the economic manufactured quantity (EMQ) with the consideration of intra-cycle shortages. *International Journal of Production Economics* 53(1): 1–11.

Jaber, M.Y., and Bonney, M., 1998. The effects of learning and forgetting on the optimal lot size quantity of intermittent production runs. *Production Planning and Control* 9(1): 20–27.

Jaber, M.Y., and Bonney, M., 1999. The economic manufacture/order quantity (EMQ/EOQ) and the learning curve: Past, present, and future. *International Journal of Production Economics* 59(1–3): 93–102.

Jaber, M.Y., and Bonney, M., 2001a. A comment on "Zhou YW and Lau H-S (1998). Optimal production lot-sizing model considering the bounded learning case and shortages backordered. J Opl Res Soc 49: 1206–1211." *Journal of Operational Research Society* 52(5): 584–590.

Jaber, M.Y., and Bonney, M., 2001b. A comment on "Zhou YW and Lau H-S (1998). Optimal production lot-sizing model considering the bounded learning case and shortages backordered. J Opl Res Soc 49: 1206–1211 — Comments on Zhou & Lau's reply." *Journal of Operational Research Society* 52(5): 591–592.

Jaber, M.Y., and Bonney, M., 2001c. Economic lot sizing with learning and continuous time discounting: Is it significant? *International Journal of Production Economics* 71(1–3): 135–143.

Jaber, M.Y., and Bonney, M., 2003. Lot sizing with learning and forgetting in set-ups and in product quality. *International Journal of Production Economics* 83(1): 95–111.

Jaber, M.Y., and Bonney, M., 2007. Economic manufacture quantity (EMQ) model with lot-size dependent learning and forgetting rates. *International Journal of Production Economics* 108(1–2): 359–367.

Jaber, M.Y., Bonney, M., and Guiffrida, A.L., 2010. Coordinating a three-level supply chain with learning-based continuous improvement. *International Journal of Production Economics* 127(1): 27–38.

Jaber, M.Y., Bonney, M., and Moualek, I., 2009. Lot sizing with learning, forgetting and entropy cost. *International Journal of Production Economics* 118(1): 19–25.

Jaber, M.Y., and El Saadany, A.M.A., 2011. An economic production and remanufacturing model with learning effects. *International Journal of Production Economics* 131(1): 115–127.

Jaber, M.Y., Goyal, S.K., and Imran, M., 2008. Economic production quantity model for items with imperfect quality subject to learning effects. *International Journal of Production Economics* 115(1): 143–150.

Jaber, M.Y., and Guiffrida, A.L., 2004. Learning curves for processes generating defects requiring reworks. *European Journal of Operational Research* 159(3): 663–672.

Jaber, M.Y., and Guiffrida, A.L., 2007. Observations on the economic order (manufacture) quantity model with learning and forgetting. *International Transactions in Operational Research* 14(2): 91–104.

Jaber, M.Y., and Guiffrida, A.L., 2008. Learning curves for imperfect production processes with reworks and process restoration interruptions. *European Journal of Operational Research* 189(1): 93–104.

Jaber, M.Y., and Khan, M., 2010. Managing yield by lot splitting in a serial production line with learning, rework and scrap. *International Journal of Production Economics* 124(1): 32–39.

Jaber, M.Y., and Kher, H.V., 2002. The dual-phase learning-forgetting model. *International Journal of Production Economics* 76(3): 229–242.

Jaber, M.Y., Kher, H.V., and Davis, D., 2003. Countering forgetting through training and deployment. *International Journal of Production Economics* 85(1): 33–46.

Jaber, M.Y., Nuwayhid, R.Y., and Rosen, M.A., 2004. Price-driven economic order systems from a thermodynamic point of view. *International Journal of Production Research* 42(24): 5167–5184.

Jaber, M.Y., and Salameh, M.K., 1995. Optimal lot sizing under learning considerations: Shortages allowed and back ordered. *Applied Mathematical Modeling* 19(5): 307–310.

Jaber, M.Y., and Sikström, S., 2004a. A numerical comparison of three potential learning and forgetting models. *International Journal of Production Economics* 92(3): 281–294.

Jaber, M.Y., and Sikström, S., 2004b. A note on: An empirical comparison of forgetting models. *IEEE Transactions on Engineering Management* 51(2): 233–234.

Jaber, M.Y., and Zolfaghari, S., 2008. Quantitative models for centralized supply chain coordination. In *Supply chains: Theory and applications*, ed. V. Kordic, 307–338, Vienna: I-Tech Education and Publishing.

Jordan, R.B., 1958. How to use the learning curve. *N.A.A. Bulletin* 39(5): 27–39.

Karni, R., 1981. Maximum part-period gain (MPG): A lot sizing procedure for unconstrained and constrained requirements planning systems. *Production and Inventory Management* 22(2): 91–98.

Karwan, K., Mazzola, J., and Morey, R., 1988. Production lot sizing under set-up and worker learning. *Naval Research Logistics* 35(2): 159–175.

Keachie, E.C., and Fontana, R.J., 1966. Effects of learning on optimal lot size. *Management Science* 13(2): B102–B108.

Kim, S.L., Banerjee, A., and Burton, J., 2008. Production and delivery policies for enhanced supply chain partnerships. *International Journal of Production Research* 46(22): 6207–6229.

King, A.M., Burgess, S.C., Ijomah, W., and McMahon, C.A., 2006. Reducing waste: Repair, recondition, remanufacture or recycle? *Sustainable Development* 14(4): 257–267.

Klastorin, T.D., and Moinzadeh, K., 1989. Production lot sizing under learning effects: An efficient solution technique. *AIIE Transactions* 21(1): 2–10.

Kopcso, D.P., and Nemitz, W.C., 1983. Learning curves and lot sizing for independent and dependent demand. *Journal of Operations Management* 4(1): 73–83.

Li, C.L., and Cheng, T.C.E., 1994. An economic production quantity model with learning and forgetting considerations. *Production and Operations Management* 3(2): 118–132.

Maity, A.K., Maity, K., Mondal, S.K., and Maiti, M., 2009. A production-recycling-inventory model with learning effect. *Optimization and Engineering* 10(3): 427–438.

Muth, E.J., and Spremann, K., 1983. Learning effect in economic lot sizing. *Management Science* 29(2): B102–B108.

Moon, I., and Choi, S., 1998. A note on lead time and distribution assumptions in continuous review inventory models. *Computers and Operations Research* 25 (11): 1007–1012.

Nanda, R., and Nam, H.K., 1992. Quantity discounts using a joint lot size model under learning effects – Single buyer case. *Computers and Industrial Engineering* 22(2): 211–221.

Nanda, R., Nam, H.K., 1993. Quantity discounts using a joint lot size model under learning effects – Multiple buyer case. *Computers and Industrial Engineering* 24(3): 487–494.

Pan, J.C.H., and Lo, M.C., 2008. The learning effect on set-up cost reduction for mixture inventory models with variable lead time. *Asia-Pacific Journal of Operational Research* 25(4): 513–529.

Porteus, E., 1985. Investing in reduced set-ups in the EOQ model. *Management Science* 31(8): 998–1010.

Pratsini, E., Camm, J.D., and Raturi, A.S., 1994. Capacitated lot sizing under set-up learning. *European Journal of Operational Research* 72(3): 545–557.

Rachamadugu, R., 1994. Performance of a myopic lot size policy with learning in set-ups. *IIE Transactions* 26(5): 85–91.

Rachamadugu, R., and Schriber, T.J., 1995. Optimal and heuristic policies for lot sizing with learning in set-ups. *Journal of Operations Management* 13(3): 229–245.

Rachamadugu, R., and Tan, L.C., 1997. Policies for lot sizing with set up learning. *International Journal of Production Economics* 48(2): 157–165.

Replogle, S., 1988. The strategic use of smaller lot sizes through a new EOQ model. *Production and Inventory Management* 29(3): 41–44.

Richter, K., 1996a. The EOQ and waste disposal model with variable set-up numbers. *European Journal of Operational Research* 95(2): 313–324.

Richter, K., 1996b. The extended EOQ repair and waste disposal model. *International Journal of Production Economics* 45(1–3): 443–448.

Rosenblatt, M.J., and Lee, H.L., 1986. Economic production cycles with imperfect production processes. *IIE Transactions* 18(1): 48–55.

Salameh, M.K., Abdul-Malak, M.U., and Jaber, M.Y., 1993. Mathematical modeling of the effect of human learning in the finite production inventory model. *Applied Mathematical Modeling* 17(11): 613–615.

Salameh, M.K., and Jaber, M.Y., 2000. Economic production quantity model for items with imperfect quality. *International Journal of Production Economics* 64(1–3): 59–64.

Schrady, D.A., 1967. A deterministic inventory model for repairable items. *Naval Research Logistics Quarterly* 14(3): 391–398.

Shiue, Y.C., 1991. An economic batch production quantity model with learning curve-dependent effects: A technical note. *International Journal of Production Economics* 25(1–3): 35–38.

Smunt, T.L., and Morton, T.E., 1985. The effect of learning on optimal lot sizes: Further developments on the single product case. *AIIE Transactions* 17(1): 33–37.

Spradlin, B.C., and Pierce, D.A., 1967. Production scheduling under learning effect by dynamic programming. *The Journal of Industrial Engineering* 18(3): 219–222.

Steedman, I., 1970. Some improvement curve theory. *International Journal of Production Research* 8(3): 189–206.

Sule, D.R., 1978. The effect of alternate periods of learning and forgetting on economic manufactured quantity. *AIIE Transactions* 10(3): 338–343.

Sule, D.R., 1981. A note on production time variation in determining EMQ under the influence of learning and forgetting. *AIIE Transactions* 13(1): 91–95.

Teyarachakul, S., Chand, S., and Ward, J., 2008. Batch sizing under learning and forgetting: Steady state characteristics for the constant demand case. *Operations Research Letters* 36(5): 589–593.

Teyarachakul, S., Chand, S., and Ward, J., 2011. Effect of learning and forgetting on batch sizes. *Production and Operations Management* 20(1): 116–128.

Thorndike, E.L., 1898. Animal intelligence: An experimental study of the associative process in animals. *The Psychological Review: Monograph Supplements* 2(4): 1–109.

Thurstone, L.L., 1919. The learning curve equation. *Psychological Monograph* 26(3): 1–51.

Towill, D.R., 1985. The use of learning curve models for prediction of batch production performance. *International Journal of Operations and Production Management* 5(2): 13–24.

Urban, T.L., 1998. Analysis of production systems when run length influences product quality. *International Journal of Production Research* 36(11): 3085–3094.

Wagner, H.M., and Whitin, T.M., 1958. Dynamic version of the economic lot size model. *Management Science* 5(1): 89–96.

Wortham, A.W., and Mayyasi, A.M., 1972. Learning considerations with economic order quantity. *AIIE Transactions* 4(1): 69–71.

Wright, T., 1936. Factors affecting the cost of airplanes. *Journal of Aeronautical Science* 3(4): 122–128.

Yelle, L.E., 1979. The learning curve: Historical review and comprehensive survey. *Decision Sciences* 10(1): 302–328.

Zhou, Y.W., and Lau, H.S., 1998. Optimal production lot-sizing model considering the bounded learning case and shortages backordered. *The Journal of the Operational Research Society* 49(11): 1206–1211.

Zhou, Y.W., and Lau, H.S., 2001. A comment on "Zhou YW and Lau H-S (1998). Optimal production lot-sizing model considering the bounded learning case and shortages backordered. J Opl Res Soc 49: 1206–1211 — Reply to Jaber and Bonney." *Journal of Operational Research Society* 52(5): 590–591.

15 Learning Effects in Inventory Models with Alternative Shipment Strategies

Christoph H. Glock

CONTENTS

INTRODUCTION

Learning effects have been analyzed in a variety of different application areas, such as the leadership of employees, purchasing decisions, or corporate strategy (see e.g., Manz and Sims 1984; De Geus 1988; Anderson and Parker 2002; Hatch and Dyer 2004; and Yelle 1979, for a review of related literature). The primary reason why the learning phenomenon has received increased attention in recent years is that learning has been identified as a source of sustainable competitive advantage by many researchers (see Hatch and Dyer 2004), which means that proactively transforming static organizations into learning organizations may help to differentiate the company from its competitors.

In the domain of production planning, researchers have studied the impact of learning processes on the development of inventory levels (e.g., Elmaghraby 1990; Jaber and Bonney 1998), the failure rate in production processes (see Jaber and Bonney 2003; Jaber et al. 2008), or the costs of machine set-ups (see Chiu et al. 2003; Jaber and Bonney 2003). The basic idea behind learning curve theory is that the performance of an individual improves in a repetitive task, leading to fewer mistakes and to faster job completion (Jaber and Bonney 1999). Seen from the perspective of inventory management, learning effects increase the production rate over time, which may lead to faster inventory build-up, an earlier start of the consumption phase and, consequently, lower inventory carrying costs (e.g., Adler and Nanda 1974). However, as has been shown by Glock (2010, 2011), increasing the production rate in a system where batch shipments are transported between subsequent production stages may increase in-process inventory and thus lead to higher total costs. By taking a closer look at the literature on learning effects in inventory models, it becomes obvious that interdependencies between the transportation strategy implemented in a production system and learning effects in the production process have not yet been studied. This is insufficient inasmuch as the timing and size of deliveries between subsequent production stages influence the relation between the production and consumption phases, which in turn impacts the learning effect. In addition, learning effects may influence the manufacturing time of a given production quantity, which may increase waiting times in the production system and consequently the in-process inventory.

To close the gap identified above, this chapter aims to analyze how the timing and size of deliveries between subsequent stages helps to take advantage of the benefits of learning effects in production. The remainder of the chapter is organized as follows: In the next section we describe the problem studied in this chapter and introduce the assumptions and definitions that will be used in the remaining parts of the chapter. Accordingly, we develop formal models for different shipment strategies and integrate a learning effect into the model formulation.

PROBLEM DESCRIPTION

In this chapter, we study the interdependencies between learning effects in production and the timing of deliveries between subsequent stages of a production system. This study is motivated by two recent papers by Glock (2010, 2011), who has shown that increasing the production rate in a system where equal-sized batches are transported between subsequent stages may increase the waiting times for some of the batches, thereupon leading to higher inventory carrying costs. We focus on a two-stage production system with a producing and a consuming stage, which is the basic building block of a broad variety of more sophisticated inventory models, and consider four basic transportation strategies that are frequently found in the literature.

In developing the proposed models, the following assumptions were made:

1. All parameters are deterministic and constant over time
2. The planning horizon is divided into J periods, and one lot is produced in each of the J periods

3. Production lots are of equal sizes
4. Learning and forgetting effects occur at the producing stage
5. The production rate exceeds the demand rate
6. Shortages are not allowed

Furthermore, the following notation was used:

A set-up costs per set-up
α_j units experience available in the production system at the beginning of period j
D total demand in the planning period
d demand rate in units per unit of time
F transportation costs per shipment
f slope of the forgetting curve
γ cost per unit of production time
h inventory carrying charges per unit per unit of time
J number of set-ups in the planning horizon
k the production count
l slope of the learning curve with $0 \le l \le 1$
m number of batch shipments per lot produced
p production rate in units per unit of time
$t_{p,i,j}$ production time of batch j of production lot i
Q production lot size with $Q = D/J$
R_j number of units that could have been produced during an interruption in period j
T_1 time required to produce the first unit
$\hat{T}_1$ first unit of the forgetting curve
T_k time required to produce the kth unit
$\hat{T}_x$ time for the xth unit of lost experience of the forgetting curve
t_c time to consume a lot
t_p time to produce a lot
$t_{w,i}$ waiting time of batch i
x unit count of the forgetting curve
IC inventory carrying costs in the planning period
PC production costs
TC total costs in the planning period
TWI time-weighted inventory
$\text{Max}[a,b]$ denotes the maximum value of a and b
$\text{Min}[a,b]$ denotes the minimum value of a and b

ALTERNATIVE MODES OF TRANSPORTING
LOTS BETWEEN PRODUCTION STAGES

The literature discusses a variety of alternative strategies for transporting lots between subsequent production stages. The mode of transport chosen in a production process determines the number of shipments and the shipment quantities, and may thus influence inventory and transfer costs in the production system. In the following sections, we discuss four basic strategies for transporting lots between a producing and a consuming stage. This will be extended to include learning effects in the later sections of this chapter.

TRANSFER OF COMPLETE LOTS (MODEL C)

One alternative to transfer a lot from a producing to a consuming stage is to wait until the lot has been completed and to ship the entire production quantity to the next stage. This strategy is often implemented where a production lot may not be split up into smaller parts due to technical reasons or where high transportation costs prohibit more than one shipment per lot. The corresponding inventory time plots for this strategy are shown in Figure 15.1a.

The inventory carrying costs in the planning period may easily be calculated as follows:

$$IC^C = \frac{Q}{2}\left(t_p + t_c\right)hJ = \frac{D^2}{2J}\left(\frac{1}{p} + \frac{1}{d}\right)h. \tag{15.1}$$

To assure comparability with the models developed in the next sections, we further consider production costs that amount to $\gamma D/p$ in the present case. By further considering set-up costs, the total cost function of this strategy amounts to:

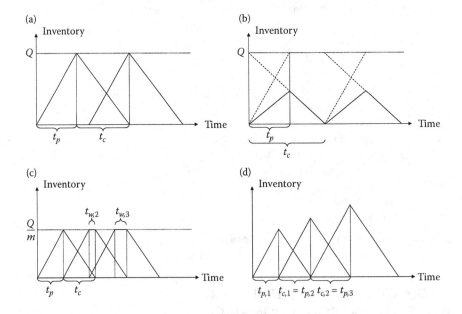

FIGURE 15.1 Alternative shipment strategies in a two-stage production system. (With kind permission from Springer Science + Business Media: *Zeitschrift für Betriebswirtschaft (Journal of Business Economics)*, Rational inefficiencies in lot-sizing models with learning effects, 79(4), 2009, 37–57, Bogaschewsky, R.W., and Glock, C.H.)

$$TC^C = \frac{D^2}{2J}\left(\frac{1}{p}+\frac{1}{d}\right)h + \gamma\frac{D}{p} + AJ. \tag{15.2}$$

IMMEDIATE PRODUCT TRANSFER (MODEL I)

Another alternative to transfer lots to the subsequent stage is to ship each (infinitesimal) unit separately. This strategy is often discussed in the context of the classical economic production quantity (EPQ) model (e.g., Silver et al. 1998; Tersine 1998) and is applicable in cases where the producing and consuming stages are located in close proximity and/or where both stages are connected with automatic transportation equipment. The corresponding inventory time plots for this case are shown in Figure 15.1b. The inventory carrying costs in the planning period may be obtained by calculating the area under the solid lines and multiplying the resulting expression with Jh. It follows that:

$$IC^I = \frac{Q}{2}\left(t_c - t_p\right)Jh = \frac{D^2}{2J}\left(\frac{1}{d}-\frac{1}{p}\right)h. \tag{15.3}$$

By comparing Equations 15.1 and 15.3, it can be seen that IC^C is always larger than IC^I, which is because consumption starts in Model C after production has been finished, whereas in Model I the production and consumption are initiated simultaneously. The total costs of this strategy are given as:

$$TC^I = \frac{D^2}{2J}\left(\frac{1}{d}-\frac{1}{p}\right)h + \gamma\frac{D}{p} + AJ. \tag{15.4}$$

EQUAL-SIZED BATCH SHIPMENTS (MODEL E)

The two shipment strategies introduced above are not mutually exclusive and can be seen as extreme points on a continuum of hybrid transportation strategies. Instead of transporting only complete batches or each manufactured unit separately to the next stage, the company may decide to aggregate several units to a batch, which is then transported to the subsequent stage. In the case where the transportation frequency is equal to 1, this strategy would be identical to the case where only complete lots are transported to the next stage; whereas in a case where the transportation frequency approaches infinity, this strategy would correspond to the case where each (infinitesimal) unit is shipped separately.

A basic transportation strategy that includes batch shipments is due to Szendrovits (1975), who assumes that successive shipments are of equal sizes. The corresponding inventory time plots for this case are illustrated in Figure 15.1c. As can be seen, the first batch is shipped to the subsequent stage directly after its completion. Due to $p > d$, the second batch is finished before the first batch is completely used up, therefore it has to be kept in stock for $t_{w,1} = t_v - t_p$ time units. As described in Glock (2010, 2011) and Bogaschewsky and Glock (2009), the time-weighted inventory in this case consists of a "regular" inventory and an inventory due to waiting times. The "regular" inventory may be obtained by calculating the area defined by the identical triangles shown in Figure 15.1c:

$$\frac{Q^2}{2m}\left(\frac{1}{p}+\frac{1}{d}\right).$$ (15.5)

Taking into account waiting times of $m-1$ sub-lots, we obtain:

$$\frac{Q}{m}\sum_{i=1}^{m-1}i\left(t_c-t_p\right)=\frac{Q^2\left(m-1\right)}{2m}\left(\frac{1}{d}-\frac{1}{p}\right).$$ (15.6)

The inventory carrying costs per unit of time may thus be calculated as follows:

$$IC^E=\frac{Q^2}{2m}\left(\frac{2-m}{p}+\frac{m}{d}\right)Jh=\frac{D^2}{2Jm}\left(\frac{2-m}{p}+\frac{m}{d}\right)h.$$ (15.7)

Equation 15.7 reduces to Equation 15.1 when $m=1$ and to Equation 15.3 when $m\to\infty$. In calculating the total cost function, it is necessary to consider transportation costs that accrue with every shipment, since otherwise it would always be optimal to set m equal to infinity. To exclude this trivial case, which has already been considered above, we formulate the total cost function of this strategy as follows:

$$TC^E=\frac{D^2}{2Jm}\left(\frac{2-m}{p}+\frac{m}{d}\right)h+\gamma\frac{D}{p}+\left(A+mF\right)J.$$ (15.8)

UNEQUAL-SIZED BATCH SHIPMENTS (MODEL U)

While transporting equal-sized batch shipments to the subsequent stage may be beneficial in case the second strategy is infeasible or too costly, it has been shown by Goyal (1977) that the process inventory may be further reduced if batches of unequal size are shipped to the subsequent stage. The corresponding inventory time plots for the case of unequal-sized batch shipments that follow a geometric series are illustrated in Figure 15.1d. As can be seen, batches 2 to m are manufactured in the consumption time of the respective precedent batch, which avoids the waiting time-related inventory. The size of the jth batch can be calculated as follows (see Goyal 1977):

$$q_j=q_1\left(p/d\right)^{j-1},\text{with }Q=\sum_{i=1}^{m}q_i=q_1\sum_{i=1}^{m}\left(p/d\right)^{i-1}.$$ (15.9)

The inventory carrying costs per unit of time take the following form (see Glock 2010):

$$IC^U = \frac{q_1^2}{2}\left(\frac{1}{p}+\frac{1}{d}\right)\frac{(p/d)^{2m}-1}{(p/d)^2-1}Jh = \frac{D^2}{2J}\left(\frac{1}{d}-\frac{1}{p}\right)\frac{\left((p/d)^m+1\right)(p/d-1)}{\left((p/d)^m-1\right)(p/d+1)}h. \quad (15.10)$$

The total costs of this strategy are thus given as:

$$TC^U = \frac{D^2}{2J}\left(\frac{1}{d}-\frac{1}{p}\right)\frac{\left((p/d)^m+1\right)(p/d-1)}{\left((p/d)^m-1\right)(p/d+1)}h+\gamma\frac{D}{p}+(A+mF)J. \quad (15.11)$$

LEARNING EFFECTS IN INVENTORY MODELS WITH BATCH SHIPMENTS

LEARNING EFFECT

In the following, we assume that learning and forgetting effects occur in the production process of the first stage. The learning effect is assumed to follow the power learning curve due to Wright (1936), which is of the form:

$$T_k = T_1 K^{-1}. \quad (15.12)$$

In this respect, T_k denotes the time to produce the kth unit, k the cumulative production quantity, T_1 the time required to produce the first unit, and l the slope of the learning curve. Furthermore, we consider the forgetting effect described by Carlson and Rowe (1976), which is of the form:

$$\hat{T}_x = \hat{T}_1 x^f, \quad (15.13)$$

where $\hat{T}_x$ equals the time for the xth unit of lost experience of the forgetting curve, x the amount of output that would have been accumulated if the production process had not been interrupted, $\hat{T}_1$ the equivalent time for the first unit of the forgetting curve, and f equals the slope of the forgetting curve.

Both effects described above have been integrated by Jaber and Bonney (1996, 1998). Because the time it takes to produce one unit at the time production stops equals the starting point of the forgetting curve, it follows from Equations 15.12 and 15.13 that:

$$\hat{T}_{1,i} = T_1 q_i^{-(l+f)}. \quad (15.14)$$

If α_i denotes the equivalent number of produced units that the production system remembers at the beginning of production run i, the time to manufacture q_i units equals (see Jaber and Bonney 1998):

$$t_{p,i} = \int_{\alpha_i}^{\alpha_i+q_i} T_1 k^{-l} dk = \frac{T_1}{1-l}\left((q_i+\alpha_i)^{1-l} - \alpha_i^{1-l}\right). \qquad (15.15)$$

Further, we assume that R_i denotes the number of units that could have been produced during the interruption $t_{f,i}$. It follows that:

$$t_{f,i} = \int_{\alpha_i+q_i}^{\alpha_i+q_i+R_i} T_1 q_i^{-(l+f)} z^f dz = \frac{T_1 q_i^{-(l+f)}}{1+f}\left((\alpha_i+q_i+R_i)^{1+f} - (\alpha_i+q_i)^{1+f}\right). \quad (15.16)$$

Solving Equation 15.16 for R_i yields:

$$R_i = \left((\alpha_i+q_i)^{1+f} + \frac{(1+f)q_i^{f+l}}{T_1} t_{f,i}\right)^{1/(1+f)} - (\alpha_i+q_i). \qquad (15.17)$$

As the last point of the forgetting curve is equivalent to the starting point of the learning curve in the following production cycle, it follows that:

$$T_1(\alpha_i+q_i)^{-(f+l)}(\alpha_i+q_i+R_i)^f = T_1\alpha_{i+1}^{-l}. \qquad (15.18)$$

Solving for α_{i+1} yields:

$$\alpha_{i+1} = \left((\alpha_i+q_i)^{-(f+l)}(\alpha_i+q_i+R_i)^f\right)^{-1/l}. \qquad (15.19)$$

Since it is not reasonable to assume that more units can be remembered than have previously been produced, feasible values for α_i are restricted to the following interval:

$$0 \le \alpha_i \le \sum_{n=1}^{j-1} q_n. \qquad (15.20)$$

Note that $\alpha_1 = 0$, since the production system encounters no prior experience in the first period.

TRANSFER OF COMPLETE LOTS (MODEL C)

We first consider the case where only complete lots are transported to the subsequent stage. As can be seen in Figure 15.2a, the production rate now increases with each unit produced due to the learning effect. Looking first at the inventory carrying costs, it is again advantageous to derive the stock in the planning period by calculating the area under the triangle-like shapes shown in Figure 15.2a. By solving

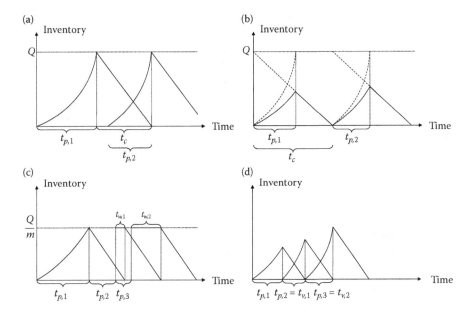

FIGURE 15.2   Alternative shipment strategies in a two-stage production system with learning effects. (With kind permission from Springer Science + Business Media: *Zeitschrift für Betriebswirtschaft (Journal of Business Economics)*, Rational inefficiencies in lot-sizing models with learning effects, 79(4), 2009, 37–57, Bogaschewsky, R.W., and Glock, C.H.)

Equation 15.15, which gives the time needed to manufacture q_i units, for q_i, we can obtain the number of units that can be produced in $t_{p,i}$ time units given α_i units of experience:

$$q_i = \left(\frac{1-l}{T_1} t_{p,i} + \alpha_i^{1-l} \right)^{1/(1-l)} - \alpha_i. \tag{15.21}$$

The time-weighted inventory per lot produced can be determined by integrating Equation 15.21 over the limits 0 and $t_{p,i}$ and by further considering the inventory at the second stage, providing:

$$TWI_i^{C,L} = \frac{T_1}{2-l} \left\{ \left[(q_i + \alpha_i)^{1-l} \right]^{\delta} - \left(\alpha_i^{1-l} \right)^{\delta} - \delta \alpha_i \left[(q_i + \alpha_i)^{1-l} - \alpha_i^{1-l} \right] \right\} + \frac{D^2}{2J^2 d}, \tag{15.22}$$

where $\delta = (2 - l)/(1 - l)$. Note that $q_i = D/J$ in this case. Inventory carrying costs in the planning period may now be formulated as follows:

$$IC^{C,L} = \sum_{i=1}^{J} TWI_i^{C,L} h. \tag{15.23}$$

Since the time to produce the total demand D is affected by the lot sizing policy, it can be assumed that the production costs themselves will also be influenced by the chosen lot sizing policy. If γ denotes the costs per unit production time, the production costs amount to:

$$PC^{C,L} = \gamma \sum_{i=1}^{J} t_{p,i}, \tag{15.24}$$

where $t_{p,i}$ is calculated according to Equation 15.15. The total cost function may be derived by further considering set-up costs:

$$TC^{C,L} = \sum_{i=1}^{J} TWI_i^{C,L} h + \gamma \sum_{i=1}^{J} t_{p,i} + AJ. \tag{15.25}$$

IMMEDIATE PRODUCT TRANSFER (MODEL I)

In case each (infinitesimal) unit is shipped separately to the next stage, inventory per lot may be calculated by subtracting the time-weighted inventory during the production phase from the time-weighted inventory during the consumption phase (cf. Figure 15.2b). It follows from Equation 15.22:

$$TWI_i^{I,L} = \frac{D^2}{2J^2 d} - \frac{T_1}{2-l} \left\{ \left[(q_i + \alpha_i)^{1-l} \right]^{\delta} - \left(\alpha_i^{1-l} \right)^{\delta} - \delta \alpha_i \left[(q_i + \alpha_i)^{1-l} - \alpha_i^{1-l} \right] \right\}. \tag{15.26}$$

Note that, despite the learning effect, $p > d$ or $1/T_1 > d$ is still a necessary condition to ensure that the demand at the second stage is satisfied without interruption. Inventory carrying costs in the planning period are thus given as:

$$IC^{I,L} = \sum_{i=1}^{J} TWI_i^{I,L} h. \tag{15.27}$$

The total cost function for this strategy is given by further considering production and set-up costs:

$$TC^{I,L} = \sum_{i=1}^{J} TWI_i^{I,L} h + \gamma \sum_{i=1}^{J} t_{p,i} + AJ. \tag{15.28}$$

EQUAL-SIZED BATCH SHIPMENTS (MODEL E)

The case where equal-sized batch shipments are transported to the subsequent stage is illustrated in Figure 15.2c. As can be seen, inventory may be differentiated into a "regular" inventory during the production and consumption phase of a batch, and an inventory due to waiting times. It is obvious that learning reduces the time it takes to build up inventory and consequently the regular inventory, but that simultaneously the inventory due to waiting times is increased.

The "regular" inventory for Q units produced in m batches can be determined by adopting Equation 15.22:

$$I^{reg} = \frac{T_1}{2-l} \sum_{j=1}^{m} \left(\left(\left(\frac{D}{Jm} + \alpha_{i,j} \right)^{1-l} \right)^{\delta} - \left(\alpha_{i,j}^{1-l} \right)^{\delta} - \delta\alpha_{i,j} \left(\left(\frac{D}{Jm} + \alpha_{i,j} \right)^{1-l} - \alpha_{i,j}^{1-l} \right) \right)$$
$$+ \frac{D^2}{2mJ^2 d},$$
(15.29)

where $\alpha_{i,j+1} = \alpha_{i,j} + D/(Jm)$ for $j = 1,...,m-1$ and $\alpha_{1,1} = 0$. $\alpha_{i,1}$ can be calculated from $\alpha_{i-1,m}$ according to Equation 15.19.

Inventory due to waiting times may be obtained in a similar way to the case without learning effects (see Bogaschewsky and Glock 2009):

$$I^{wait} = \frac{Q}{m} \sum_{i=1}^{m-1} \left(it_c - \sum_{j=2}^{i+1} t_{p,i,j} \right) = \frac{D}{Jm} \left(\frac{(m-1)D}{2Jd} - \sum_{i=1}^{m-1} \sum_{j=2}^{i+1} t_{p,i,j} \right),$$
(15.28)

where $t_{p,i,j}$ is the manufacturing time of the jth batch of production lot i (cf. Equation 15.15). The time-weighted inventory per lot is now given as the sum of Equations 15.27 and 15.28:

$$TWI_i^{E,L} = \frac{T_1}{2-l} \sum_{j=1}^{m} \left(\left(\left(\frac{D}{Jm} + \alpha_{i,j} \right)^{1-l} \right)^{\delta} - \left(\alpha_{i,j}^{1-l} \right)^{\delta} - \delta\alpha_{i,j} \left(\left(\frac{D}{Jm} + \alpha_{i,j} \right)^{1-l} - \alpha_{i,j}^{1-l} \right) \right)$$
$$+ \frac{D}{Jm} \left(\frac{mD}{2Jd} - \sum_{i=1}^{m-1} \sum_{j=2}^{i+1} t_{p,i,j} \right).$$
(15.29)

Inventory carrying costs may now be calculated as follows:

$$IC^{E,L} = \sum_{i=1}^{J} TWI_i^{E,L} h.$$
(15.30)

The total cost function can thus be expressed as:

$$TC^{E,L} = \sum_{i=1}^{J} TWI_i^{E,L} h + \gamma \sum_{i=1}^{J} \sum_{j=1}^{m} t_{p,i,j} + (A + mF)J.$$
(15.31)

UNEQUAL-SIZED BATCH SHIPMENTS (MODEL U)

The case where unequal-sized batch shipments are transported to the subsequent stage is illustrated in Figure 15.2d. As can be seen, no inventory due to waiting times emerges in this case. As described above, the time-weighted inventory for the jth batch of production lot i may be calculated as follows:

$$TWI_{i,j}^{U,L} = \frac{T_1}{2-l}\left(\left(\left(q_{i,j}+\alpha_{i,j}\right)^{1-l}\right)^{\delta} - \left(\alpha_{i,j}^{1-l}\right)^{\delta} - \delta\alpha_{i,j}\left(\left(q_{i,j}+\alpha_{i,j}\right)^{1-l} - \alpha_{i,j}^{1-l}\right)\right) + \frac{q_{i,j}^2}{2d}.$$ (15.32)

With the help of Equation 15.21, $q_{i,j}$ can be expressed as:

$$q_{i,j+1} = \left(\frac{1-l}{T_1}t_{c,j} + \alpha_{i,j}^{1-l}\right)^{1/(1-l)} - \alpha_{i,j},$$ (15.33)

where $t_{c,j}$ denotes the consumption time of batch j; that is, $t_{c,j} = q_{i,j}/d$. Further, we note that $\alpha_{i,j+1} = \alpha_{i,j} + q_{i,j}$ for $j = 1,...,m-1$ and $\alpha_{1,1} = 0$. Again, $\alpha_{i,1}$ can be calculated from $\alpha_{i-1,m}$ according to Equation 15.19. Inventory carrying costs in the planning period are thus given as:

$$IC^{U,L} = \sum_{i=1}^{J}\sum_{j=1}^{m} TWI_{i,j}^{U,L}h.$$ (15.34)

The total cost function may now be formulated as follows:

$$TC^{U,L} = \sum_{i=1}^{J}\sum_{j=1}^{m} TWI_{i,j}^{U,L}h + \gamma\sum_{i=1}^{J}\sum_{j=1}^{m} t_{p,i,j} + (A+mF)J.$$ (15.35)

NUMERICAL STUDIES

Due to the complexity of the total cost functions 15.25, 15.28, 15.31, and 15.35, it is difficult to formally prove convexity in J and m. However, numerical studies indicated that the cost functions are quasi-convex in both decision variables, wherefore we applied a two-dimensional search algorithm that successively increased both decision variables until an increase in either J or m led to an increase in the total costs. The best solution found so far was taken as the optimal solution. To confirm our results, we used the NMinimize-function of the software-package Mathematica 7.0 (Wolfram Research, Inc), a function that contains several methods for solving constrained and unconstrained global optimization problems, such as genetic algorithms, simulated annealing, or the simplex method (see Champion 2002). For each instance we tested, the results of our enumeration algorithm and the results derived by Mathematica were identical.

To illustrate how learning effects may impact the total costs under different shipment strategies, we considered the test problem shown in Table 15.1.

First, we analyzed how the learning effect influences the time-weighted inventory in the models developed above. If we abstract from learning (i.e., if we assume that $l = 0$), it can be seen in Figure 15.3 that transferring each unit separately to the

TABLE 15.1

Test Problem Used for Computational Experimentation

D	p	d	h	A	F	γ	f
1000	150	100	0.5	200	20	2	0.1

next stage leads to the lowest time-weighted inventory for fixed values of J and m, whereas shipping only complete lots results in the highest inventory. If the learning rate is gradually increased, time-weighted inventory increases for Models I, E, and U, but leads to a reduction in inventory for Model C. Further, it becomes obvious that for a high learning rate, the time-weighted inventory is almost identical for all four transportation strategies. Note that we only considered values for the learning rate in the range $0 \leq l \leq 0.9$ since the model used to describe the learning effect in this chapter becomes invalid for the special case $l = 1$ (see Jaber and Guiffrida 2007).

These results may be explained as follows: In case only complete lots are transferred to the second stage, it can be seen in Figure 15.1 that the inventory consists of an inventory during the production and an inventory during the consumption phase. If the learning rate is increased, the output per unit time of the production system increases, which leads to a shorter production time for a lot of size Q. In the hypothetical case of a production rate that approaches infinity, the inventory has to be carried solely during the consumption phase t_c. In this case, the time-weighted inventory equals $D^2/(2Jd)$, which takes on a value of 2500 in the example introduced above. In case each unit is shipped immediately to the next stage, the inventory during the production phase increases according to the difference between the production rate p and the demand rate d. Consequently, if learning occurs and the production rate is increased, the inventory increases at a higher rate. In the hypothetical case where the production rate approaches infinity, the lot is available immediately and the Models C and I are identical.

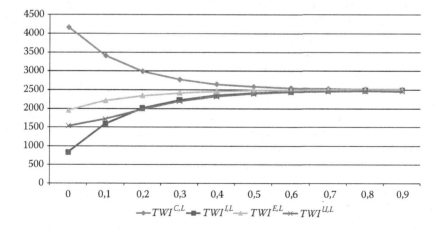

FIGURE 15.3   Time-weighted inventory for different learning rates and $J = 2$ and $m = 3$.

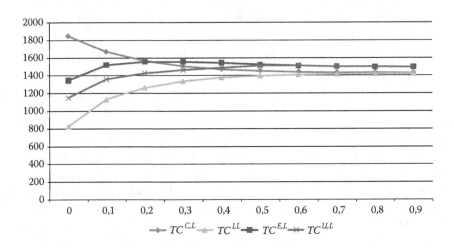

FIGURE 15.4   Total costs for alternative shipment strategies and different learning rates.

If equal-sized batches are transported to the second stage, the inventory consists of a "regular" inventory (i.e., an inventory during the production and consumption time of a batch), and an inventory due to waiting times. If learning occurs and the production rate increases, the regular inventory is reduced and the inventory during waiting times is increased, as has been outlined above. In the hypothetical case where the production rate approaches infinity, the production lot is again available immediately and Model E is identical to Models C and I. Finally, in case unequal-sized batches are shipped to the subsequent stage, a similar effect occurs than the one described above. However, due to the fact that batches are of unequal sizes, an increase in the production rate may be balanced by reducing the size of the first shipment, which reduces the minimum inventory in the system. If the production rate takes on a very high level, the whole production lot is again available immediately, in which case Model U leads to similar results than Models C, I, and E.

Figure 15.4 shows the development of the total costs for alternative values of l. It can be seen that for Model C, an increase in the learning rate leads to a reduction in the total costs. By contrast, if each unit is shipped separately to the next stage, an increase in the learning rate results in an increase in the total costs. For high values of l, Models C and I lead to identical costs.

As to Models E and U, Figure 15.4 illustrates that an increase in the learning rate first increases and then slightly reduces the total costs. The reduction is due to a decrease in production costs, which overcompensates the additional inventory carrying costs. The difference between the total costs of Models C and I on the one hand, and Models E and U on the other hand, is due to the fact that transportation costs have not been considered in the formulation of Models C and I.

CONCLUSIONS

In this chapter, we have analyzed the impact of learning effects on the total costs of a two-stage production system under different transportation strategies. It has been shown that learning effects on the producing stage may increase in-process inventory and thus lead to higher total costs. The results are congruent with the logic of the theory of constraints, which suggests that increasing the capacity of a machine supplying to a bottleneck increases waiting times and inventory in front of the bottleneck (see Goldratt and Cox 1989). In such a case, it is beneficial to synchronize the production rates of the bottleneck and the machines supplying to it in order to avoid excessive inventory.

When interpreting the results derived in this chapter, it has to be considered that the problems identified may be due to the specific definition of the problem domain and the implicit restrictions set by defining this domain (see Bogaschewsky and Glock 2009). The learning effects we considered in our model result in higher costs (in three out of four models) because they increase the production rate and thus induce an additional inventory in front of the consuming stage. However, if the production materials that are kept in stock in the raw material inventory are taken into account, it becomes obvious that faster production leads to a faster transformation of input materials into semi-finished or finished products. Thus, the inventory is transferred from the raw material storage location to the sales warehouse, and no additional inventory emerges. Therefore, higher inventory holding costs in the sales area (or after finishing production of any amount of items) would be compensated by lower inventory holding costs regarding the material stock, if all the materials needed for production are on stock no matter when production exactly starts. Obviously, problems arise only in a case where the production materials are delivered (and paid for) according to the actual work-in-process.

Further, we note that mechanisms exist that are appropriate to reduce the problems described in this paper in part. For example, in order to avoid excessive inventory, the machine could be switched off after a certain period of time in order to wait until the inventory in front of the consuming stage has been depleted (see Bogaschewsky and Glock 2009; Szendrovits 1987). This might result in a partial loss of experience, but could simultaneously reduce inventory in the system.

The models presented in this chapter may be used to study more complex production systems, for example, multi-stage or integrated inventory systems. Further, alternative formulations for the learning effect could be used to analyze how other learning curves—e.g., in combination with a plateau—impact inventory levels and the total costs in production systems.

REFERENCES

Adler, G.L., and Nanda, R.,1974. The effects of learning on optimal lot size determination – Single product case. *AIIE Transactions* 6(1): 14–20.

Anderson, E.G., Jr., and Parker, G.G., 2002. The effect of learning on the make/buy decision. *Production and Operations Management* 11(3): 313–339.

Bogaschewsky, R.W., and Glock, C.H., 2009. Rational inefficiencies in lot-sizing models with learning effects. *Zeitschrift für Betriebswirtschaft (Journal of Business Economics)* 79(4): 37–57.

Carlson, J.G.H., and Rowe, R.G., 1976. How much does forgetting cost? *Industrial Engineering* 8(9): 40–47.

Champion, B., 2002. *Numerical optimization in Mathematica: An insider's view of Minimize.* Proceedings of the 2002 World Multiconference on Systemics, Cybernetics, and Informatics. Orlando 2002, may be accessed online at http://library.wolfram.com/infocenter/Conferences/4311/

Chiu, H.N., Chen, H.M., and Weng, L.C., 2003. Deterministic time-varying demand lot-sizing models with learning and forgetting in set-ups and production. *Production and Operations Management* 12(1): 120–127.

De Geus, A.P.,1988. Planning as learning. *Harvard Business Review* 66(1): 70–74.

Elmaghraby, S.E., 1990. Economic manufacturing quantities under conditions of learning and forgetting (EMQ/LaF). *Production Planning & Control* 1(4): 196–208.

Glock, C. H., 2010. Batch sizing with controllable production rates. *International Journal of Production Research* 48(20): 5925–5942.

Glock, C.H., 2011. Batch sizing with controllable production rates in a multi-stage production system. *International Journal of Production Research* (doi: 10.1080/00207543.2010.528058).

Goldratt, E.M., and Cox, J., 1989. *The goal* (revised edition), Aldershot: Gower.

Goyal, S.K., 1977. Determination of optimum production quantity for a two-stage production system. *Operations Research Quarterly* 28(4): 865–870.

Hatch, N.W., and Dyer, J.H., 2004. Human capital and learning as a source of sustainable competitive advantage. *Strategic Management Journal* 25(12): 1155–1178.

Jaber, M.Y., and Bonney, M., 1996. Production breaks and the learning curve: The forgetting phenomenon. *Applied Mathematical Modeling* 20(2): 162–169.

Jaber, M. Y., and Bonney, M., 1998.The effects of learning and forgetting on the optimal lot size quantity of intermittent production runs. *Production Planning and Control* 9(1): 20–27.

Jaber, M.Y., and Bonney, M.,1999. The economic manufacture/order quantity (EMQ/EOQ) and the learning curve: Past, present, and future. *International Journal of Production Economics* 59(1–3): 93–102.

Jaber, M.Y., and Bonney, M., 2003. Lot sizing with learning and forgetting in set-ups and in product quality. *International Journal of Production Economics* 83(1): 95–111.

Jaber, M.Y., Goyal, S.K., and Imran, M., 2008. Economic production quantity model for items with imperfect quality subject to learning effects. *International Journal of Production Economics* 115(1): 143–150.

Jaber, M.Y., and Guiffrida, A.L., 2007. Observations on the economic manufacture quantity model with learning and forgetting. *International Transactions in Operational Research* 14(2): 91–104.

Manz, C.C., and Sims H.P., Jr., 1984. The potential for "groupthink" autonomous work groups. *Human Relations* 35(9): 773–784.

Silver, E.A., Pyke, D. F., and Peterson, R., 1998. *Inventory management and production planning and scheduling,* 3rd ed., New York: Wiley.

Szendrovits, A.Z., 1975. Manufacturing cycle time determination for a multi-stage economic production quantity model. *Management Science* 22(3): 298–308.

Szendrovits, A.Z., 1987. An inventory model for interrupted multi-stage production. *International Journal of Production Research* 25(1): 129–143.

Tersine, R.J., 1998. *Principles of inventory and materials management,* 4th ed., Upper Saddle River: Prentice Hall

Wright, T.P., 1936. Factors affecting the cost of airplanes. *Journal of the Aeronautical Sciences* 3(4): 122–128.

Yelle, L.E., 1979. The learning curve: Historical review and comprehensive survey. *Decision Sciences* 10(2): 302–328.

16 Steady-State Characteristics under Processing-Time Learning and Forgetting

Sunantha Teyarachakul

CONTENTS

INTRODUCTION AND MAJOR ASSUMPTIONS

This chapter covers the long-term characteristics of batch production times in the repetitive work environment over an infinite horizon in which the demand rate is constant and the production of a constant batch size of q units takes place at regular intervals (such as once every Monday). The model that we are discussing differs from the traditional inventory models [i.e., economic order quantity (EOQ), economic manufacturing quantity (EMQ)] in that the production rate is no longer a constant; instead, it is influenced by workers learning while processing units and their forgetting during the break between two successive batches. When returning to produce a batch of the same product, the worker is then relearning.

Early work in this area (Sule 1978; Axsäter and Elmaghraby 1981; and Elmaghraby 1990) has reported some common observations; as the number of lots produced increases, the batch production time converges to a unique value.

Sule (1978, 338) explains it well: "In the steady state condition [...] the drop in productivity due to forgetting would be equal to the increase in productivity due to learning during the manufacture of Q units." This long-term characteristic implies that an operator starts every new batch with the same experience (skill) level due to the forgetting effect canceling out the learning effect within a batch. Teyarachakul et al. (2008) named this characteristic as the *single-point convergence* of the batch production time.

More recent work (Teyarachakul 2003; Teyarachakul et al. 2008) found that the single-point convergence is not the sole possibility. Instead, in the long term, all of the odd batch numbers require the same longer production time than the even batch numbers do, or vice versa. Specifically in the steady state, every other batch starts with the same high experience level, while the other batch begins with the same low experience level; therefore, the amount of time to complete each batch alternates between two different values. This characteristic is referred to as the *alternating convergence* in batch production times by Teyarachakul et al. (2008).

This chapter is devoted to the study of the long-term characteristics that were found to exist in power forgetting functions with a variable y-intercept and a fixed slope (Sule 1978; Axsäter and Elmaghraby 1981), those with a fixed y-intercept and a fixed slope (Elmaghraby 1990), and also in an exponential forgetting curve (Globerson and Levin 1987; Teyarachakul 2003). Later in the chapter, the generalization of the long-term characteristics is presented. The results—which are not specific to particular forgetting or learning functions—are based on a list of sensible characteristics of learning and forgetting.

To summarize, the major common assumptions being made by scholars who analyze similar problems in this area are as follows: (1) infinite horizon, (2) constant demand rate of d, (3) constant lot size of q, and (4) the use of an original learning curve as a relearning curve.

Additionally, many papers in this area have implicitly assumed the equal-spaced production cycles with the length $T = q/d$ time periods. Each cycle starts when a batch production begins and ends when the stored inventory is depleted (zero inventory policy).

The learning function in Globerson and Levin's (1987) convergence study is Wright's (1936) learning curve, whereas the learning curve in Sule's (1978) and Elmaghraby's (1990) studies has a slight modification to the unit of productivity measurement. They used the production rate (unit/time) as a function of continuous production time, while Globerson and Levin (1987) used the unit production time (time/unit) as a function of the cumulative number of units of uninterrupted production.

We next present a brief review for one of the most widely used learning curves—Wright's learning curve (1936). Its most important feature is that as the number of units produced doubles, the unit production time declines by a constant percentage, say $(1-\delta)100$ percentage. In sections "Steady-State Characteristics: The Case of Power Forgetting Functions" and "Steady-State Characteristics: The Case of an Exponential Forgetting Function," it will be used to capture workers' learning or the experience gained in the convergence analysis under the cases of function-specific forgetting curves. Specifications of the continuous version of Wright's learning curve (Teyarachakul et al. 2008) are the following:

$$T(x) = T_0 x^{-m},$$

where $T(x)$ is the instantaneous per-unit production time at the instant when the xth unit starts, $x \geq 1$; T_0 is $T(1)$, the initial instantaneous per-unit production time by an operator who has no prior experience or learning; m is $(-\log(\delta)/\log(2))$ and $0 < m < 1$; and $\delta = T(2x)/T(x)$.

Note that the function $T(x)$ is continuous, thus x is allowed to take non-integer values. The interpretation for $T(x)$ was given by Teyarachakul et al. (2008, 589) that "if the work were to continue at the rate it was at the start of the xth unit, then it would take $T(x)$ amount of time to complete one unit." The $T(x)$ could be easily converted into the production rate $R(x)$ by letting $R(x) = 1/T(x)$. To simplify our presentation we use the unit production time $T(x)$, which corresponds to an experience level x, in the place of the instantaneous per-unit production time at the instant when the xth unit starts.

As illustrated in some earlier chapters, a learn-forget cycle occurs in a production cycle, within which there are: (1) a batch production interval with the length of $p(a) = \int_a^{a+q} T_0 x^{-m} dx$, where a here represents the experience level when the batch starts, measured as the cumulative number of units produced in a learning curve; and (2) an interruption with the duration of $I(a) = T - p(a)$ time periods. Consider a cycle n with the cycle initial per-unit time of $T(\alpha_n)$, the batch production time of $p(\alpha_n)$, and an interruption interval of $I(\alpha_n)$. The worker will then start the next batch with an experience level α_{n+1}, or the unit production time $T(\alpha_{n+1})$, on the *original learning curve*.

Note that we assume that $\alpha_n \geq 1 \ \forall \ n$, and the experience level of 1 is defined as the initial per-unit production time by an operator who has no prior learning, which is the minimal possible experience level. Therefore, after a worker loses all of their learning, their experience level will become equivalent to 1. We further assume that the experience level that is remaining at the start of the next batch cannot be infinite. Specifically, $\lim_{\alpha_n \to \infty} \alpha_{n+1}(\alpha_n) < \infty$, and similarly, $\lim_{\alpha_n \to \infty} \alpha_{n+2}(\alpha_n) < \infty$. Teyarachakul et al. (2009, 7) address the interpretation of this assumption: "Intuitively, after the break and losing part of his/her skill level due to forgetting, an operator cannot start the production with the perfect skill level (experience level) of infinity."

For clarity and consistency of the models presented in this chapter, the forgetting rate and learning rate are related to the performance time of a unit, rather than the number of units per time, for reasons similar to those addressed by Globerson and Levin's (1987, 89), "an interruption shows itself in performance time and not in number of units. [...] performance time can be used as a 'memory' for previous interruptions."

To proceed in our analysis of the long-term characteristics under both cases of function-specific and of generalization, notations here are used for modeling workers' forgetting:

$T(S)$ = the production time for the next unit if there were no forgetting during
   the interruption;

I = the interruption duration; and

$F(T(S), I)$ = the production time of the first unit after an interruption.

The main purpose of many forgetting functions is to provide an expression or the computation of the time per-unit produced when the next cycle resumes. The forgetting function $F(T(S), I)$ captures the feature that the subsequence drop in productivity at the beginning of the next cycle depends on the interruption duration I periods and the performance level when the interruption occurs, $T(S)$.

This chapter is organized as follows. The next two sections provide the analysis of steady-state characteristics for the cases of power forgetting functions ("Steady-State Characteristics: The Case of Power Forgetting Functions" section) and exponential forgetting functions ("Steady-State Characteristics: The Case of an Exponential Forgetting Function" section). Section "Steady-State Characteristics: The Generalization Case" examines the convergence properties of the experience level under the generalized case or the non-function-specific case. Section "A Discussion on Form of the Optimal Policy" discusses form of the optimal policy (FOOP) and considers alternatives to the best equal-spaced and constant lot size policy. Section "Concluding Remarks" closes the paper by providing concluding remarks.

STEADY-STATE CHARACTERISTICS: THE CASE
OF POWER FORGETTING FUNCTIONS

Two power forgetting functions (Sule 1978; Axsäter and Elmaghraby 1981; Elmaghraby 1990) were reported to provide the convergence in the batch production time under certain conditions.

POWER FUNCTION WITH A VARIABLE Y-INTERCEPT AND A FIXED SLOPE

A good representative of the power function with a variable y-intercept and a fixed slope is the variable regression to variable forgetting (VRVF) function, introduced by Carlson and Rowe (1976) and used in the analysis of Sule's convergence results (1978).

$$\hat{T}(t) = F_0 t^c,$$

where
   $\hat{T}(t)$ = the unit production time at time t of lost experience on the forgetting curve
   F_0 = the equivalent time for the first unit on the forgetting curve, which varies
      from cycle to cycle
   c = the known fixed slope of the forgetting curve, where c could be defined as
      $(-\log(\beta)/\log(2))$, β is $\hat{T}(t)/\hat{T}(2t)$, and $0 < c < 1$

Via the use of this forgetting curve, the computation procedures for $F(T(S), I(\alpha_n))$ and α_{n+1} given a learn-forget cycle n with an experience level α_n at the start of the cycle, can be described as follows.

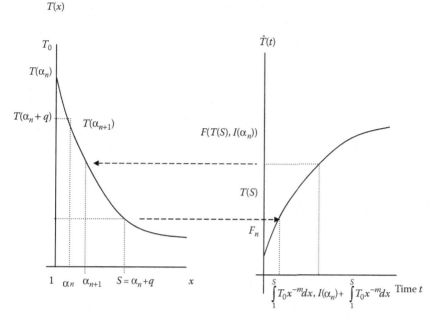

Learning effect within batch n, $T(\alpha_n+q) < T(\alpha_n)$, and $T(\alpha_{n+1})$, the starting of relearning in batch $n+1$.

Forgetting effect in batch n with $F(T(S), I(\alpha_n)) > T(S = \alpha_n+q)$

FIGURE 16.1 A learn-forget cycle n using Wright's learning curve and VRIF forgetting function.

1. Find an experience level at the start of the interruption, $S = \alpha_n + q$, uses the learning curve to compute the cycle n production time $p(\alpha_n) = \int_{\alpha_n}^{\alpha_n + q} T_0 x^{-m} dx = T_0/1 - m((\alpha_n + q)^{1-m} - \alpha_n^{1-m})$, and calculate the cycle n interruption duration $I(\alpha_n) = T - p(\alpha_n)$.
2. As illustrated in Figure 16.1, recompute F_0 for cycle n, called as F_n, by letting $T(S) = \hat{T}\left(\int_1^S T_0 x^{-m} dx\right)$. Thus, $F_n = T_0^{1-c} S^{-m}(1 - m/S^{1-m} - 1)^c$.
3. Find $F(T(S), I(\alpha_n))$, the next unit production time after an interruption of $I(\alpha_n)$ time periods, and its corresponding experience level, α_{n+1}, by equating $F(\cdot)$ with $\hat{T}\left(\int_1^S T_0 x^{-m} dx + I(\alpha_n)\right)$ and $T(\alpha_{n+1})$ with $F(\cdot)$.

Therefore, $F(T(S), I(\alpha_n)) = F_n(I + T_0 \cdot S^{1-m}/1 - m - T_0/1 - m)^c$, and $\alpha_{n+1} = (F(T(S), I)/T_0)^{(-1/m)}$

Example 16.1

Suppose the values of the problem parameters are: the learning rate δ of 85%; the forgetting rate β of 80%; the initial unit production time T_0 of 0.09 day; the batch size of 14 units; and the demand rate of 11 units per day. The table below shows the convergence behavior of the system.

Cycle n	Starting Experience Level α_n	End-of-Production Phase Experience Level $S = \alpha_n + q$	Cycle Production Time $p(\alpha_n)$	Cycle Interruption Duration $I(\alpha_n) = (q/d) - p(\alpha_n)$	y-Intercept on the Forgetting Curve F_n	First-Unit Production Time for the Next Cycle $F(T(S), I(\alpha_n))$
1	1.00	15.00	0.8253	0.4475	0.0513	0.0554
2	8.28	22.28	0.6790	0.5937	0.0419	0.0502
3	12.62	26.62	0.6374	0.6353	0.0383	0.0477
4	15.66	29.66	0.6153	0.6575	0.0363	0.0462
5	17.86	31.86	0.6015	0.6712	0.0350	0.0453
:	:	:	:	:	:	:
29	24.87	38.87	0.5666	0.7061	0.0317	0.0428
30	24.88	38.88	0.5666	0.7062	0.0317	0.0428
31	24.88	38.88	0.5666	0.7062	0.0317	0.0428

As shown in the table above, the power forgetting function variable regression to invariant forgetting (VRIF) allows the convergence of batch production time to a unique value (the case of single-point convergence). This result is justified by the mathematical proof, originated by Axsäter and Elmaghraby (1981), with our modifications, to fit it with the chapter content.

Proof of the convergence results

To claim the convergence result of Sule (1978) to exist, we need the assumption that

$$\frac{T_0}{(1-m)c}(\alpha_n + q - 1) > \frac{q}{d}\alpha_n^m + \frac{T_0}{(1-m)}\left(\alpha_n - \alpha_n^m\right) \quad \forall \alpha_n.$$

We will prove the result by showing that as $n \to \infty$, $\alpha_n \to \alpha$, where α, a unique positive value, represents an experience level in the steady state. First, we combine all the equations mentioned above in the procedure into one function, α_{n+1}. Then, we show some properties of the function necessary to obtain the convergence result of α_n to a single value, α. The function α_{n+1} in α_n can be illustrated as follows. Let A be $T_0/1-m$ to simplify our presentation.

$$\alpha_{n+1} = (\alpha_n + q)\left\{\left[\frac{1}{(\alpha_n+q)^{1-m}-1}\right]\bullet\left[\frac{1}{A}\right]\right\}^{\frac{-c}{m}}\left\{\frac{q}{d}+A\left(\alpha_n^{1-m}-1\right)\right\}^{\frac{-1}{m}}.$$

If $\alpha_n < (>) \alpha_{n+1}$, then $A^c\left(1+\frac{q}{\alpha_n}\right)^m > (<)\left[\frac{1}{(\alpha_n+q)^{1-m}-1}\right]^c\left\{\frac{q}{d}+A\alpha_n^{1-m}-A\right\}.$

We observe that $A^c(1 + q/\alpha_n)^m$, referred to as the left-hand side (LHS), decreases in α_n and $\lim_{\alpha_n \to \infty} A^c(1 + q/\alpha_n)^m = A^c$.

$$\frac{\partial \left[\dfrac{1}{(\alpha_n + q)^{1-m} - 1} \right]^c \left\{ \dfrac{q}{d} + A\alpha_n^{1-m} - A \right\}}{\partial \alpha_n} = \left((\alpha_n + q)^{1-m} - 1 \right)^c (1 - m)$$

$$\left\{ A\alpha_n^{-m} - \frac{c\left(\dfrac{q}{d} + A(\alpha_n^{1-m} - 1) \right)}{\alpha_n + q - 1} \right\}.$$

The condition of $A/c\,(\alpha_n + q - 1) > (q/d)\alpha_n^m + A(\alpha_n - \alpha_n^m)\ \forall \alpha_n$ ensures the positive derivative. Thus, $[1/(\alpha_n + q)^{1-m} - 1]^c \{q/d + A\alpha_n^{1-m} - A\}$, referred to as the right-hand side (RHS), increases in α_n with the limit of ∞ as $\alpha_n \to \infty$. At $\alpha_n = \alpha_{n+1}$, the LHS and RHS of the inequality are equal or intercepted. Based on the characteristics of the LHS and RHS, there is, at most, one interception point or equivalently, at most, one value of α_n (say α) such that $\alpha_n = \alpha_{n+1}\ (= \alpha)$ for $\alpha_n \geq 1$. We next show that such an interception point ($\alpha \geq 1$) exists by contradiction.

Suppose the interception is at some point of α less than one. Then, for some $\alpha_n \geq 1$, $\alpha_{n+1} < \alpha_n$ and therefore the sequence of $\{\alpha_n\}$ is decreasing with no lower bound in the range of $\alpha_n \geq 1$. Thus, for some n such that $\alpha_n = 1$, then $\alpha_{n+1} < 1$. This contradicts our assumption that $\alpha_n \geq 1$. We can conclude that α exists and that it is unique with a value no less than 1.

Note that all examples provided by Sule (1978) and Axsäter and Elmaghraby (1981) show the case of the sequence $\{\alpha_n\}$ either increasing or decreasing as n increases. We therefore do not consider the case of non-monotonic sequence $\{\alpha_n\}$. Our proof here is aimed towards establishing the point in regard to Sule's assumption of the single-point convergence for the given sequences.

The sequence $\{\alpha_n\}$ is monotonic and infinite with a limit at α (a finite value). Thus, it must converge to the single point α. This completes our proof for Sule's report on the convergence results.

FUNCTION WITH A FIXED *Y*-INTERCEPT AND A FIXED SLOPE

The function that we will consider next has been known as variable regression to invariant forgetting (VRIF). It has the same functional form as VRVF except that VRIF's *y*-intercept of the function is fixed. This feature results in a unique forgetting function for all production cycles. It is justified by Elmaghraby's (1990) claim that F_0 (intercept) and c (slope) of the forgetting curve are system-dependent, similar to T_0 (intercept) and m (slope) of the learning curve. The value of F_0 is assumed to be determined from the first cycle in the same way as VRVF is. The constant

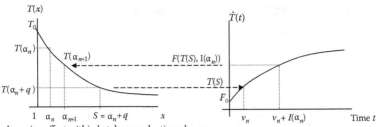

Learning effect within batch-n production phrase, $T(\alpha_n + q) < T(\alpha_n)$, and the experience level α_{n+1} to begin batch $n+1$.

Forgetting effect in Batch n, $F(T(S), I(\alpha_n)) > T(S = \alpha_n + q)$

FIGURE 16.2 A learn-forget cycle n using Wright's learning curve and VRIF function.

y-intercept F_0 gives rise to have the variable v map the performance level at the end-of-production phase onto the forgetting curve, in the similar way that α_{n+1} relates the productivity at the end-of-interruption phase in cycle n onto the learning curve. See Figure 16.2 for an illustration.

Thus, by applying VRIF to the learn-forget cycle n with an experience level α_n to start the cycle, the computation steps for the performance time of the next cycle are illustrated as follows:

1. Find F_0 from cycle 1, in the same way that we do for the case of the VRVF model by allowing $T(1+q) = T(\int_1^{1+q} T_0 x^{-m} dx)$, and $v_1 = \int_1^{1+q} T_0 x^{-m} dx$.

   Hence, $F_0 = T_0^{1-c}(1+q)^{-m}(1 - m/(1+q)^{1-m} - 1)^c$.

2. For each cycle (such as cycle n), compute S and $I(\alpha_n)$ using the same approach as in the case of VRVF.
3. Map the unit time at the end of cycle n production interval to time per-unit of the forgetting curve by letting $T(S) = \hat{T}(v_n)$ to determine the values of v_n for $n = 2, 3, \dots$. $v_n = (T_0/F_0)^{1/c}(\alpha_n + q)^{-m/c}$.
4. Find the performance unit time at the end-of-interruption phase, $F(T(S), I(\alpha_n))$, and the corresponding experience level α_{n+1} when the next batch resumes.

$$F(T(S), I(\alpha_n)) = F_0(v_n + I(\alpha_n))^c; \quad \alpha_{n+1} = \left(\frac{F(T(S), I(\alpha_n))}{T_0}\right)^{\frac{-1}{m}}.$$

Example 16.2

Suppose the values of the problem parameters are: the learning rate δ of 85%; the forgetting rate β of 80%; the initial unit production time T_0 of 0.09

day; a fixed batch size of 14 units; and a constant demand rate of 11 units per day. The table below shows the convergence behavior of the system with $F_0 = 0.0513$.

Cycle n	Starting Experience Level α_n	End-of-Production-Phase Experience Level $S = \alpha_n + q$	Cycle Production Time $p(\alpha_n)$	Cycle Interruption Duration $I(\alpha_n) = (q/d) - p(\alpha_n)$	End-of-Production-Phase Performance Level on the Forgetting Curve V_n	First-Unit Production Time for the Next Cycle $F(T(S), I(\alpha_n))$
1	1.00	15.00	0.8252	0.4475	0.8253	0.0554
2	8.28	22.28	0.6790	0.5937	0.6188	0.0545
3	8.84	22.84	0.6726	0.6001	0.6075	0.0545
4	8.89	22.89	0.6721	0.6006	0.6065	0.0545
5	8.90	22.90	0.6720	0.6007	0.6064	0.0545
6	8.90	22.90	0.6720	0.6007	0.6064	0.0545
7	8.90	22.90	0.6720	0.6007	0.6064	0.0545
8	8.90	22.90	0.6720	0.6007	0.6064	0.0545

Proof of the convergence results

Recall the following expressions from the steps mentioned above:

$$F_0 = T_0^{1-c}(1+q)^{-m}\left(\frac{1-m}{(1+q)^{1-m}-1}\right)^c ; v_n = \left(\frac{T_0}{F_0}\right)^{1/c}(\alpha_n + q)^{-m/c};$$

$$F\left(T(S), I(\alpha_n)\right) = F_0(v_n + I(\alpha_n))^c ; \alpha_{n+1} = \left(\frac{F(T(S), I)}{T_0}\right)^{-1/m}.$$

All of these equations are combined to form a single equation α_{n+1} as a function of α_n, which relates learning and forgetting within the cycle n to the next cycle, cycle $n + 1$.

$$\alpha_{n+1}\left[\frac{T_0}{1-m}\left((1+q)^{1-m}-1\right)\right]^{c/m}(1+q)$$

$$\left[\frac{T_0}{1-m}\left((1+q)^{1-m}-1\right)(1+q)^{m/c}(\alpha_n + q)^{-m/c} + \frac{q}{d} - \frac{T_0}{1-m}\left((\alpha_n + q)^{1-m} - \alpha_n^{1-m}\right)\right]^{-c/m}.$$

We would like to prove the convergence property; that is, as $n \to \infty$, $\alpha_n \to \alpha$. The sequence α_n is recursive and thus, if $\alpha_{n+1} = \alpha_n$, such α_{n+1} is α. Otherwise $\alpha_n < (>) \alpha_{n+1}$, then

$$\left(\frac{\alpha_n}{1+q}\right)^{-m/c}\left[\frac{T_0}{1-m}\left((1+q)^{1-m}-1\right)\right] >$$

$$(<)\frac{T_0}{1-m}\left[\left((1+q)^{1-m}-1\right)\left(\frac{\alpha_n+q}{1+q}\right)^{-m/c}+\left(\alpha_n^{1-m}-(\alpha_n+q)^{1-m}\right)\right]+\frac{q}{d},$$

which is equivalent to

$$\left[\frac{T_0}{1-m}\left((1+q)^{1-m}-1\right)\right](1+q)^{m/c}\left[\alpha_n^{-m/c}-(\alpha_n+q)^{-m/c}\right] >$$

$$(<)\frac{q}{d}+\frac{T_0}{1-m}\left(\alpha_n^{1-m}-(\alpha_n+q)^{1-m}\right)$$

Let us call the left-hand side of the above in equality as LHS and similarly, the right-hand side of it as RHS.

Since $d(\alpha_n^{-m/c}-(\alpha_n+q)^{-m/c})/d\alpha_n = (-m/c)(\alpha_n^{-m/c-1}-(\alpha_n+q)^{-m/c-1}) < 0$, the LHS is decreasing in α_n. On the other side, $d(\alpha_n^{1-m}-(\alpha_n+q)^{1-m})/d\alpha_n = (1-m)(\alpha_n^{-m}-(\alpha_n+q)^{-m}) > 0$. The RHS is increasing in α_n. We can conclude, here, that there is, at most, one interception between the LHS and RHS; the corresponding value of α_n at the point of interception has the property that $\alpha_n = \alpha_{n+1} = \alpha$. Additionally, for $\alpha_n < \alpha$, $\alpha_n < \alpha_{n+1}$; for $\alpha_n > \alpha$, $\alpha_n > \alpha_{n+1}$. Next we will show that such an interception must exist.

Recall an assumption of $\alpha_{n+1} \geq 1$ for all n thus, at $\alpha_n = 1$, α_{n+1} remains no less than 1. So, LHS($\alpha_n = 1$) $\geq$ RHS($\alpha_n = 1$). As $\alpha_n \to \infty$, LHS $\to 0$ and RHS $\to q/d$ and LHS < RHS. Hence, there is a unique α, where $1 < \alpha < \infty$ such that LHS(α) = RHS(α).

Similar to our proof of Sule's convergence result, we intend to establish the point in regard to Elmaghraby's report of the single-point convergence for the given sequences, which do not include the case of $\alpha_n < \alpha_{n+1}$ and $\alpha_{n+1} > \alpha_{n+2}$ for any n. The sequence $\{\alpha_n\}$ is monotonic and infinite with a limit at α. Thus, the sequence must converge to the α.

STEADY-STATE CHARACTERISTICS: THE CASE OF AN EXPONENTIAL FORGETTING FUNCTION

The exponential forgetting function we are about to discuss next was originated by Globerson and Levin (1987) with the characteristics of:

1. $F(T(S), 0) = T(S)$: the performance time of the next unit is $T(S)$ if there was no interruption and therefore no forgetting.
2. $\lim_{I \to \infty} F(T(S), I) = T_0$: all previous learning could be forgotten if the interruption duration is sufficiently long.

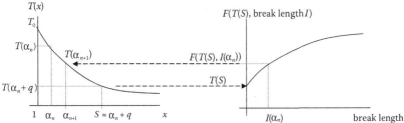

Learning effect within batch-n production phrase, $T(\alpha_n+q) < T(\alpha_n)$, and the experience level α_{n+1} to begin batch $n+1$ ($T(\alpha_{n+1})$). Forgetting effect in Batch n, $F(T(S),I(\alpha_n)) > T(S = \alpha_n+q)$

FIGURE 16.3   A learn-forget cycle n using Wright's learning curve and an exponent.

3. $\partial F(T(S),I)/\partial I \geq 0$: the performance time of the next unit after the break is monotonously non-decreasing in interruption duration.
4. $\partial F(T(S),I)/\partial T(S) \geq 0$: given the fixed break length, the performance time of the next unit after the break is non-decreasing with the unit time prior to the break.
5. $\lim_{I \to \infty} \partial F(T(S),I)/\partial I = 0$: if the interruption duration is very long, there is no more to forget since everything learned prior to the break has been forgotten.

The specifications of the exponential forgetting function can be described as $F(T(S), I) = T_0 - (T_0 - T(S))e^{-bI}$,

Where b is the forgetting parameter, and $F(T(S), I)$ increases with b and as $b \to \infty$, $F(T(S), I) \to T_0$ or the entire amount learned was lost. The function illustrates a rapid initial decrease in productivity, followed by a gradual leveling off.

The computation procedures are explained below when using the exponential forgetting function together with the continuous version of Wright's learning curve in the production cycle n (Figure 16.3).

a. Compute S and I in the same way as in the application of VRVF, given the cycle n start with an experience level of α_n.
b. Then calculate $F(T(S), I(\alpha_n))$ using the equation above and its corresponding experience level, α_{n+1}: $F(T(S), I(\alpha_n)) = T_0 - T_0(1 - S^{-m})e^{-bI(\alpha_n)}; \alpha_{n+1} = (T_0/1 - (1 - S^{-m})e^{-bI(\alpha_n)})^{1/m}$.

Recently, Teyarachakul et al. (2008) specified two possible types of convergences of the sequence $\{\alpha_n\}$; namely, the single-point convergence and the alternating convergence.

1. Single-point convergence: $\{\alpha_n\} \to \alpha$ as $n \to \infty$.
   In the steady state, every batch starts the production with the same level of learning of α. See Figure 16.4.
2. Alternating convergence

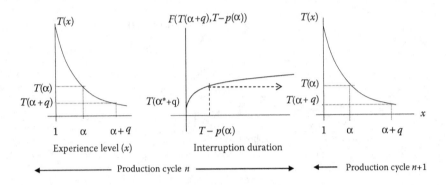

FIGURE 16.4 Learn-forget-relearn cycles with the steady state experience level α.

In the steady state, every other batch starts with an experience level α^-, while the other batch begins with an experience level α^+, where $\alpha^- < \alpha < \alpha^+$. The characteristic of α^- and α^+ will be discussed later. The sequence $\{\alpha_n\}$ is decomposed into two subsequences: (1) $\{\alpha_{2n-1}\} = \{\alpha_1, \alpha_3, \alpha_5, \ldots\}$; and (2) $\{\alpha_{2n}\} = \{\alpha_2, \alpha_4, \alpha_6, \ldots\}$. One of the subsequence convergence to α^- and the other convergences to α^+. Thus, as $n \to \infty$, $\{\alpha_n\} = \{\alpha_1, \alpha_2, \alpha_3, \ldots, \alpha^-, \alpha^+, \alpha^-, \alpha^+, \alpha^-, \alpha^+, \ldots\}$. See Figure 16.5 for an illustration.

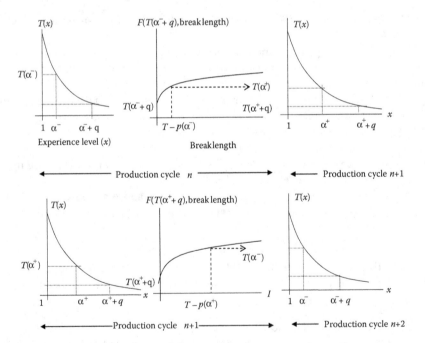

FIGURE 16.5 The steady state production cycles with experience levels α^- and α^+, alternating from one cycle to the next.

Example 16.3: Single-Point Convergence

The problem parameters are supposed to be: the learning rate δ of 75%; the forgetting parameter b of 0.5; the initial unit production time T_0 of 0.20 day; the batch size of 20 units; and the demand rate of 10 units per day. The table below shows the behaviors of the single-point convergence in the example.

Cycle n	Starting Experience Level α_n	End-of-Production-Phase Experience Level $S = \alpha_n + q$	Cycle Production Time $p(\alpha_n)$	Cycle Interruption Duration $I(\alpha_n) = (q/d) - p(\alpha_n)$	First-Unit Production Time for the Next Cycle $F(T(S), I(\alpha_n))$
1	1.000	21.000	1.687	0.313	0.077
2	9.884	29.884	1.189	0.811	0.099
3	5.411	25.411	1.351	0.649	0.093
4	6.297	26.297	1.312	0.688	0.095
5	6.051	26.051	1.322	0.678	0.094
6	6.114	26.114	1.319	0.681	0.094
7	6.098	26.098	1.320	0.680	0.094
8	6.102	26.102	1.320	0.680	0.094
9	6.101	26.101	1.320	0.680	0.094
10	6.101	26.101	1.320	0.680	0.094

Example 16.4: Alternating Convergence

Let assume the following problem parameters: the learning rate δ of 75%; the forgetting parameter b of 0.5; the initial unit production time T_0 of 1.50 day; the batch size of 20 units; and the demand rate of 1.575 units per day. The table below shows the behaviors of the alternating convergence of the problem.

Cycle n	Starting Experience Level α_n	End-of-Production-Phase Experience Level $S = \alpha_n + q$	Cycle Production Time $p(\alpha_n)$	Cycle Interruption Duration $I(\alpha_n) = (q/d) - p(\alpha_n)$	First-Unit Production Time for the Next Cycle $F(T(S), I(\alpha_n))$
1	1.000	21.000	12.656	0.044	0.448
2	18.430	38.430	7.573	5.127	1.410
3	1.161	21.161	12.490	0.210	0.530
4	12.257	32.257	8.456	4.244	1.363
5	1.260	21.260	12.394	0.306	0.575
6	10.095	30.095	8.870	3.830	1.333

(continued)

Cycle n	Starting Experience Level α_n	End-of-Production-Phase Experience Level $S = \alpha_n + q$	Cycle Production Time $p(\alpha_n)$	Cycle Interruption Duration $I(\alpha_n) = (q/d) - p(\alpha_n)$	First-Unit Production Time for the Next Cycle $F(T(S), I(\alpha_n))$
7	1.329	21.329	12.330	0.370	0.603
8	8.973	28.973	9.117	3.583	1.312
:	:	:	:	:	:
156	5.995	25.995	9.933	2.767	1.221
157	1.641	21.641	12.064	0.636	0.713
158	5.995	25.995	9.933	2.767	1.221
159	1.641	21.641	12.064	0.636	0.713
:	:	:	:	:	:
1002	5.995	25.995	9.933	2.767	1.221
1003	1.641	21.641	12.064	0.636	0.713

Proof of the convergence results

The proof is separated into two parts (results), according to two different categories of convergence, each of which occurs under a set of specific conditions. The
approaches that were similar to the earlier proofs are not applicable to the case of
the exponential forgetting function we are now considering. The rigorous but complicated proofs of both types of convergence have been suggested by Teyarachakul
et al. (2008); we are here proposing a simplified version of them. The following
general observations and claims will be used in the main proof of the convergence
results.

Observation 1. $F(T(S), I(\alpha_n)) = T(\alpha_n) - \ell(\alpha_n) + f(\alpha_n) = T(\alpha_{n+1})$,

where $\ell(\alpha_n) = T_0(\alpha_n^{-m} - (\alpha_n + q)^{-m})$; $f(\alpha_n) = (T_0 - T(\alpha_n + q))(1 - e^{-bI})$.

We can view that $\ell(\alpha_n)$ and $f(\alpha_n)$ are, within the batch n, the amount of learning
in the production phase and the amount forgetting in the break phase, respectively.

Observation 2. For any given positive value of q, $\ell(\alpha_n)$ is strictly decreasing in α_n
whereas $f(\alpha_n)$ is strictly increasing in α_n.

Observation 3. By combining all of the expressions in the computation procedures which have been recently mentioned, the function α_{n+1} in
α_n can be written as:

$$\alpha_{n+1}(\alpha_n) = \left(1 - \left(1 - (\alpha_n + q)^{-m}\right)e^{-b\left(\frac{q}{d} - \frac{T_0}{1-m}\left((\alpha_n + q)^{1-m} - \alpha_n^{1-m}\right)\right)} \right)^{-1/m}.$$

To proceed in our analysis, we defined $\alpha_{n+2}(\alpha_n)$ as $\alpha_{n+1}(\alpha_{n+1}(\alpha_n))$.

Observation 4. As $\alpha_n \rightarrow \infty$, $\alpha_{n+1}(\alpha_n) \rightarrow c_1$ and $\alpha_{n+2}(\alpha_n) \rightarrow c_2$, where both c_1 and $c_2 < \infty$. This observation can be easily verified mathematically.

Claim 16.1

There is one and only one value of α_n, called as α, such that $\alpha_{n+1}(\alpha_n) = \alpha_n$.

Proof. Recall that $T(\alpha_n) - \ell(\alpha_n) + f(\alpha_n) = T(\alpha_{n+1})$ and if $\alpha_n = \alpha$, $\alpha_n = \alpha_{n+1}$. Therefore, when $\alpha_n = \alpha$, $\ell(\alpha_n) = f(\alpha_n)$. Because $\ell(\alpha_n)$ is strictly decreasing and $f(\alpha_n)$ is strictly increasing with α_n; there is, at most, one value of α_n (i.e., α) such that $\ell(\alpha_n) = f(\alpha_n)$. The next step is to show the existence of α.

Recall that $\alpha_n \geq 1$. For $\alpha_n = 1$, $\alpha_{n+1} \geq 1$ and $\ell(\alpha_n = 1) \geq f(\alpha_n = 1)$. If $\ell(\alpha_n = 1) = f(\alpha_n = 1)$, then α is 1; otherwise $\ell(\alpha_n = 1) > f(\alpha_n = 1)$. As $\alpha_n \rightarrow \infty$, $\ell(\alpha_n) \rightarrow 0$ and $f(\alpha_n) \rightarrow T_0(1 - e^{-bq/d})$. Both functions are continuous and strictly monotone. Thus, there must exist α, where $1 \leq \alpha \leq \infty$, such that $\ell(\alpha) = f(\alpha)$. We can conclude that α exists, and that it is unique.

Note that the property of α implies that α is a fixed point of α_{n+1} (i.e., $\alpha_{n+1} = \alpha_n = \alpha$). Similarly, if $\hat{\alpha}$ is a fixed point of $\alpha_{n+2}(\alpha_n)$, then $\alpha_{n+2}(\alpha_n) = \alpha_n = \hat{\alpha}$. □

Claim 16.2

$\alpha_{n+1} > \alpha_n$ iff $\alpha_n < \alpha$; and $\alpha_{n+1} < \alpha_n$ iff $\alpha_n > \alpha$.

Proof. $\alpha_{n+1} > \alpha_n$ is equivalent to $T(\alpha_{n+1}) < T(\alpha_n)$, where the learning function $T(x)$ is continuous and strictly decreasing with x.

Recall that $T(\alpha_{n+1}) = T(\alpha_n + q) + (1 - e^{-bl})(T(1) - T(\alpha_n + q))$. Thus, $T(\alpha_{n+1}) < T(\alpha_n)$ gives $T(\alpha_n + q) + (1 - e^{-bl})(T(1) - T(\alpha_n + q)) < T(\alpha_n)$, which can be rewritten as $(1 - e^{-bl})(T(1) - T(\alpha_n + q)) < T(\alpha_n) - T(\alpha_n + q)$.

Notice that the LHS of the above inequality is $f(\alpha_n)$, and its RHS is $\ell(\alpha_n)$. Thus, $f(\alpha_n) < \ell(\alpha_n)$; this occurs only when $\alpha_n < \alpha$ by Observation 2 and the fact that $f(\alpha) = \ell(\alpha)$. Similarly, $\alpha_{n+1} < \alpha_n$ iff $\alpha_n > \alpha$. □

Claim 16.3

If $\alpha_{n+2}(\alpha_n)$ has exactly three fixed points (say, α^-, α and α^+, where $\alpha^- < \alpha < \alpha^+$) with $d\alpha_{n+2}(\alpha_n)/d(\alpha_n)|_{\alpha_n = \alpha} > 1$, then (1) $\alpha_{n+2} > \alpha_n$ for $\alpha_n < \alpha^-$ and for $\alpha < \alpha_n < \alpha^+$, and (2) $\alpha_{n+2} < \alpha_n$ for $\alpha^- < \alpha_n < \alpha$ and for $\alpha^+ < \alpha_n$.

Proof. The properties of three fixed points of $\alpha_{n+2}(\alpha_n)$, together with $|d\alpha_{n+2}(\alpha_n)/d(\alpha_n)|_{\alpha_n = \alpha} > 1$, have formed three different switching points, where $\alpha_{n+2}(\alpha_n) > (<) \alpha_n$ changes to $\alpha_{n+2}(\alpha_n) < (>) \alpha_n$. At the lowest possible value of $\alpha_n = 1$, $\alpha_{n+2}(\alpha_n) \geq \alpha_n$, and similarly at the probably highest value of $\alpha_n = \infty$, $\alpha_{n+2}(\alpha_n) \leq \alpha_n$ (see

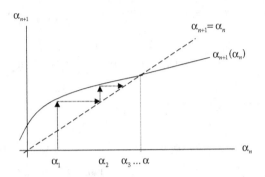

FIGURE 16.6 $\{\alpha_n\} \to \alpha$ when $\partial\alpha_{n+1}/\partial\alpha_n \geq 0 \ \forall \ \alpha_n$ and $\alpha_1 < \alpha$.

Observation 4). Note that $\alpha_{n+2}(\alpha_n)$ is continuous. Thus, it must be the case that $\alpha_{n+2}(\alpha_n) > \alpha_n$ for $\alpha_n < \alpha^-$ and for $\alpha < \alpha_n < \alpha^+$, and $\alpha_{n+2}(\alpha_n) < \alpha_n$ for $\alpha^- < \alpha_n < \alpha$ and for $\alpha^+ < \alpha_n$. □

Result 16.1 (Type 1 Convergence): As $n \to \infty, \{\alpha_n\} \to \alpha$

16.1   If $d\alpha_{n+1}(\alpha_n)/d\alpha_n \geq 0 \quad \forall \alpha_n$, or

16.2   If $d\alpha_{n+1}(\alpha_n)/d\alpha_n < 0 \quad \forall \alpha_n$ and if $\alpha_{n+2}(\alpha_n)$ has one and only one fixed point, α.

Proof. We will start with the proof for Result 16.1. $d\alpha_{n+1}(\alpha_n)/d\alpha_n \geq 0 \ \forall \alpha_n$ implies that: for $\alpha_n < \alpha$, $\alpha_{n+1} < \alpha$; otherwise, $\alpha_{n+1} > \alpha$. This fact is then combined with Claim 16.2 to obtain the result that if $\alpha_n < \alpha$, $\alpha_n < \alpha_{n+1} < \alpha$ and, similarly, if $\alpha_{n+1} < \alpha$, $\alpha_{n+1} < \alpha_{n+2} < \alpha$, and so on. Thus, if $\alpha_1 < \alpha$, the infinite sequence $\{\alpha_n\}$ is increasing and bounded above by α. Therefore, it must convergence. We will prove that $\{\alpha_n\}$ converges to its fixed point by contradiction. Any infinite, increasing, and bounded sequence must converge to its minimum upper bound, and we know that α is one upper bound of $\{\alpha_n\}$ if $\alpha_1 < \alpha$. Suppose the $\{\alpha_n\}$ converges to some α', where $\alpha' < \alpha$. Then, by letting $\alpha_n = \alpha'$, we obtain $\alpha_{n+1} > \alpha'$. This contradicts the assumption that α' is the minimum upper bound of the $\{\alpha_n\}$. Thus, $\{\alpha_n\}$ does not converge to α'; it must converge to α. Similarly, it is true for the case of $\alpha_1 > \alpha$; if $\alpha_1 > \alpha$, $\alpha_1 > \alpha_2 > \alpha_3 > \cdots > \alpha$, the sequence $\{\alpha_n\}$ is increasing and infinite; therefore, it must converge to α. See Figure 16.6 for an illustration for the convergence when $\alpha_1 < \alpha$. Next, we will prove Result 16.2.

In Result 16.2, $d\alpha_{n+1}(\alpha_n)/d\alpha_n < 0 \ \forall \alpha_n$ implies that: (1) $d\alpha_{n+2}(\alpha_n)/d\alpha_n > 0 \ \forall \alpha_n$, which can be verified using chain rule; hence, for $\alpha_n < (>)\alpha$, $\alpha_{n+2} < (>)\alpha$; (2) for $\alpha_n < (>)\alpha$, $\alpha_{n+1} > (<)\alpha$. The condition that $\alpha_{n+2}(\alpha_n)$ has a unique fixed point α as well as Observation 4 ($\lim\alpha_n \to \infty$ $\alpha_{n+2}(\alpha_n) < \infty$.) and $\alpha_{n+2}(\alpha_n = 1) \geq 1$ implies that $\alpha_{n+2} > \alpha_n$ for all $\alpha_n < \alpha$, and $\alpha_{n+2} < \alpha_n$ for all $\alpha_n > \alpha$.

Let us consider the case $\alpha_n < \alpha$, the implications from the above paragraph results in $\alpha_n < \alpha_{n+2} < \alpha_{n+4} < \cdots < \alpha$ and $\alpha < \cdots < \alpha_{n+5} < \alpha_{n+3} < \alpha_{n+1}$. Both infinite sequences are monotone and bounded by α. They must converge, and the convergence must be to the fixed point of α_{n+2} (α_n), α (see Figure 16.7.) A similar result holds true for the case $\alpha_n > \alpha$. Thus, $\{\alpha_n\} \to \alpha$ as $n \to \infty$ □

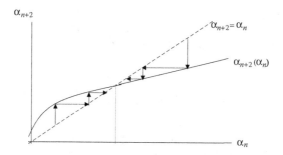

FIGURE 16.7 $\{\alpha_n\} \to \alpha$ when $d\alpha_{n+1}(\alpha_n)/d\alpha_n < 0 \ \forall \ \alpha_n$ and if $\alpha_{n+2}(\alpha_n)$ has the unique fixed point.

Result 16.2 *(Type 2 Convergence):* One of the subsequence, $\{\alpha_{2n-1}\}$ and $\{\alpha_{2n}\}$, converges to α^- and the other converges to α^+ if (a) $d\alpha_{n+1}(\alpha_n)/d\alpha_n < 0 \quad \forall \alpha_n$, and if (b) $\alpha_{n+2}(\alpha_n)$ has three fixed points, including α^-, α, and α^+, where $\alpha^- < \alpha < \alpha^+$, with $\partial\alpha_{n+2}(\alpha_n)/\partial\alpha_n|_{\alpha_n=\alpha} > 1\ldots$

Proof. Similarly to parts of the Result 16.2 proof, $d\alpha_{n+1}(\alpha_n)/d\alpha_n < 0 \ \forall \alpha_n$ gives:
(1) $\forall \alpha_n < (>)\hat{\alpha}$, $\alpha_{n+2} < (>)\hat{\alpha}$, where $\hat{\alpha}$ is a fixed point of $\alpha_{n+2}(\alpha_n)$; and (2) $\forall \alpha_n < (>) \alpha$, $\alpha_{n+1} > (<)\alpha$. The implication is that no elements in $\{\alpha_{2n-1}\}$ or in $\{\alpha_{2n}\}$ cross any fixed points of α_{n+2} (boundedness of the subsequence); additionally, the entire subsequences $\{\alpha_{2n-1}\}$ and $\{\alpha_{2n}\}$ are located on the opposite sides of α. By conditions (b) in Result 16.2, the range that α_n can take value on is divided into four intervals, including: (1) $\alpha_n < \alpha^-$; (2) $\alpha^- < \alpha_n < \alpha$; (3) $\alpha < \alpha_n < \alpha^+$; and (4) $\alpha^+ < \alpha_n$. Without loss of generality, we start the sequence $\{\alpha_n\}$ with α_1, which is possibly located in any one of the four intervals. Note that we will combine the analysis of intervals (1) and (3), and similarly we investigate the intervals (2) and (4) together.

Intervals (1) and (3): For $\alpha_1 < \alpha^-$, $\alpha_1 < \alpha_3 < \alpha^-$ by the above implication and Claim 16.3. Thus, the infinite subsequence $\{\alpha_{2n-1}\}$ is monotonically increasing and bounded above by α^-: $\alpha_1 < \alpha_3 < \alpha_5 < \ldots < \alpha^-$. As a result, it must converge to a fixed point of $\alpha_{n+2} (\alpha_n)$. There is only one fixed point in Interval 1, α^-. $\{\alpha_{2n-1}\}$ must converge to α^-.

Similar to the case of $\alpha_1 < \alpha^-$, for $\alpha < \alpha_1 < \alpha^+$, $\alpha_1 < \alpha_3 < \alpha_5 < \ldots < \alpha^+$. Thus, $\{\alpha_{2n-1}\}$ this series must converge to α^+. Note that in Interval 3, although there are two fixed points of $\alpha_{n+2} (\alpha_n)$, α and α^+, the subsequence moves away from α, and moves towards α^+. So, it must converge to α^+, not to α.

Intervals (2) and (4): For $\alpha^- < \alpha_1 < \alpha$, $\alpha^- < \alpha_3 < \alpha_1 < \alpha$ by the above implication and Claim 3. Thus, the infinite subsequence $\{\alpha_{2n-1}\}$ is monotonically decreasing and bounded below by α^-. It must converge to α^-. For $\alpha^+ < \alpha_1$, $\alpha^+ < \alpha_3 < \alpha_1$. So $\alpha^+ < \cdots < \alpha_5 < \alpha_3 < \alpha_1$. $\{\alpha_{2n-1}\}$ must converge to α^+.

Next, we claim that one of the subsequences, $\{\alpha_{2n-1}\}$ and $\{\alpha_{2n}\}$, converges to α^- and the other converges to α^+; that is $\{\alpha_n\} = \{\alpha_1,\alpha_2,\alpha_3,\ldots,\alpha^-,\alpha^+,\alpha^-,\ldots\}$. This is based on the fact that $\alpha_{n+1} (\alpha_n = \alpha^-) = \alpha^+$ and vise versa. It is verified by α_{n+2} $(\alpha_n = \alpha^-) = \alpha^- \Leftrightarrow \alpha_{n+1} (\alpha_{n+2}(\alpha_n = \alpha^-)) = \alpha_{n+2}(\alpha_{n+1}(\alpha^-)) = \alpha_{n+1}(\alpha^-)$; therefore $\alpha_{n+1}(\alpha^-)$ is another fixed point and it must be α^+. Once the subsequence $\{\alpha_{2n-1}\}$ or $\{\alpha_{2n}\}$ converges to α^- or α^+, the sequence $\{\alpha_n\}$ alternates between α^- and α^+.

Note that for type 2 convergence to occur, we have to exclude the initial value of $\alpha_n = \alpha$.   □

STEADY-STATE CHARACTERISTICS: THE GENERALIZATION CASE

The main purpose of our analysis in this section is to generalize convergence results beyond specific forgetting functions. We list the *fundamental characteristics* of learning and forgetting functions which lead to the convergence results given previously for the specific forgetting functions discussed in the sections "Steady-State Characteristics: The Case of Power Forgetting Functions" and "Steady-State Characteristics: The Case of an Exponential Forgetting Function". Below are the *fundamental characteristics*.

> The learning functions are such that $\partial T(x)/\partial x < 0$, $\partial^2 T(x)/\partial x^2 > 0$ and $\lim_{x \to \infty} T(x) = 0$. That is, as the operator gains more experience by producing more units over time, the unit production time declines in a decreasing manner. At most, the reduction of per-unit production time is $T(1)$.
>
> The forgetting functions are such that $(\partial(F[T(S),I] - T(S))/\partial(T(1) - T(S)))$ > 0, $\lim_{I \to \infty} F[T(S),I] = T(1)$, and $(\partial F(T(S),I)/\partial I) \geq 0$. These functions have properties and the associated interpretations as follows: (1) the amount of forgetting increases in the amount of learning; that is, workers forget more as there is more to forget; (2) the amount of forgetting does not exceed the amount of learning; that is, workers can forget at most what they have learned, not more; and (3) the amount forgotten is increasing in the interruption duration I; that is as workers take longer break, they lose more of their learning and therefore the amount forgotten increases.

The general procedures for computations are: (1) compute $T(S) = T(x_n + q)$ and $I = T - p(\alpha_n)$; (2) find the time for the next unit when the production resumes, $F(T(S), I)$, using a forgetting function and let $T(\alpha_{n+1}) = F(T(S), I)$; and (3) use a learning function to figure out the value of α_{n+1}.

Proof of the convergence results

Some observations and claims from section "Steady-State Characteristics: The Case of an Exponential Forgetting Function" are applicable or similar to the generalization case that we are now considering. Specifically:

Observation 1. $F(T(S), I(\alpha_n)) = T(\alpha_n) - \ell(\alpha_n) + f(\alpha_n) = T(\alpha_{n+1})$.
Observation 2. For any given positive value of q, (1) $\ell(\alpha_n)$ is strictly decreasing in α_n, and (2) $f(\alpha_n)$ is strictly increasing in α_n.

The characteristic of $\ell(\alpha_n)$, strictly decreasing in the value of α_n, is easily to verify by the fact that $\ell(x) = T(x) - T(x + q)$, $\partial T(x)/\partial x < 0$, and $\partial^2 T(x)/\partial x^2 > 0$. The characteristic of $f(\alpha_n)$, strictly increasing in the value of α_n, is verified as follows.
Recall that $f(x) = F[T(x + q), T - p(x)] - T(x + q)$.

Thus, $\dfrac{\partial f(x)}{\partial x} = \dfrac{\partial\left(F\left[T(x+q),T-p(x)\right]-T(x+q)\right)}{\partial(T(1)-T(x+q))} \bullet \dfrac{\partial(T(1)-T(x+q))}{\partial x}.$

Since, $\dfrac{\partial(T(1)-T(x+q))}{\partial x} = -\dfrac{\partial T(x+q)}{\partial x} = -\dfrac{\partial T(x+q)}{\partial(x+q)} \bullet \dfrac{\partial(x+q)}{\partial x} > 0,$ and

$\dfrac{\partial(F[T(\alpha+q),I]-T(\alpha+q))}{\partial(T(1)-T(\alpha+q))} > 0,$ by a *fundamental characteristic*.

Claim 16.4

There is one and only one value of α_n, referred to as α, such that $\alpha_{n+1}(\alpha_n) = \alpha_n$.

Proof. It is similar to the proof for Claim 16.1 in the section "Steady-State Characteristics: The Case of an Exponential Forgetting Function" with a minor modification that as $\alpha_n \to \infty$, $\ell(\alpha_n) \to 0$ and $f(\alpha_n) \to F[0,T]$ and $F[0,T] > 0$. ☐

Claim 16.5

$\alpha_{n+1} > \alpha_n$ iff $\alpha_n < \alpha$; and $\alpha_{n+1} < \alpha_n$ iff $\alpha_n > \alpha$.

Proof. From Observations 1 and 2, together with Claim 16.4, we obtain for $\alpha_n < \alpha$, $\ell(\alpha_n) > f(\alpha_n)$. So, $T(\alpha_{n+1}) < T(\alpha_n)$ (recall: $T(\alpha_{n+1}) = T(\alpha_n) - \ell(\alpha_n) + f(\alpha_n))$. Thus, $\alpha_{n+1} > \alpha_n$.
Similar proof can be used to obtain the result of $\alpha_{n+1} < \alpha_n$ iff $\alpha_n > \alpha$. ☐

Claim 16.6

If $\alpha_{n+2}(\alpha_n)$ has exactly three fixed points (say, α^-, α and α^+, where $\alpha^- < \alpha < \alpha^+$) with $d\alpha_{n+2}(\alpha_n)/d(\alpha_n)|_{\alpha_n=\alpha} > 1$ then (1) $\alpha_{n+2} > \alpha_n$ for $\alpha_n < \alpha^-$ and for $\alpha < \alpha_n < \alpha^+$; and (2) $\alpha_{n+2} < \alpha_n$ for $\alpha^- < \alpha_n < \alpha$ and for $\alpha^+ < \alpha_n$.

Proof. Similar to the proof for Claim 16.3 in the section "Steady-State Characteristics: The Case of an Exponential Forgetting Function" ☐

Claim 16.7

If $\alpha_{n+2}(\alpha_n)$ has only one fixed point, α, then $\alpha_n < \alpha$ iff $\alpha_{n+2} > \alpha_n$, and $\alpha_n > \alpha$ iff $\alpha_{n+2} < \alpha_n$.

Proof. We know the properties of $\alpha_{n+2}(\alpha_n)$ at the two extreme points of α_n: $\alpha_{n+2}(\alpha_n = 1) \geq \alpha_n$ and $\lim_{\alpha_n \to \infty}\alpha_{n+2}(\alpha_n) < \alpha_n$. By combining such properties with the condition of only one fixed point, α, and the continuity of $\alpha_{n+2}(\alpha_n)$, it must be the case that $\alpha_n < \alpha$ iff $\alpha_{n+2} > \alpha_n$, and $\alpha_n > \alpha$ iff $\alpha_{n+2} < \alpha_n$.

The next claim investigates the impact of the relationship between the production time at the start of the next batch $F[T(\alpha_n + q), T - p(\alpha_n)]$ and the worker experience level α_n at the start of the current batch upon the locations of $\alpha_{n+2}(\alpha_n)$ and $\alpha_{n+1}(\alpha_n)$. $\square$

Claim 16.8

16.8.1 If $\partial F[T(\alpha_n + q), T - p(\alpha_n)]/\partial\alpha_n \leq 0$ for all values of α_n, then $\alpha_n < (>)$ $\alpha \Rightarrow \alpha_{n+1}(\alpha_n) \leq (\geq)\ \alpha$.

16.8.2 If $\partial F[T(\alpha_n + q), T - p(\alpha_n)]/\partial\alpha_n > 0$ for all values of α_n, then (a) $\alpha_n < (>)$ $\alpha \Rightarrow \alpha_{n+1}(\alpha_n) > (<)\ \alpha$; and (b) $\alpha_n < (>)\ \alpha' \Rightarrow \alpha_{n+2}(\alpha_n) < (>)\ \alpha'$, where α' is a fixed point of $\alpha_{n+2}(\alpha_n)$.

Proof. We observe that $\partial F[T(\alpha_n + q), T - p(\alpha_n)]/\partial\alpha_n \leq (>) 0\ \forall \alpha_n$ iff $\partial\alpha_{n+1}(\alpha_n)/\partial\alpha_n \geq (<)\ 0$ $\forall\alpha_n$, justified by the fact that $(\partial F[T(\alpha_n + q), T - p(\alpha_n)]/\partial\alpha_n) = (\partial T(\alpha_{n+1}(\alpha_n))/\partial\alpha_n) = (\partial T$ $(\alpha_{n+1}(\alpha_n))/\partial\alpha_{n+1}(\alpha_n)) \cdot (\partial\alpha_{n+1}(\alpha_n)/\partial\alpha_n)$ and $\partial T(x)/\partial x < 0 \forall x$. Note that $\partial\alpha_{n+1}(\alpha_n)/\partial\alpha_n < 0$ iff $\partial\alpha_{n+2}(\alpha_n)/\partial\alpha_n > 0$ by chain rule.

If $\partial\alpha_{n+1}(\alpha_n)/\partial\alpha_n \geq 0$, $\alpha_n < (>)\ \alpha \Rightarrow \alpha_{n+1}(\alpha_n) \leq (\geq)\ \alpha$. Similarly, if $\partial\alpha_{n+1}(\alpha_n)/\partial\alpha_n < 0$, then $\alpha_n < (>)\ \alpha \Rightarrow \alpha_{n+1}(\alpha_n) > (<)\ \alpha$ and $\alpha_n < (>)\ \alpha' \Rightarrow \alpha_{n+2}(\alpha_n) < (>)\ \alpha'$. $\square$

We now have enough observations and claims to prove the existence of the convergence results.

Result 16.3 *(Type 1 Convergence):* As $n \to \infty$, $\{\alpha_n\} \to \alpha$

16.3.1 If $\partial F[T(\alpha_n + q), T - p(\alpha_n)]/\partial\alpha_n \leq 0\ \forall\alpha_n$, or

16.3.2 If $\partial F[T(\alpha_n + q), T - p(\alpha_n)]/\partial\alpha_n$ and if $\alpha_{n+2}(\alpha_n)$ has one and only one fixed point, α_n.

Proof. By Claims 16.5 and 16.8.1, if $\partial F[T(\alpha_n + q), T - p(\alpha_n)]/\partial\alpha_n \leq 0\ \forall\alpha_n$, for $\alpha_1 < \alpha$, $\alpha_1 < \alpha_2 < \alpha_3 < \cdots < \alpha$; the opposite relationship of α_n holds for $\alpha_1 > \alpha$. Thus, the infinite sequence $\{\alpha_n\}$ is monotone with the closest bound of α. Hence, it must converge to α (Result 16.3). Note that the convergence Result 16.3 is such that all elements in the sequence $\{\alpha_n\}$ are either $\leq \alpha$ or $\geq \alpha$. It is referred to as "one-sided convergence" (Teyarachakul 2003).

Consider the proof for Result 16.3.2. As previously mentioned, $\partial F[T(\alpha_n + q),$ $T - p(\alpha_n)]/\partial\alpha_n > 0 \Leftrightarrow d\alpha_{n+1}(\alpha_n)/d\alpha_n > 0$, which implies that if $\alpha_n < (>)\alpha$, $\alpha_{n+2} < (>)\alpha$. That is, no elements in the subsequence $\{\alpha_{2n-1}\}$ or $\{\alpha_{2n-1}\}$ cross over the fixed point α and hence they are bounded by α. Claim 16.7 ($\alpha_{n+2} > \alpha_n$ for all $\alpha_n < \alpha$, and $\alpha_{n+2} < \alpha_n$ for all $\alpha_n > \alpha$) results in the monotone for both subsequences. Both infinite subsequences must converge to the one and only one fixed point of $\alpha_{n+2}(\alpha_n)$, α. Note that all elements in one subsequence are $\leq \alpha$, and those in the other are $\geq \alpha$. This characteristic is referred to as double-sided convergence (Teyarachakul 2003).

Result 16.4 *(Type 2 Convergence):* One of the subsequence, $\{\alpha_{2n-1}\}$ and $\{\alpha_{2n}\}$, converges to α^- and the other converges to α^+ if (a) $\partial F[T(\alpha_n + q), T - p(\alpha_n)]/\partial\alpha_n > 0\ \forall\alpha_n$;

and if (b) $\alpha_{n+2}(\alpha_n)$ has three fixed points, including α^-, α, and α^+, where $\alpha^- < \alpha < \alpha^+$, with $\partial\alpha_{n+2}(\alpha_n)/\partial\alpha_n|_{\alpha_n=\alpha} > 1$.

Proof. To prove Result 16.4, we need the boundedness and monotonicity for the subsequences, which are ensured to exist by conditions (a) and (b) of the result. Note that one subsequence contains in $(1, \alpha)$ if, and only if, the other is in (α, ∞). Similar to the proof of Result 16.2 in the section "Steady-State Characteristics: The Case of an Exponential Forgetting Function", if $\alpha_1 < \alpha$, then (a) all elements in the subsequence $\{\alpha_{2n-1}\} < \alpha$ and converges to α^-; and (b) the entire subsequence $\{\alpha_{2n}\} > \alpha$ and converges to α^+. Its negation is also true. This completes the proof. ☐

DISCUSSION ON FORM OF THE OPTIMAL POLICY (FOOP)

The analysis relevant to form of the optimal policy (FOOP) has gained little prior attention. To the best of our knowledge, only the work of Teyarachakul et al. (2008) considers non-ESCLS production policies that may have lower costs than the best ESCLS policy, where ESCLS (equal-spaced and constant lot size) is defined as a policy with characteristics that the time interval between the start times for two successive batches is constant and equal to q/d and that the lot size q is fixed. They consider the problems with parameters for which $\{\alpha_n\}$ converges to α.

Similarly, the steady state costs associated with ESCLS policies have received little attention. For example, cost expressions and cost analysis are not included in the study by Globerson and Levin (1987). Sule (1978) assumes the steady state batch production time is a linear function of lot size; however, this assumption was shown to be invalid by Axsäter and Elmaghraby (1981). Teyarachakul et al. (2008) demonstrate that the average per-period total cost could be non-convex or non-unimodel. It is assumed to include: (1) the average setup cost per period, decreasing convex in q; (2) an average inventory holding cost per period, linearly increasing in q; and (3) an average production labor cost per period, incorporating the long-term influence of learning and forgetting.

We will continue their FOOP analysis and restrict our analysis in this section to Wright's learning curve and Globerson and Levin's exponential forgetting curve. The model of Elmaghraby (1990) allows lot sizes to be unequal; however, they provided no discussion on whether unequal lot sizes could have lower costs than the best fixed lot size.

We now consider alternatives to an ESCLS policy that may lead to lower costs. Example 16.5 shows that a policy of alternating batch sizes (q_1 and q_2) can have lower costs than the best constant batch size policy, based on an assumption that each batch starts when inventory reaches zero. In Example 16.5, the alternating batch policy "converges" to a pair of alternating experience levels. That is, the subsequence of the experience levels at the start of the batches of size q_1 converge to some α_1 and the subsequence of size q_2 converge to some α_1. We used these limits to evaluate the average costs; the function specification is provided by Teyarachakul et al. (2008). Further, the parameters of Example 16.5 (as well

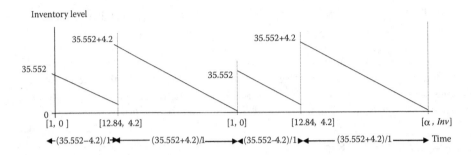

FIGURE 16.8 Inventory level under unequal-spaced policy with $q^* = 35.552$ units.

as Example 16.6 below) were chosen so that the average per-period total cost is unimodal in q. Thus, the optimal constant lot size q^* uniquely satisfies the first order condition.

Example 16.5

$T_0 = 1.00$ day, $m = 0.05$, $b = 1.20$, $d = 1.00$ unit per day, the constant inventory holding cost = $0.33 per unit per day, wage rate = $92.00 per day, the setup cost = $3.00 per setup. This problem has its optimal fixed lot size at $q^* = 6.08$ units, with $\alpha = 3.20$ and the associate average total cost = $85.64. The alternating lot sizes of $(q_1 = 4, q_2 = 20)$ gives the average total cost = $85.33.

We also questioned whether costs can be lowered by relaxing the ESCLS assumption that the batches (with constant q) are equally spaced. Under the constant-demand-rate assumption, it is easy to see that for any ESCLS policy, costs are lowest when each batch starts exactly when the inventory hits zero. Relaxing the equal-spaced requirement means that we consider policies where at least some batches start before the inventory reaches zero. Example 16.6 shows that a fixed lot size policy using unequal-spaced batches can have lower costs than the best ESCLS policy.

Example 16.6

$T_0 = 1$ day, $m = 0.05$, $b = 1.2$, $d = 1$ unit per day, the constant inventory holding cost = $0.33 per unit per day, wage rate = $150.00 per day, the setup cost = $3.00 per setup. The least cost ESCLS policy is for $q^* = 35.552$, with $\alpha = 1.016$ and the associate average total cost = $137.10 per period. An unequal-spaced policy, with the same value of q, but with every other batch starting when inventory (*Inv*) equals 4.2 units gives an average cost of $135.64 per period. Figure 16.8 illustrates the steady state condition of such an unequal-spaced policy.

We can conclude that unequal-spaced production cycles could improve cost and a constant batch size policy may not be optimal. That is, the optimal "traditional EOQ-type" policies could be sub-optimal in the steady state when there is learning and forgetting in processing time.

CONCLUDING REMARKS

This chapter presents the convergence characteristics of batch production times (equivalently, workers' skill levels) in an environment where learning and forgetting in production time occurs intermittently. That is, workers learn while producing units within a batch, forget during the interruption in productions, and relearn when returning to the production of a product.

This chapter provides the collective analysis of convergence in batch production time for a fixed batch size under constant demand. It was first observed by Sule (1978) in his numerical study: as the number of lots produced increased, the batch production time converged to a unique value. Sule (1978) used the variable regression to variable forgetting (VRVF) model in which a y-intercept, the equivalent time for the first unit on the forgetting curve, is variable and a forgetting slope is fixed. Elmaghraby (1990) developed the variable regression to invariant forgetting (VRIF) model, in which both the y-intercept and the forgetting slope are fixed, and provided convergence results.

Globerson and Levin (1987) introduced an exponential forgetting function of which the amount forgotten is bounded by the amount learned. His convergence results have been observed in numerical examples. The mathematical proof of Globerson and Levin's (1987) convergence characteristics were later provided by Teyarachakul et al. (2008).

Studies (Sule 1978; Globerson and Levin 1987; Elmaghraby 1990) prior to 2000 found that the batch production time converges to a unique value as the number of batches produced goes to infinity. A new type of convergence was found in Teyarachakul et al.'s (2008) numerical examples in which the batch times "converged" to two alternating values. The generalization case (the convergence analysis with no restriction to the specific functions) is reported in this chapter.

Teyarachakul et al. (2008) also considered the steady state average cost per time function. We then analyzed the forms of the production policy that improve this cost over the best ESCLS policy. In particular, costs may be lowered by using non-constant lot sizes, or starting batches before the inventory hits zero.

Potential research in this area includes: (1) the development of a more efficient algorithm to find the optimal fixed batch size; (2) an analysis of optimal policies under forgetting functions other than Globerson and Levin's (1987) exponential function; (3) more detailed specification of policies that could lower costs more than the best ESCLS policy; for example, under a constant lot size policy, what is the best timing of batch start times? For a non-constant lot size policy, can an alternating cycle of more than two sizes lead to further improvements?

REFERENCES

Axsäter, S., and Elmaghraby, S., 1981. A note on EMQ under learning and forgetting. *AIIE Transactions* 13(1): 86–89.

Carlson, J.G., and Rowe, A.J., 1976. How much does forgetting cost? *Industrial Engineering* 8(9): 40–47.

Elmaghraby, S.E., 1990. Economic manufacturing quantities under conditions of learning and forgetting (EMQ/LaF). *Production Planning and Control* 1(4): 196–208.

Globerson, S., and Levin, N., 1987. Technical paper: Incorporating forgetting into learning curves. *International Journal of Operations and Production Management* 7(4): 80–94.

Sule, D.R., 1978. The effect of alternate periods of learning and forgetting on economic manufacturing quantity. *AIIE Transactions* 10(3): 338–343.

Teyarachakul, S., 2003. The impact of learning and forgetting on production batch size. Thesis, Purdue University.

Teyarachakul, S., Chand, S., and Ward, J., 2008. Batch sizing under learning and forgetting: Steady state characteristics for the constant demand case. *Operations Research Letters* 36(5): 589–593.

Teyarachakul, S., Comez, D., and Tarakci, H., 2009. "Steady-State Skill Levels of Workers in Learning and Forgetting Environments: The Fixed-Point Property Approach," working paper.

Wright, T.P., 1936. Factors affecting the cost of airplanes. *Journal of Aeronautical Sciences* 3(4): 122–128.

17 Job Scheduling in Customized Assembly Lines Affected by Workers' Learning

Michel J. Anzanello and Flavio S. Fogliatto

CONTENTS

INTRODUCTION

A company's adoption of a mass customization (MC) production strategy implies a large variety of product models being produced in small lot sizes. That requires high flexibility of productive resources (e.g., workers) in order to enable the fast adaptation to the new model to be produced (Da Silveira et al. 2001). These productive characteristics may cause losses in manual-based operations, where workers need to adapt to the features of the new model. Job scheduling becomes particularly difficult since the time needed for lot completion under the learning process is often unknown. Hence, such production environments could potentially benefit from the integration of learning curve (LC) modeling and scheduling techniques.

Learning curves (LCs) are non-linear regression models that associate workers' performance, usually described in units produced per time interval, to task characteristics. LC modelling enables a detailed description of the learning profile of workers (or teams of workers), depicting how their efficiency improves as an operation is continuously repeated (Uzumeri and Nembhard 1998).

Job scheduling is a major issue in the manufacturing and services industries; it aims at allocating jobs to resources by optimizing an objective (Pinedo 2008). However, it seems that understanding workers' learning impacts on the scheduling framework has only recently become a research issue. The seminal work of Biskup (1999) presented an analysis on the effect of learning on the position of jobs in a single machine. Later, Mosheiov and Sidney (2003) integrated LCs with different parameters for each job to programming formulations aimed at optimizing objectives such as flow-time and makespan on a single machine, and flow-time in unrelated parallel machines.

Jobs' processing times in Mosheiov and Sidney (2003) are generated under a uniform distribution. However, statistical distributions may not be appropriate to estimate processing times that follow a functional pattern as in LCs. The need for a method to estimate workers' job-completion times becomes evident if better scheduling schemes are desired.

This chapter addresses the scheduling problem of minimizing the completion time when learning effects are to be considered. We first use a hyperbolic LC, as proposed by Anzanello and Fogliatto (2007), to quantitatively evaluate workers' adaptation to a given set of jobs with different complexities. Workers' learning profiles are then used to estimate the processing times of new jobs. In our approach, each worker is considered as an unrelated parallel machine, since the speed of job execution differs independently between workers. Thus, machine, worker, and team of workers will be treated as synonyms throughout this chapter. In addition, lot and job will also have the same meaning. This leads to an $R_m \parallel \Sigma C$ problem in the standard scheduling representation, where R_m denotes an unrelated parallel machine environment with m teams of workers, and C_j is the completion time of job j.

In the next step of the proposed method we generate four simple heuristics by combining modified stages of existing heuristics for the unrelated parallel machines problem, and then identify the one with the best performance. The proposed heuristics are deployed in three stages. In Stage 1 an initial job order is defined by testing two distinct rules. In Stage 2 we decide on the jobs to be performed by each of the I teams, so that the workload balance among teams is maintained; two rules are tested here. In Stage 3, the jobs assigned to each team are sequenced with the aim of minimizing completion times, yielding $m \, 1 \parallel \Sigma C$ problems.

The resulting heuristics are tested by simulating jobs with different lot sizes and complexity. The estimated completion times are then compared with optimal schedules for two teams of workers obtained by complete enumeration, following two criteria: (i) the deviation of the proposed heuristics' objective function values from their optimal value, and (ii) workload unbalance among teams of workers. The recommended heuristic leads to a 4.9% average deviation from optimality, and satisfactorily balances the workload among teams. That heuristic is then applied to a real shoe manufacturing application consisting of three teams and 90 jobs of distinct size and complexity.

There are two main contributions here. First, we systematize the use of LCs to precisely estimate the processing time required by the different teams of workers to complete a job, depending on its size and complexity. Second, we combine and test modified stages of scheduling heuristics from the literature in order to minimize the completion times in unrelated parallel machine environments. The proposed approach captures workers' learning effects by means of the processing times, leading to more realistic scheduling schemes in customized manufacturing applications.

We now provide a brief review of LC models, and the fundamentals of scheduling in unrelated parallel machines.

BACKGROUND

LEARNING CURVES (LCs)

LCs are mathematical representations of a worker's performance when he or she is repeatedly exposed to a manual task or operation. As repetitions take place workers require less time to perform a task, either due to familiarity with the task and the tools required to perform it, or because shortcuts to task completion are discovered (Wright 1936; Teplitz 1991). There are several LC models proposed in the literature, most notably power models such as Wright's, and hyperbolic models.

Wright's model, arguably the best known LC function in the literature due to its simplicity and efficiency in describing empirical data, is given by:

$$t = U_1 z^b, \tag{17.1}$$

where z represents the number of units produced, t denotes the average accumulated time or cost to produce z units, U_1 is the time or cost to produce the first unit, and b is the curve's slope ($-1 \le b \le 0$).

The hyperbolic LC model allows a more precise description of the learning process if compared to Wright's model. The three-parameter hyperbolic model reported in Mazur and Hastie (1978) is given by:

$$y = k\left(\frac{x+p}{x+p+r}\right), \tag{17.2}$$

such that $p + r > 0$. In Equation 17.2, y describes workers' performance in terms of units produced after x time units of cumulative practice ($y \ge 0$ and $x \ge 0$), k gives the upper limit of y ($k \ge 0$), p denotes previous experience in the task given in time units ($p \ge 0$), and r is the operation time demanded to reach $k/2$, which is half the maximum performance.

The hyperbolic LC model enables a better understanding of workers' learning profiles, potentially optimizing the assignment of jobs to workers (Uzumeri and Nembhard 1998). In Anzanello and Fogliatto (2007) jobs are assigned to workers according to the parameters of the hyperbolic LC, such that teams with a higher final performance receive longer jobs and fast learners receive jobs with smaller lot sizes. However, no additional effort was devoted to the scheduling of jobs to teams in that study.

SCHEDULING IN UNRELATED PARALLEL MACHINES

Scheduling in parallel machines has received increasing attention in recent years, as reported by Cheng and Sin (1990) and Pinedo (2008). A subclass of this problem is the unrelated parallel machines problem in which the processing time in each machine depends only on that machine; that is, machines are independent (Pinedo 2008). Yu et al. (2002) state that the unrelated parallel machine problem is one of the hardest in scheduling theory, and thus justifies the large number of methods used. In fact, most unrelated parallel machine problems are NP (Non-Deterministic Polynomial-Time)-hard, demanding exponential time for obtaining a solution.

Many heuristics have been proposed to solve the scheduling problem in unrelated parallel machines. Mokotoff and Jimeno (2002) suggested several heuristics using partial enumeration aimed at minimizing the makespan (i.e., the competition time of the last job). Chen and Wu (2006) developed a heuristic to minimize the total lateness of secondary operations related to the main job, as set-up procedures, resource availability, and process restrictions. In a similar way, Kim et al. (2009) developed a heuristic focused on the minimization of completion times in scenarios characterized by precedence restrictions, where a task is to promptly begin after the completion of the previous task. Further approaches focused on the minimization of completion times in unrelated parallel machines are reported by Suresh and Chaudhuri (1994) and Randhawa and Kuo (1997), while a comprehensive survey on scheduling methods using tabu search, genetic algorithm, and simulated annealing has also been reported (Jungwattanakit et al. 2009).

Research on the impact of the learning process on scheduling problems is still incipient. In Biskup (1999), workers' learning is assumed as a function of the job position in the schedule in single machine applications; the proposed method aimed at minimizing the weighted completion time and flow-time under a common due date. That method was extended by Mosheiov (2001a, b) to scenarios comprised of several machines as well as parallel identical machines, using an LC with identical parameters for all jobs. In a further study, Mosheiov and Sidney (2003) evaluated how distinct learning patterns affect job sequence using LCs with different parameters for each job. Such LC parameters were inputted into integer programming formulations in order to minimize objective functions such as flow-time and makespan on a single machine; the method was also tested in unrelated parallel machines.

METHOD

The method enables the scheduling of manual-based jobs in highly customized production environments characterized by small sized lots. In such manufacturing environments, workers' learning rate and final performance dictate the time required to complete a job, thereby affecting job scheduling. The method we propose consists of two steps, each comprised of a number of operational stages.

In the first step we follow the procedure reported in Anzanello and Fogliatto (2007); namely, we identify the relevant product models (jobs) and describe those models using classification variables. Product models are grouped in homogeneous families through cluster analysis on those variables. Families are then assigned to predefined assembly lines, and the performance data on teams of workers performing one or more bottleneck operations are collected. We then use the hyperbolic LC on each combination of family model and worker team. The area under the curve defines the processing time of each job.

In the second step we apply new heuristics for job scheduling based on the processing times obtained from the LC analysis in the first step. The set of worker teams is treated as a set of unrelated parallel machines; that is valid since the processing times of the teams are not related. Next we modify and integrate the stages of scheduling heuristics suggested by Adamopoulos and Pappis (1998), Bank and Werner (2001), and Pinedo (2008) for the minimization of completion time, and generate four new choices of heuristics. These heuristics are also expected to balance the workload among the teams. Finally, we compare the results from the heuristics with the optimal schedule obtained by the enumeration of scenarios comprised of two teams of workers.

There are three assumptions in the proposed heuristics: (1) all jobs are available for processing at time zero, (2) teams do not process two or more jobs simultaneously, and (3) pre-emption and job splitting are not allowed at any time.

STEP 1 – ESTIMATION OF JOB PROCESSING TIME

Select teams of workers from which learning data will be collected. We recommend choosing teams comprised of workers familiar with the operations to be analyzed, as well as teams with low turnover. Teams of workers are denoted by $i = 1, ..., I$.

The next stage consists of selecting product models for analysis. Products with demand for customization, reflected in small lot sizes, are the natural choice. We describe product models in terms of their relevant characteristics, such as the physical aspects of the product and the complexity of its manufacturing operations, which may be objectively or subjectively assessed. A clustering analysis on product models is performed using such characteristics as clustering variables; we aim at creating model families from which learning data will be collected. The clustering procedure allows us to extend the LC data collected from a specific model to others in the same family (Jobson 1992; Hair et al. 1995). Model families are denoted by $f = 1, ..., F$.

LC data is collected from teams performing bottleneck manufacturing operations in each model family; we understand bottleneck operations as manual procedures that demand extra learning time and worker ability. All combinations of i and f must be sampled, and replications are recommended. Performance data for each combination are to be collected from the beginning of the operation, and should last until no significant modifications are perceived in the data being collected. Performance data can be collected by counting the number of units processed in each time interval.

Performance data collected from the process are analyzed using the three-parameter hyperbolic model presented in Equation 17.2. The hyperbolic model is selected based on its superior performance in empirical studies, as reported by Nembhard and Uzumeri (2000), and Anzanello and Fogliatto (2007); however, other models may also be tested. Parameter estimates for the LC may be obtained through non-linear regression routines available in most statistical packages. Modeling procedures use the performance data as the dependent variable (y), and the accumulated

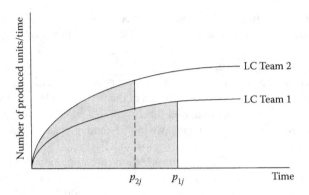

FIGURE 17.1   Job processing times.

operation time as the independent variable (x). Thus, associated to a given family f, there will be a set of parameters k_{if}, p_{if}, and r_{if} estimated using performance data from the ith team of workers. We then average the LC parameters from replications generating parameters $\bar{k}_{if}$, $\bar{p}_{if}$ and $\bar{r}_{if}$. Parameter $\bar{k}_{if}$ is later adjusted to represent the final performance per minute of the operation analyzed. We then construct f sets of graphs consisting of i LCs per set; those curves represent the performance profile of each team when processing a given product family.

Finally, we use the previously generated graphs to estimate the time required by team i to perform job j (i.e., the processing times p_{ij}). The number of units processed in a given time interval corresponds to the area under each LC, as illustrated in Figure 17.1. Hence, the processing time p required to complete a lot comprised of Q units may be estimated by integrating each LC from 0 to p, until an area equivalent to Q is obtained. We repeat this procedure for each team. It is important to state that p_{ij} refers to the time to process the entire job comprised of Q units, and not a single unit.

STEP 2 – SCHEDULING LEARNING-DEPENDENT JOBS

There are three stages in the heuristics proposed here: (1) define an initial order for job distribution, (2) assign jobs to teams in a balanced way, and (3) sequence the subset of jobs assigned to each team in view of the objective function to be optimized. These stages are explained in detail in the following sections.

Stage 1 – Define order for job distribution

We initially order the set of N jobs according to two priority rules:

1. *Decreasing absolute difference between jobs' processing times.* Originally suggested by Adamopoulos and Pappis (1998), this rule finds the two teams (say A and B) with the smallest processing times (p_{ij}, $i = A, B$) and calculates their difference, as in Equation 17.3. This procedure is repeated for each job j. Jobs are then listed in decreasing order of D_j and assigned to teams.

$$D_j = |\, p_{Aj} - p_{Bj}\,|. \tag{17.3}$$

2. *Increasing absolute difference between jobs' processing times rule.* Similar to rule (1), but now jobs are listed in increasing order of D_j and assigned to teams.

Stage 2 – Assign jobs to teams

In this stage we decide which team will process each job following the order established in Stage 1. A main challenge here is to assign jobs ensuring a balanced workload distribution among the I teams, avoiding idleness. We test two rules:

1. *Cumulative processing time.* This is a modification of Bank and Werner's (2001) distribution rule for balancing the total processing time in each team i. A job l is tested in each candidate team according to Equation 17.4 and is assigned to the team with the smallest C_i. This procedure is repeated until all jobs are assigned to teams.

$$C_i = \sum_{j=1}^{Z_i} p_{ij} + p_{il} \quad i = 1,\ldots,I, \tag{17.4}$$

where Z_i is the number of jobs already allocated to team i.

2. *Cumulative processing time and number of jobs.* This rule, suggested by Adamopoulos and Pappis (1998), assigns jobs to teams monitoring both the cumulative processing time and the cumulative number of jobs already assigned to each team. The rule is comprised of two phases: a regular (primary) assignment and a secondary assignment.

We start the regular assignment phase by determining $\lfloor H \rfloor = N/I$, where H is an upper bound on the number of jobs to be assigned to each team in the regular phase of the assignment. The first job is assigned to the team requiring the smallest p_{ij} to complete the job. This procedure is repeated for all jobs, always considering the upper bound H. In case H is exceeded upon assignment of a job to a team, the job will be temporarily assigned to a dummy (non-existent) team. Once the regular assignment is concluded, the secondary assignment starts by assigning jobs from the dummy team to real teams based in the cumulative processing time. The processing time p_{ij} of the first remaining job is added to the cumulative processing time of the jobs that have already been assigned to each team, and the job is assigned to the team with the smallest cumulative processing time. This procedure is repeated until the list of jobs in the dummy team is empty.

Stage 3 – Sequence jobs inside each team

The m resulting $1 \| \sum C$ scheduling problems have their jobs scheduled according to the shortest processing time (SPT) rule. The processing times of jobs inserted in each team are organized in increasing order $(p_1 \le p_2 \le \ldots \le p_n)$, where p_j is the processing time of job j (Pinedo 2008). Note that there is only one sub-index in the processing time, since each team is considered separately. For convenience, the resulting heuristics will be labeled Hu, $u = 1,\ldots,4$, as in Table 17.1.

TABLE 17.1

Summary of Proposed Heuristics

Heuristic	Stage 1 – Define Order for Job Distribution	Stage 2 – Assign Jobs to Teams	Stage 3 – Sequence Jobs Inside Each Team
H1	Decreasing Absolute Difference between Jobs' Processing Times	Cumulative Processing Time	
H2	Decreasing Absolute Difference between Jobs' Processing Times	Cumulative Processing Time and Number of Jobs	Minimization of completion time
H3	Increasing Absolute Difference between Jobs' Processing Times	Cumulative Processing Time	
H4	Increasing Absolute Difference between Jobs' Processing Times	Cumulative Processing Time and Number of Jobs	

The proposed heuristics and the optimal schedule are compared by analyzing two criteria:

1. Relative deviation between objective functions—The heuristic objective function value ($OB_{heuristic}$) and the optimal objective function value ($OB_{opt.sch.}$) are compared as in Equation 17.5, where $OB_{opt.sch.}$ stands for the minimum objective function value among all enumerations of jobs and teams.

$$error = \frac{OB_{heuristic} - OB_{opt.sch.}}{OB_{opt.sch.}}. \tag{17.5}$$

2. Average workload unbalance among teams generated by the heuristics— This criterion is calculated as follows. Suppose the cumulative processing time of jobs that have been assigned to Team 1 equals 500 minutes, while for Team 2 it is 550 minutes. In this case there is a workload unbalance of 9.09% [=1−(500/550)]. The smaller the workload unbalance generated by a heuristic the better. One should note that workload balance issues are not considered in the enumeration when searching for the optimal solution; however, it is known that the minimal objective function happens in a job sequence that balances the workload among teams.

CASE EXAMPLE

We applied the method in a shoe manufacturing plant in the south of Brazil. Shoe producers have faced decreasing lot sizes in the past decade, forcing their mass production configuration to adapt to an increasingly customized market. Shoes are assembled through different stages; independent of the type of shoe produced, the sewing stage is the bottleneck operation, being highly dependent on workers' manual skills. All analyses that follow were performed in Matlab® 7.4.

Twenty shoe models are considered in this study. Models were characterized with respect to manufacturing complexity through the following clustering variables:

overall complexity, parts complexity (deployed into four categories), and the number of parts in the model. These variables were subjectively assessed by company experts using a three-point scale, where three denotes the highest complexity or number of parts. An additional variable, type of shoe, was used to enhance the clustering procedure. It presented two levels: (1) for shoes and sandals, and (2) for boots, which tend to be more complex in terms of assembly. We then applied a k-means cluster analysis on the 20 models, yielding three complexity families: easy, medium, and difficult.

We selected three teams of workers (Teams 1, 2, and 3), each comprised of approximately 40 supervised workers organized in an assembly line from which performance data were collected. Models from the three families were directed to teams in a balanced way. Performance data were collected as number of pairs produced in 10-minute intervals, and adjusted to the hyperbolic LC model using Matlab® 7.4. We averaged the parameters k_{if}, p_{if}, and r_{if} since there were several replicated observations on a same shoe family, and then adjusted parameter $\bar{k}_{if}$ to represent the production performance in units per minute for the integration procedure presented next, with the results given in Table 17.2.

LC graphs were generated using the parameters $\bar{k}_{if}$, $\bar{p}_{if}$, and $\bar{r}_{if}$, and then grouped according to shoe family. Three graphs were obtained, corresponding to the complexity families easy, medium, and difficult, and in each graph there were three average LCs, one for each team analyzed. Areas under the graphs enabled the estimation of processing times for each team; processing times were then used in the scheduling heuristics.

We now compare the performance of the proposed heuristics with the optimal schedule through simulation. Since the optimal schedule is obtained by complete enumeration of all possible combinations of jobs in each team, we consider only 10 jobs and two teams. Job sizes (in units) were assumed to follow a $N\sim(\mu,\sigma^2)$ distribution. We tested three levels of job size following shoe manufacturing experts' opinions: $N\sim(500,100)$, $N\sim(300,75)$ and $N\sim(150,25)$. We randomly inserted each job into a complexity family by means of a uniform distribution in the interval [1,3], where 1 indicates an "easy" family. Table 17.3 depicts an example of the processing times required by Teams 1 and 2 to complete each of the 10 jobs generated according to $N\sim(500,100)$.

We repeated each heuristic 200 times for each job size distribution. The first numerical column of Table 17.4 depicts the average deviation estimated by Equation 17.5. The second and third numerical columns display the workload unbalance among teams generated by the optimal solution and by the heuristics, respectively. These values are the average of the three lot size distributions.

Deviations with respect to the optimum objective function value generated by the heuristics range from 4.9% to 9.6%; that may be considered satisfactory given the simplicity of the heuristics tested. We recommend heuristic H1 since it leads to the minimum deviation, with low workload unbalance.

We then evaluated the influence of job size in the heuristic deviations; our testing is given in Table 17.5. There are no significant differences or trends in deviations as the job size changes. Heuristic H1 performs better overall tested distributions.

Finally, heuristic H1 is applied to a shoe manufacturing process. We considered 90 jobs of different complexities and sizes to be scheduled, as presented in Appendix 17.1. Three teams of workers are considered. Table 17.6 depicts the

TABLE 17.2
LC Parameters for Shoe Families

	Difficult			Medium			Easy		
	Team 1	Team 2	Team 3	Team 1	Team 2	Team 3	Team 1	Team 2	Team 3
$\bar{k}$ (units/min)	0.94	1.11	1.57	1.62	1.34	2.66	1.19	1.30	1.26
$\bar{p}$ (min)	77.9	21.1	34.1	15.9	14.4	16.1	80.3	62.9	51.5
$\bar{r}$ (min)	68.7	50.9	97.3	46.9	69.8	38.0	145.9	122.5	66.6

TABLE 17.3
Job Process Time Estimated by LCs

Family	Lot Size (Units)	Processing Time Team 1 (Hours)	Processing Time Team 2 (Hours)
Difficult	457	8.7	7.2
Difficult	333	6.6	5.6
Easy	513	9.4	8.7
Difficult	529	9.9	8.1
Medium	385	6.8	3.4
Easy	619	11.0	10.2
Easy	550	9.3	9.1
Medium	496	8.4	4.2
Difficult	533	9.9	8.2
Difficult	517	9.7	8.0

TABLE 17.4
Performance of Suggested Heuristics Considering Deviation and Workload Unbalance (Average of all Lot Size Distributions)

Heuristic	Deviation	Workload Unbalance Optimal Solution	Workload Unbalance Heuristic
H1	4.9	2.8	9.1
H2	9.6	2.9	25.5
H3	6.8	2.9	12.2
H4	7.9	3.4	18.8

TABLE 17.5
Deviation under Different Lot Size Distributions

Heuristic	Lot Size Distribution (in Units)		
	$N\sim(150,25)$	$N\sim(300,75)$	$N\sim(500,100)$
H1	3.5	6.6	4.8
H2	8.8	11.6	8.3
H3	5.4	7.0	8.2
H4	6.9	8.5	8.4

TABLE 17.6
Recommended Job Sequence and Teams' Occupancy Times

Team	Job Sequence																			Occupancy Time (%)
Team 1	28	34	75	22	6	25	57	65	30	7	59	66	72	3	56	19	13	87	81	75
	41	18	47	73	63	58	84													
Team 2	32	67	8	21	14	76	43	82	78	10	15	70	11	2	64	71	12	24	74	82
	53	27	85	68	45	9	77	55												
Team 3	52	89	33	42	62	39	23	29	50	16	1	79	4	36	83	69	80	31	44	100
	54	86	48	90	88	60	5	49	37	46	17	35	61	20	38	26	51	40		

recommended job sequence for each team, as well as the occupancy time. Here we decided to measure the workload in terms of occupancy time because it is more intuitive than workload unbalance when there are more than two teams. A large number of jobs were assigned to Team 3 due to its higher final performance and faster learning rate, expressed by the parameters k and r, respectively, in Table 17.2. In addition, the recommended heuristic leads to a satisfactory balance between teams' occupancy times.

CONCLUSION

We proposed a method to schedule jobs in highly customized applications where workers' learning takes place. The method integrates LCs to new scheduling heuristics aimed at minimizing completion time. LCs enabled the estimation of the processing time required by teams of workers to complete jobs of different sizes and complexity. These times were inputted in four choices of heuristics developed based on modifications of existing heuristics for the unrelated parallel machine problem.

The recommended heuristic determined in a simulation study yielded an average deviance of 4.9% compared to the optimal schedule, and led to a satisfactory balance of workload among teams. When applied to a shoe manufacturing case study the recommended heuristic prioritized job allocation to the fastest team and led to remarkable balance in the teams' workloads.

Future research includes the analysis of more complex scheduling problems where workers' learning takes place, including the job shop problem. We will also explore extensions by introducing uncertainty in the processing times estimated by the LCs, and evaluate their effects in the scheduling problem.

REFERENCES

Adamopoulos, G., and Pappis, C., 1998. Scheduling under a common due-date on parallel unrelated machines. *European Journal of Operational Research* 105(3): 494–501.

Anzanello, M., and Fogliatto, F., 2007. Learning curve modeling of work assignment in mass customized assembly lines. *International Journal of Production Research* 45(13): 2919–2938.

Bank, J., and Werner, F., 2001. Heuristic algorithms for unrelated parallel machine scheduling with a common due date, release dates, and linear earliness and tardiness penalties. *Mathematical and Computer Modeling* 33(4–5): 363–383.

Biskup, D., 1999. Single-machine scheduling with learning considerations. *European Journal of Operational Research* 115(1): 173–178.

Cheng, T.C.E., and Sin, C.C.S., 1990. A state-of-the-art review of parallel-machine scheduling research. *European Journal of Operational Research* 47(3): 271–292.

Chen, J., and Wu, T., 2006. Total tardiness minimization on unrelated parallel machine scheduling with auxiliary equipment constraints. *Omega* 34(1): 81–89.

Da Silveira, G., Boreinstein, D., and Fogliatto, F., 2001. Mass customization: Literature review and research direction. *International Journal of Production Economics* 72(1): 1–13.

Hair, J., Anderson, R., Tatham, R., and Black, W., 1995. *Multivariate data analysis with readings.* New Jersey: Prentice-Hall Inc.

Jungwattanakit, J., Reodecha, M., Chaovalitwongse, P., and Werner, F., 2009. A comparison of scheduling algorithms for flexible flow shop problems with unrelated parallel machines, setup times, and dual criteria. *Computers and Operations Research* 36(2): 358–378.

Kim, E., Sung, C., and Lee, I., 2009. Scheduling of parallel machines to minimize total completion time subject to s-precedence constraints. *Computers and Operations Research* 36(3): 698–710.

Mazur, J. E., and Hastie, R., 1978. Learning as accumulation: A re-examination of the learning curve. *Psychological Bulletin* 85(6): 1256–1274.

Mokotoff, E., and Jimeno, J., 2002. Heuristics based on partial enumeration for the unrelated parallel processor scheduling problem. *Annals of Operations Research* 117(1–4): 133–150.

Mosheiov, G., 2001a. Scheduling problems with learning effect. *European Journal of Operational Research* 132(3): 687–693.

Mosheiov, G., 2001b. Parallel machine scheduling with learning effect. *Journal of the Operational Research Society* 52(10): 391–399.

Mosheiov, G., and Sidney, J., 2003. Scheduling with general job-dependent learning curves. *European Journal of Operational Research* 147: 665–670.

Nembhard, D., and Uzumeri, M. 2000. An individual-based description of learning within an organization. *IEEE Transactions on Engineering Management* 47(3): 370–378.

Pinedo, M., 2008. Scheduling, theory, algorithms and systems. New York: Springer.

Suresh, V., and Chaudhuri, D., 1994. Minimizing maximum tardiness for unrelated parallel machines. *International Journal of Production Economics* 34(2): 223–229.

Teplitz, C. 1991. *The learning curve deskbook: A reference guide to theory, calculations and applications*. New York: Quorum Books.

Uzumeri, M., and Nembhard, D., 1998. A population of learners: A new way to measure organizational learning. *Journal of Operations Management* 16(5): 515–528.

Wright, T.P., 1936. Factors affecting the cost of airplanes. *Journal of the Aeronautical Sciences* 3(2): 122–128.

Yu, L., Shih, H., Pfund, M., Carlyle, W., and Fowler, J., 2002. Scheduling of unrelated parallel machines: An application to PWB manufacturing. *IIE Transactions* 34(11): 921–931.

APPENDIX 17.1 SIZE AND COMPLEXITY OF SHOE LOTS

Lot	Family	Lot size (units)	Lot	Family	Lot size (units)
1	Medium	460	46	Easy	525
2	Difficult	345	47	Easy	450
3	Difficult	390	48	Difficult	430
4	Difficult	640	49	Easy	590
5	Easy	630	50	Medium	485
6	Medium	500	51	Easy	360
7	Medium	545	52	Medium	730
8	Medium	530	53	Easy	500
9	Easy	380	54	Difficult	465
10	Difficult	505	55	Easy	130
11	Difficult	425	56	Difficult	335
12	Easy	585	57	Difficult	620
13	Easy	610	58	Easy	360
14	Medium	440	59	Difficult	510
15	Difficult	475	60	Easy	760
16	Medium	485	61	Easy	455
17	Easy	500	62	Medium	545
18	Easy	480	63	Easy	420
19	Easy	700	64	Easy	820
20	Easy	445	65	Difficult	580
21	Difficult	720	66	Difficult	475
22	Medium	535	67	Difficult	855
23	Medium	530	68	Easy	420
24	Easy	580	69	Difficult	560
25	Difficult	715	70	Difficult	445
26	Easy	400	71	Easy	620
27	Easy	460	72	Difficult	440
28	Medium	720	73	Easy	440
29	Medium	495	74	Easy	520
30	Medium	345	75	Medium	565
31	Difficult	535	76	Difficult	605
32	Medium	715	77	Easy	350
33	Medium	615	78	Difficult	545
34	Medium	580	79	Medium	460
35	Easy	490	80	Difficult	545
36	Difficult	600	81	Easy	540
37	Easy	580	82	Difficult	550
38	Easy	425	83	Medium	365
39	Medium	540	84	Easy	275
40	Easy	330	85	Easy	445
41	Easy	505	86	Difficult	465
42	Medium	575	87	Easy	580
43	Difficult	585	88	Difficult	300
44	Difficult	500	89	Medium	620
45	Easy	410	90	Difficult	390

18 Industrial Work Measurement and Improvement through Multivariate Learning Curves

Adedeji B. Badiru

CONTENTS

INTRODUCTION

Industrial innovation and advancement owe their sustainable foundation to some measurement scale (Badiru 2008). We must measure work before we can improve it as part of the evaluation of human performance (Wilson and Corlett 2005). This conveys the importance of work measurement in industry, business, and government. Productivity is the battle cry of industry, but what is not measured cannot be improved. Thus, work measurement should be a key strategy for productivity measurement. The following quote provides an inspirational foundation for the pursuit of work measurement activities, both from the standpoint of research and that of applications:

> "No man can efficiently direct work about which he knows nothing."
> (Col Thurman H. Bane, Dayton, Ohio 1919)

Aft (2010) suggests that the academic research of work measurement has real-world values and practical applications for industry. He summarizes that the standards obtained through work measurement help provide essential information for the successful management of an organization. Such information includes the following:

- Data for scheduling
- Data for staffing
- Data for line balancing
- Data for materials requirement planning
- Data for wage payment
- Data for costing
- Data for employee evaluation

In each of these data types, the effect of learning in the presence of multiple factors should be taken into consideration for operational decision-making purposes.

EFFECT OF LEARNING

"The illiterate of the twenty-first century will not be those who cannot read and write, but those who cannot learn, unlearn, and relearn."

(Alvin Toffler)

Manufacturing progress functions, also known as "learning curves" or "experience curves," are a major topic of interest in operations analysis and cost estimation. Learning, in the context of manufacturing operations management, refers to the improved productivity obtained from the repetition of an operation. Several research studies have confirmed that human performance improves with reinforcement or with frequent repetitions (Carr 1946; Conley 1970; Hirchman 1964; Yelle 1979; Belkaoui 1986). The reduction of operation processing times, achieved through learning curve effects, can directly translate to cost savings for manufacturers and improved morale for employees (Badiru 1988). Learning curves are essential for setting production goals, monitoring progress, reducing waste, and improving efficiency (Knecht 1974; Richardson 1978). The applications of learning curves extend well beyond conventional productivity analysis. For example, Dada and Srikanth (1990) examine the impact of accumulated production on monopoly pricing decisions.

Typical learning curves present the relationship between cumulative average production cost per unit and cumulative production volume based on the effect of learning. For example, an early study by Wright (1936) disclosed an 80% learning effect, which indicates that a given operation is subject to a 20% productivity improvement each time the production quantity doubles. This productivity improvement phenomenon is illustrated later in this paper. With information about expected future productivity levels, a learning curve can serve as a predictive tool for obtaining the time estimates for tasks that are repeated within a production cycle (Chase and Aquilano 1981). Manufacturing progress functions are applicable to all aspects of manufacturing planning and control.

This chapter presents an analytic framework for extending univariate manufacturing progress functions to multivariate models. Multivariate models are significant for cost analysis because they can help account for the multivariate influences on learning and, thus, facilitate a wider scope of learning curve analysis for manufacturing operations. A specific example of a bivariate manufacturing progress function is presented later in this chapter. Since the production time and cost are inherently related, both terms are often used interchangeably as the basis for manufacturing progress function analysis. For consistency, cost is used as the basis for discussion in this chapter and the terms "manufacturing progress function" and "learning curve" are used interchangeably.

In many manufacturing operations, tangible, intangible, quantitative, and qualitative factors intermingle to compound the productivity analysis problem. Consequently, a more comprehensive evaluation methodology, such as a multivariate learning curve, can be used for productivity analysis. Some of the specific analyses that can benefit from the results of multivariate learning curve analysis include cost estimation, work design and simplification, breakeven analysis, manpower scheduling, make or buy decision-making, production planning, budgeting and resource allocation, and management of productivity-improvement programs.

UNIVARIATE LEARNING CURVE MODELS

The conventional univariate learning curve model presents several limitations in practice. Since the first formal publication of learning curve theory by Wright (1936), there have been numerous alternative propositions concerning the geometry and functional forms of learning curves (Baloff 1971; Jewell 1984; Kopcso and Nemitz 1983; Smunt 1986; Towill and Kaloo 1978; Yelle 1983). Some of the classical models are listed below (Asher 1956; DeJong 1957; Levy 1965; Glover 1966; Pegels 1969; Knecht 1974; Yelle 1976; Waller and Dwyer 1981):

- The log-linear model
- The S-curve
- The Stanford-B model
- DeJong's learning formula
- Levy's adaptation function
- Glover's learning formula
- Pegel's exponential function
- Knecht's upturn model
- Yelle's product model
- The multiplicative power model

LOG-LINEAR MODEL

The log-linear model (Wright 1936) is often referred to as the "conventional learning curve" model. This model states that the improvement in productivity is constant (constant slope) as the output increases. There are two basic forms of the log-linear model: (i) the average cost model and (ii) the unit cost model. The average cost model is more popular than the unit cost model. It specifies the relationship between the cumulative average cost per unit and the cumulative production. The relationship indicates that cumulative cost per unit will decrease by a constant percentage as the cumulative production volume doubles. The model is expressed as:

$$C(x) = C_1 x^b,$$

where
  $C(x)$ = cumulative average cost of producing x units
  C_1 = cost of the first unit
  x = cumulative production count
  b = the learning curve exponent (i.e., the constant slope of the learning curve on log-log paper).

When linear graph paper is used, the log-linear learning curve is a hyperbola of the form shown in Figure 18.1.

On log-log paper, the model is represented by the following straight line equation:

$$\log C_x = \log C_1 + b \log x,$$

where b is the constant slope of the line. The expression for the learning rate percent, p, is derived by considering two production levels where one level is double the other. For example, given the two levels x_1 and x_2 (where $x_2 = 2x_1$), we have the following expressions:

$$C_{x_1} = C_1 (x_1)^b,$$

$$C_{x_2} = C_1 (2x_1)^b.$$

The percent productivity gain is then computed as:

$$p = \frac{C_1 (2x_1)^b}{C_1 (x_1)^b} = 2^b.$$

In general, the learning curve exponent can be computed as:

$$b = \frac{\log C_{x_1} - \log C_{x_2}}{\log x_1 - \log x_2},$$

where:
  x_1 = first production level
  x_2 = second production level
  C_{x_1} = cumulative average cost per unit at the first production level
  C_{x_2} = cumulative average cost per unit at the second production level.

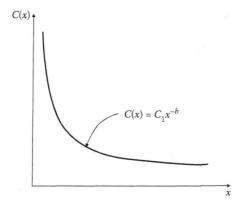

FIGURE 18.1 Log-linear learning curve model.

Expression for total cost: Using the basic cumulative average cost function, the total cost of producing x units is computed as:

$$TC_x = (x)(C_x)$$

$$= C_1 x^{(b+1)}.$$

Expression for unit cost: The unit cost of producing the xth unit is given by:

$$UC_x = C_1 x^{(b+1)} - C_1 (x-1)^{(b+1)}.$$

Expressing for marginal cost: The marginal cost of producing the xth unit is given by:

$$MC_x = \frac{d[TC_x]}{dx}$$

$$= (b+1)C_1 x^b.$$

MULTIVARIATE LEARNING CURVE MODEL

The learning rate of employees is often influenced by other factors that may be within the control of the organization. The conventional learning curve model is developed as a function of the production level only. Nonetheless, there are other factors apart from the cumulative production volume that can influence how fast, how far, and how well a production operator learns within a given time horizon. The overall effect of learning may be influenced by several factors including skill level, level of experience, level of prior training, amount of concurrent training, design changes, methods improvement, material substitutions, changes in tolerance levels, complexity of the task, degree of external interference affecting the task and/or operator, level of competence with available tools, and prior experience with related job functions (transfer of skill).

MODEL FORMULATION

Conway and Schultz (1959) first suggested the need for a multivariate generalized progress function. They point out that there are other factors that influence the cost in learning curve analysis. For example, they present a hypothetical response surface relating the cost to the production rate and cumulative production volume. This paper extends that idea to the development of an analytical model. To account for the multivariate influence on learning, a model is developed here to facilitate a wider scope of learning curve analysis for manufacturing operations. The general form of the hypothesized model is given as

$$C_x = K \prod_{i=1}^{n} C_i x_i^{b_i},$$

where
    C_x = cumulative average cost per unit for a given set of factor values
    K = model parameter (i.e., cost of first unit of the product)
    x = vector of specific values of independent variables (factors)
    x_i = specific value of the ith factor
    n = number of factors in the model
    C_i = coefficient for the ith factor
    b_i = learning exponent for the ith factor.

Figure 18.2 shows a generic graphical representation of a two-factor learning curve model based on the aforementioned formulation.

BIVARIATE MODEL

This section presents a computational experience with a learning curve model containing two independent variables: (i) cumulative production (x_1) and (ii) cumulative training time (x_2). The inclusion of training time as a second variable in the model is reasonable because many factors that influence learning can be expressed as training-dependent variables. With the two-factor model, the expected cost of production can be estimated on the basis of cumulative production and cumulative training time.

    Training time in this illustration refers to the amount of time explicitly dedicated to training an operator for his or her job functions. Previous learning curve models have considered the increase in productivity due to increases in production volumes, but not much consideration has been given to the fact that the production level itself is time-dependent and is influenced by other factors. Carlson (1973)

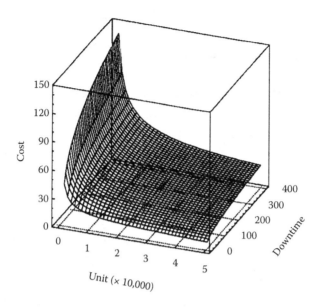

FIGURE 18.2   Response surface for bivariate learning curve model.

pointed out that multi-slope and curve-linear learning curve models have not received the recognition they deserve, probably due to the straight-line syndrome of the log-linear model. This paper addresses that long-standing neglect. For a two-factor model, the mathematical expression for the learning curve is hypothesized to be given by:

$$C_{x_1 x_2} = K c_1 x_1^{b_1} C_2 x_2^{b_2},$$

where
   C_x = cumulative average cost per unit for a given set of factor values
   K = intrinsic constant
   x_1 = specific value of first factor
   x_2 = specific value of second factor
   c_i = coefficient for the ith factor
   b_i = learning exponent for the ith factor.

The aforementioned model is based on a parametric cost model presented by Waller and Dwyer (1981). A set of real data compiled from a local manufacturing company is shown in Table 18.1. The data set is used to illustrate the procedure for fitting and diagnostically evaluating a multivariate learning curve function. Two data replicates are used for each of the 10 combinations of cost and time values. Observations are recorded for the number of units representing double production volumes. Data replicates were obtained by recording the observations from identical setups of the process being studied. The process involved the assembly of electronic

TABLE 18.1

Multivariate Learning Curve Data

Treatment Number	Observation Number	Cumulative Average Cost ($)	Cumulative Production (units)	Cumulative Training Time (hours)
1	1	120	10	11
	2	140	10	8
2	3	95	20	54
	4	125	20	25
3	5	80	40	100
	6	75	40	80
4	7	65	80	220
	8	50	80	150
5	9	55	160	410
	10	40	160	500
6	11	40	320	660
	12	38	320	600
7	13	32	640	810
	14	36	640	750
8	15	25	1280	890
	16	25	1280	800
9	17	20	2560	990
	18	24	2560	900
10	19	19	5120	1155
	20	25	5120	1000

components in communication equipment. The cost and time values were rounded off to the nearest whole numbers in accordance with strategic planning reports used in the company.

The two-factor model is represented in logarithmic scale to facilitate the curve-fitting procedure

$$\log C_x = \left[\log K + \log(c_1 c_2) \right] + b_1 \log x_1 + b_2 \log x_2$$

$$= \log a + b_1 \log x_1 + b_2 \log x_2,$$

where a represents the combined constant in the model such that

$$a = (K)(c_1)(c_2).$$

Using statistical analysis software, the following model was fitted for the data:

$$\log C_x = 5.70 - 0.21(\log x_t) - 0.13(\log x_2),$$

which transforms into the multiplicative model

$$C_x = 298.88 x_1^{-0.21} x_2^{-0.13},$$

where:

$a = 298.88$ (i.e., $\log(a) = 5.70$)
C_x = cumulative average cost per unit
x_1 = cumulative production units
x_2 = cumulative training time in hours.

Since $a = 298.88$, we have $(Kc_1c_2) = 298.88$. If two of the constants K, c_1, or c_2 are known, then the third can be computed. The constants may be determined empirically from the analysis of historical data.

Figure 18.3 shows the response surface for the multiplicative form of the fitted bivariate learning curve example. A visual inspection of the response surface plot indicates that the cumulative average cost per unit is sensitive to changes in both independent variables – cumulative units and training time. It does appear, however, that the sensitivity to cumulative units is greater than the sensitivity to training time. Diagnostic statistical analyses indicate that the model is a good fit for the data. The 95% confidence intervals for the parameters in the model are shown in Table 18.2. Figure 18.4 shows a plot of the predicated and observed cost values, with 95% confidence intervals for the predictions.

The result of the analysis of variance for the full regression is presented in Table 18.3. The P-value of 0.0000 in the table indicates that we have a highly significant regression fit. The R-squared value of 0.966962 indicates that almost 97% of the variabilities in the cumulative average cost are explained by the terms in the model. Table 18.4 shows the breakdown of the model component of the sum of squares. Based on the low P-values shown in the table, it is concluded that units and training time contributed significantly to the multiple regression model. It is also noted, based on the sum of squares, that production units account for most of the fit in this particular bivariate learning curve model.

The correlation matrix for the estimates of the coefficients in the model is shown in Table 18.5. It is evident that units and training time are negatively correlated and that the constant is positively correlated with production units, while it is negatively correlated with training time. The strong negative correlation (-0.9189) between units and training time confirms that the training time decreases as production volume increases (i.e., learning curve effect). Variables that are statistically independent will have an expected correlation of zero. As expected, Table 18.5 does not indicate any zero correlations. A residual plot of the modeling indicates a fairly random distribution of the residuals; thus suggesting a good fit. The following additional results are obtained for the residual analysis: residual average = 4.66294E-16, coefficient of skewness = 0.140909, coefficient of kurtosis = -0.557022, and Durbin-Watson statistic = 1.21472. A normal probability plot of the residuals was also analyzed. The fairly straight line fit in the plot indicates that the residuals are approximately normally distributed; an essential condition for a good regression model. A perfectly normal distribution would plot as a straight line in a normal probability plot.

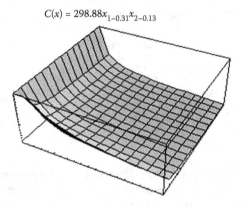

$$C(x) = 298.88x_{1-0.31}x_{2-0.13}$$

FIGURE 18.3 Bivariate learning curve example.

Given a production level of 1750 units and a cumulative training time of 600 hours, the multivariate model indicates an estimated cumulative average cost per unit as shown in the following:

$$C_{(1750,600)} = (298.88)(1750^{-0.21})(600^{-0.13})$$

$$= 27.12.$$

Similarly, a production level of 3500 units and a training time of 950 hours yields the following cumulative average cost per unit:

$$C_{(3500,950)} = (298.88)(3500^{-0.21})(950^{-0.13})$$

$$= 22.08.$$

An important application of the bivariate model involves addressing problems such as the following:

Situation: The standards department of a manufacturing plant has set a target average cost per unit of $12.75 to be achieved after 1000 hours of training. Find the cumulative units that must be produced to achieve the target cost.

TABLE 18.2
95% Confidence Interval for Model Parameters

Parameter	Estimates	Lower Limit	Upper Limit
$\log(a)$	5,7024	5,4717	5,9331
b_1	−0.2093	−0.2826	−0.1359
b_2	−0.1321	−0.2269	−0.0373

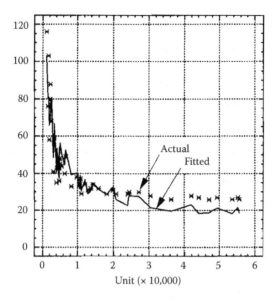

FIGURE 18.4 Cost-unit predicted data versus actual data (on normalized axes).

Analysis: Using the bivariate model previously fitted, the following relationship is obtained:

$$\$12.75 = (298.88)\left(X^{-0.21}\right)\left(1000^{-0.31}\right),$$

which yields the required cumulative production units of $X = 46.444$ units.

Based on the large number of cumulative production units that are required to achieve the expected standard cost, the standards department may want to review the cost standard. The standard of $12.75 may not be achievable if there is a limited market demand for the particular product being considered. The flat surface of the learning curve model as units and training time increase implies that more units will be needed to achieve additional cost improvements. Thus, even though

TABLE 18.3
ANOVA for the Full Regression of the Learning Curve Model

Source	Sum of Squares	D.F.	Mean Square	F-Ratio	P-Value
Model	7.41394	2	3.70697	248.778	0.0000
Error	0.253312	17	0.00149007		
Total	7.66725	19			

R-squared $= 0.966962$; R-adjusted (adjusted for degree of freedom [D.F.]) $= 0.963075$; Standard error of estimate $= 0.122069$.

TABLE 18.4

Further ANOVA for the Variables in the Model Fitted

Source	Sum of Squares	D.F.	Mean Square	F-Ratio	P-Value
LOG (units)	7.28516187	1	7.2851619	488.91	0.000
LOG (time)	0.12877864	1	0.1287786	8.64	0.0092
Model	7.41394052	2			

an average cost of $22.08 can be obtained at a cumulative production level of 3500 units, it takes several thousand additional units to bring the average cost down to $12.75 per unit.

COMPARISON WITH UNIVARIATE MODEL

The bivariate model permits us to simultaneously analyze the effects of production level and training time on the average cost per unit. If a univariate function is used with the conventional log-linear model, we would be limited to only a relationship between the average cost and cumulative production. This would deprive us of the full range of information that we could have obtained about the process productivity. This fact is evidenced by using only the second and third columns of the data in Table 18.1 as shown in the following. The univariate log-linear model is represented as:

$$C_x = C_1 x^b,$$

where the variables are as defined previously. The second and third columns of Table 18.1 are fitted to the aforementioned model to yield:

$$C_x = 240.03 X^{(-0.3032)},$$

with $R^2 = 95.02\%$. Thus, the learning exponent, b, is equal to -0.3032 and the corresponding percent learning, p, is 81.05%. This indicates that a productivity gain of 18.95% is expected whenever the cumulative production level doubles. This is consistent with Wright's (1936) observation in his pioneer learning curve experiment.

TABLE 18.5

Correlation Matrix for Coefficient Estimates

Source	Constant	LOG (units)	LOG (time)
Constant	1.0000	0.3654	−0.6895
LOG (units)	0.3654	1.0000	−0.9189
LOG (time)	−0.6895	−0.9189	1.0000

Figure 18.5 shows a plot of the fitted univariate model. Even though we obtained a good fit for the univariate model, the information on training time is lost. For example, the univariate model indicates that the cumulative average cost would be $33.84 per unit when cumulative production reaches 640 units. By contrast, the bivariate model shows that the cumulative average cost is $32.22 per unit at that cumulative production level, with a cumulative training time of 810 hours. This lower cost is due to the additional effect of training time. Thus, the bivariate model provides a more accurate picture of the interactions between the factors associated with the process.

POTENTIAL APPLICATIONS

A multivariate learning curve model can change the conventional analysis of the effect of learning into a more robust decision-making tool for operations management. As pointed out by Smith (1989), learning curves can be used to plan manpower needs; set labor standards; establish prices; analyze, make, or buy decisions; review wage incentive payments; evaluate organizational efficiency; develop quantity sales discounts; evaluate employee training needs; evaluate capital equipment alternatives; predict future production unit costs; create production delivery schedules; and develop work design strategies. The parameters that affect learning rates are broad in scope. Unfortunately, practitioners do not fully comprehend the causes of, and influences upon, learning rates. Learning is typically influenced by many factors in a manufacturing operation. Such factors may include:

Governmental Factors:

- Industry standards
- Educational facilities
- Regulations
- Economic viability
- Social programs

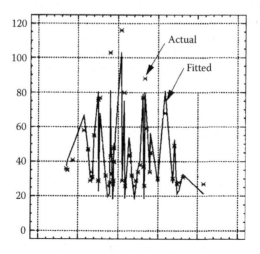

FIGURE 18.5   Fitted values of a single-variable model with 95% C.I.

- Employment support services
- Financial and trade supports

Organizational Factors:
- Management awareness
- Training programs
- Incentive programs
- Employee turnover rate
- Cost accounting systems
- Labor standards
- Quality objectives
- Pool of employees
- Departmental interfaces
- Market and competition pressures
- Obsolescence of machinery and equipment
- Long-range versus short-range goals

Product-Based Factors:
- Product specifications
- Raw material characteristics
- Environmental impacts of production
- Delivery schedules
- Number and types of constituent operations
- Product diversity
- Assembly requirements
- Design maturity
- Work design and simplification
- Level of automation

Multivariate learning curve models offer a robust tool through which several of the aforementioned factors may be quantitatively evaluated simultaneously. Decisions made on the basis of multivariate analyses are generally more reliable than decisions based on single-factor analyses. The following sections present brief outlines of specific examples of the potential applications of multivariate learning curve models.

Design of training programs

As shown in the bivariate model, the need for employee training can be assessed with the aid of a multivariate learning curve model. Projected productivity gains are more likely to be met if more of the factors influencing productivity can be included in the conventional analysis.

Manufacturing economic analysis

The results of multivariate learning curve analysis are important for various types of economic analysis in the manufacturing environment. The declining state of manufacturing in many countries has been a subject of much discussion in recent years. A reliable methodology for the cost analysis of manufacturing technology for specific operations is essential to the full exploitation of the recent advances in the available technology. Manufacturing economic analysis is the process of evaluating manufacturing operations on a cost basis. In manufacturing systems, many tangible, intangible, quantitative, and qualitative factors intermingle to compound the cost analysis problem. Consequently, a more comprehensive evaluation methodology, such as a multivariate learning curve, can be very useful.

Breakeven analysis

The conventional breakeven analysis assumes that the variable cost per unit is constant. On the contrary, learning curve analysis recognizes the potential reduction in the variable cost per unit due to the effect of learning. Owing to the multiple factors involved in manufacturing, multivariate learning curve models should be investigated and adopted for breakeven cost analysis.

Make or buy decision-making

Make or buy decisions can be enhanced by considering the effect of learning on items that are manufactured in-house. Make or buy analysis involves a choice between the cost of producing an item and the cost of purchasing it. Multivariate learning curves can provide data for determining the accurate cost of producing an item. A make or buy analysis can be coupled with breakeven analysis to determine when to make or buy a product.

Manpower scheduling

A consideration of the effect of learning in the manufacturing environment can lead to a more accurate analysis of manpower requirements and the accompanying schedules. In integrated production, where parts move sequentially from one production station to another, the effect of multivariate learning curves can become even more applicable. The allocation of resources during production scheduling should not be made without due consideration of the effects of learning (Liao 1979).

Production planning

The overall production planning process can benefit from multivariate learning curve models. Preproduction planning analysis of the effect of multivariate learning curves can identify areas where better and more detailed planning may be needed. The more preproduction planning that is done, the more potential there is for achieving the productivity gains that are produced by the effects of learning.

Labor estimating

Carlson (1973) showed that the validity of log-linear learning curves may be suspect in many labor analysis problems. For manufacturing activities involving operations in different stations, several factors interact to determine the learning rate of workers. Multivariate curves can be of use in developing accurate labor standards in such cases. Multivariate learning curve analysis can complement conventional work measurement studies.

Budgeting and resource allocation

Budgeting or capital rationing is a significant effort in any manufacturing operation. Multivariate learning curve analysis can provide a management guide for the allocation of resources to production operations on a more equitable basis. The effects of learning can be particularly useful in zero-base budgeting policies (Badiru 1988). Other decision scenarios where a multivariate learning curve analysis could be of use include: bidding (Yelle 1979), inventory analysis, productivity improvement programs (Towill and Kaloo 1978), goal setting (Richardson 1978), and lot sizing (Kopcso and Nemitz 1983).

IMPACT OF MULTIVARIATE LEARNING CURVES

The learning curve phenomenon has been of interest to researchers and practitioners for many years. The variety of situations to which learning curves are applicable has necessitated the development of various functional forms for the curves. The conventional view of learning curves considers only one factor at a time as the major influence on productivity improvement. However, in today's integrated manufacturing environment, it is obvious that several factors interact to activate and perpetrate productivity improvement. This paper has presented a general approach for developing multivariate learning curve models. Multivariate models are useful for various types of analyses in a manufacturing environment. Such analyses include manufacturing economic analysis, breakeven analysis, make or buy decision-making, manpower scheduling, production planning, budgeting, resource allocation, labor estimating, and cost optimization.

Multivariate learning curves can generate estimates about expected cost, productivity, process capability, work load composition, system response time, and so on. Such estimates can be valuable to decision makers for manufacturing process improvement and work simplification functions. The availability of reliable learning curve estimates can enhance the communication interface between different groups in an organization. Multivariate learning curves can provide mechanisms for the effective cost-based implementation of design systems, facilitate the systems integration of production plans, improve supervisory interfaces, enhance design processes, and provide cost information to strengthen the engineering and finance interface. The importance of the productivity of human resources and the workforce is becoming a major concern in many industries. Thus, a multivariate learning curve model can make a significant contribution toward the more effective utilization of available manpower resources.

The multiplicative model used in the illustrative example presented in this paper is just one of several possible expressions that can be investigated for multivariate learning curve models. Further research and detailed experiments with alternate functional forms are necessary to formalize the proposed methodology. It is anticipated that such further studies will be of interest to both practitioners and researchers.

CONCLUSION

The fast-changing operating environments that we face nowadays necessitate interdisciplinary improvement projects that may involve physicists, material industrial engineer scientists, chemists, engineers, bio-scientists, and a host of other disciplines.

They may all use different measurement scales and communicate in different jargons, however, the common thread of work measurement permeates the collective improvement efforts. We should not shy away from work measurement because of the inglorious past falsely associated with it. It should be embraced as the forerunner and foundation for contemporary approaches to pursuing improvement in industrial operations. Work measurement, under whatever names we choose to attach to it, must remain a core competency of industrial engineering. The measurement of productivity, human performance, and resource consumption are essential components of achieving organizational goals and increasing profitability. Work rate analysis, centered on work measurement, helps identify areas where operational efficiency and improvement can be pursued. In many production settings, workers encounter production breaks that require an analysis of the impact on production output. Whether production breaks are standard and scheduled or non-standard and unscheduled, the workstation is subject to work rate slowdown (ramp-down) and work rate pickup (ramp-up), respectively, before and after a break. These impacts are subtle and are hardly noticed unless a formal engineered work measurement is put in place. In the contemporary practice of industrial engineering, practitioners, researchers, students, and policy-makers must embrace and promote industrial work measurement as a tool of productivity improvement.

REFERENCES

Aft, L., 2010. Don't abandon your work measurement cap. *Industrial Engineering* 42(3): 37–40.

Asher, H., 1956. *Cost-quantity relationships in the airframe industry*. Report No. R-291. The Rand Corporation, July 1, 1956, Santa Monica, CA.

Badiru, A.B., 1988. *Project management in manufacturing and high-technology operations*. New York: Wiley.

Badiru, A.B., 2008. Long live work measurement. *Industrial Engineer* 40(3): 24.

Baloff, N., 1971. Extension of the learning curve – Some empirical results. *Operational Research Quarterly* 22(4): 329–340.

Belkaoui, A., 1986. *The learning curve*. Westport: Quorum Books.

Carlson, J.G.H., 1973. Cubic learning curves: Precision tool for labor estimating. *Manufacturing Engineering and Management* 71(5): 22–25.

Carr, G.W., 1946. Peacetime cost estimating requires new learning curves. *Aviation* 45(4): 76–77.

Chase, R.B., and Aquilano, N.J., 1981. *Production and operations management*. Irwin: Homewood.

Conley, P., 1970. Experience curves as a planning tool. *IEEE Spectrum* 7(6): 63–68.

Conway, R., and Schultz, A., 1959. The manufacturing progress function. *Journal of Industrial Engineering* 10(1): 39–53.

Dada, M., and Srikanth, K.N., 1990. Monopolistic pricing and the learning curve: An algorithmic approach. *Operations Research* 38(4): 656–666.

DeJong, J.R., 1957. The effects of increasing skill on cycle time and its consequences for time standards. *Ergonomics* 1(1): 51–60.

Glover, J.H., 1966. Manufacturing progress functions: An alternative model and its comparison with existing functions. *International Journal of Production Research* 4(4): 279–300.

Hirchmann, W.B., 1964. Learning curve. *Chemical Engineering* 71(7): 95–100.

Jewell, W.S., 1984. A generalized framework for learning curve reliability growth models. *Operations Research* 32(3): 547–558.

Knecht, G., 1974. Costing, technological growth and generalized learning curves. *Operations Research Quarterly* 25(3): 487–491.

Kopcso, D., and Nemitz, W., 1983. Learning curves and lot sizing for independent and dependent demand. *Journal of Operations Management* 4(1): 73–83.

Levy, F., 1965. Adaptation in the production process. *Management Science* 11(6): 136–154.

Liao, W.M., 1979. Effects of learning on resource allocation decisions. *Decision Sciences* 10(1): 116–125.

Pegels, C., 1969. On startup or learning curves: An expanded view. *AIIE Transactions* 1(3): 216–222.

Richardson, W.J., 1978. Use of learning curves to set goals and monitor progress in cost-reduction programs. Proceedings of 1978 IIE Spring Conference, pp. 235–239.

Smith, J., 1989. *Learning curve for cost control.* Norcross: Industrial Engineering and Management Press.

Smunt, T.L., 1986. A comparison of learning curve analysis and moving average ratio analysis for detailed operational planning. *Decision Sciences* 17(4): 475–495.

Towill, D.R., and Kaloo, U., 1978. Productivity drift in extended learning curves. *Omega* 6(4): 295–304.

Waller, E.W., and Dwyer, T.J., 1981. Alternative techniques for use in parametric cost analysis. *Concepts – Journal of Defense Systems Acquisition Management* 4(2): 48–59.

Wilson, J.R., and Corlett, N., eds. 2005. *Evaluation of human work* (3rd Edition). Taylor and Francis: New York.

Wright, T.P., 1936. Factors affecting the cost of airplanes. *Journal of the Aeronautical Sciences* 3(2): 122–128.

Yelle, L.E., 1976. Estimating learning curves for potential products. *Industrial Marketing Management* 5(2–3): 147–154.

Yelle, L.E., 1979. The learning curve: Historical review and comprehensive survey. *Decision Sciences* 10(2): 302–328.

Yelle, L.E., 1983. Adding life cycles to learning curves. *Long Range Planning* 16(6): 82–87.

19 Do Professional Services Learn, Sustain, and Transfer Knowledge?

Tonya Boone, Ram Ganeshan, and Robert L. Hicks

CONTENTS

INTRODUCTION

Organizational knowledge management has become a critical competitive capability for many companies and recent research has confirmed that organizational knowledge can provide an inimitable competitive advantage (Adler 1990; Zander and Kogut 1995). Three skills are critical to knowledge management capability: (1) acquiring new knowledge, (2) storing and sustaining the acquired knowledge, and (3) disseminating the knowledge within the organization to all potential users (Cohen and Levinthal 1990). This research draws on organizational learning curve theory to examine these knowledge management capabilities in a professional service organization.

Professional services provide a new business context to study the acquisition, retention, and transfer of knowledge. Professional services (i.e., consultants, engineers, lawyers, etc.) use their knowledge base to develop business solutions to solve client problems. Solutions are often tailored to clients' needs and the skill set needed for any given project is determined by the project context and complexity.* The success of professional service firms depends on how well they manage the knowledge that they accrue, and how efficiently this accrued knowledge is shared across the firm in order to solve clients' problems.

Over the past few years, we have been exploring and drawing upon organizational learning curve theory to examine two critical knowledge management capabilities: (i) knowledge retention (or alternatively depreciation) and (ii) diffusion in a professional service organization. In this chapter we summarize some of our past results on retention and depreciation (Boone et al. 2008) and provide new ones on diffusion (Boone et al. 2010).

We use organizational learning curve theory since it is now a crucial and essential tool to measuring and tracking productivity. Learning curves measure the knowledge acquired from production experience via experience-based productivity improvements (Yelle 1979). They provide a mechanism for evaluating knowledge sustainability by measuring the rate at which knowledge depreciates. Past research has found that knowledge erodes, or depreciates, over time and varies considerably among organizations (Bailey 1989; Darr et al. 1995). Higher levels of knowledge depreciation appear to be due to several factors, many of which are distinguishing characteristics of professional services. For example, professional service work is typically project-based and is characterized by high levels of process variety, which has been associated with faster erosion of knowledge (Jaber and Bonney 1997). The models in the "Firm-Level Learning Curve" section examine the issues related to learning and depreciation in a professional service firm.

Knowledge diffusion, or intra-firm knowledge transfer, can be evaluated using "spillover models," which evaluate the association between different units of knowledge gains. Knowledge transferred into an organization unit enables that unit to improve its own performance by leveraging the production experience of others within the organization (Jarmin 1994). When knowledge created elsewhere is assimilated into a unit, the result is an association between an organization's performance and the production experience of others. The models in the "Inter-Departmental Knowledge Transfer" section examine such cross-functional knowledge transfer. Past work indicates that knowledge diffuses more easily between departments that are similar in terms of skills, experiences, and function (see Lane and Lubatkin 1998). An imperative of knowledge management, however, is that the relevant knowledge is distributed throughout an organization to those requiring that knowledge, irrespective of the functional or departmental boundaries.

CONTEXT AND RESEARCH QUESTIONS

The context for this research project is an architecture-engineering (A/E) firm that primarily provides facilities engineering services to its clients. The firm has expertise in four disciplines organized into departments or units: architectural, electrical,

---

* Prior research has largely focused on manufacturing or mass-service environments where much of the learning is either substantiated in technology (such as process improvements) or via repetition of the same task. See, e.g., Argote and Epple 1990.

mechanical, and civil/structural engineering. This firm also designs technical drawings and estimates the costs for systems such as power generation and distribution systems, Heating, Venting, and Air Conditioning (HVAC), and for mechanical systems.

The organization operates in a project process environment, creating drawings, technical specifications, and cost estimates to meet customer requirements. Many of the projects require input from more than one department. In such cases, a multidisciplinary project team collaborates on the project. Much of the organization's work is collaborative, involving workers from mutiple departments.

Workers typically work on more than one project at any given time, although each project may be at a different stage of development. There is considerable variability among projects, although some project tasks, such as cost estimation and site visits, are similar across projects; indeed, all draw on common bodies of knowledge—for example, architectural and engineering rules seldom change, although the designs may.

The firm provided data for all projects started between 1992 and 2000. The data included pages of design output, labor hours, project start and completion dates, and a list of the participating engineers from different departments for every project. The analysis focuses on the relationship between productivity improvements and knowledge depreciation and the transfer during that time.

Based on this data, we provide summaries of two areas of enquiry that we have explored in the past three years:

- At the firm level, is it possible to detect a learning curve in this professional service organization? How well does this firm retain any knowledge that is accumulated?[*]
- At the department, unit, or discipline (these terms are used interchangeably) level, how successful are these units in accruing productivity gains through experience? For example, does an electrical project take less time than a mechanical project of the same complexity? Why? To answer this we will investigate if each of the departments are able to transfer the knowledge accumulated from other disciplines into solutions to the projects at hand.[†]

BACKGROUND LITERATURE

Wright (1936) was the first to document the relationship between production experience and productivity improvements. He observed that the production time of airplanes decreased at a predictable rate. According to this model, each doubling of cumulative production results in a constant reduction in the unit production time. As an organization gains production experience, represented by the cumulative number of units produced, it is able to produce individual units faster and/or at a lower cost.

Most learning curve research has been conducted in manufacturing organizations. Prior to Boone et al. (2008), learning curves that were found in service businesses typically involved contexts with relatively stable processes or in-service back rooms (Argote and Epple 1990; Dutton and Thomas 1984; Yelle 1979).

---

[*] We originally published these results in Boone et al. (2008).
[†] Our results are based on the working paper by Boone et al. (2010).

However, the distinguishing characteristics of services—intangibility, heterogeneity, simultaneous production and consumption, and the customer's involvement in the service creation process—may affect the association between the production experience and productivity improvements (Lovelock 1992; Morris and Johnston 1986).

Professional services have the highest levels of labor intensity, customization, and customer interaction (Schmenner 1986). Professional service organizations must contend with more variable worker skill levels, experience, and knowledge. Much of the work within professional service organizations is carried out via project type processes—one-of-a-kind, long-term, resource-intensive operations, frequently involving several workers from several departments. The defining characteristics of professional services and project processes (i.e., high labor intensity, product variability, process variability, relatively small production volumes) can potentially endanger experience-based productivity improvements. However, on the other hand, professional service workers are also highly trained, requiring certifications over time to carry out their professional service work. Consequently, as they develop more domain experience, they become more capable of tackling client problems in a quicker and more efficient way. Since most of the work product—in this case CAD designs—are stored electronically, the ability to codify past work and the ability to retrieve it for current client work also significantly impacts upon how knowledge is retained in the firm.

The proportion of human labor has been shown to affect the relationship between productivity and production experience in manufacturing organizations. Human labor has demonstrated a greater capacity for learning than machine labor (Hirschmann 1964a, 1964b). Production processes that use higher proportions of human versus machine labor typically have steeper learning curves and plateau more slowly than less labor-intensive processes (Yelle 1979). This is especially true for professional services where a part of the knowledge is tacit and "belongs" to the worker. The rates of learning will depend, in part, on how well firms are able to harness this tacit knowledge.

Sufficient project volume is a significant source of experience-based productivity improvements. The more often a task is repeated, the better the worker performs that task and the more productive the organization becomes. This implies that a task must be completed some number of times in order for there to be appreciable improvements.

KNOWLEDGE DEPRECIATION

Experience-based productivity improvements depend in part on repeated actions. Productivity improves because a worker repeats some task until he or she gets better at it. Professional service project processes are, by their nature, highly variable. Each successive product, or project, is likely to be different from the next, although there may be similarities and commonalities among projects. In order for productivity improvements to be realized, both the worker and the organization must be able to benefit from past projects that are meaningfully different from the one at hand.

Past empirical studies suggest that, all else being equal, less process variability results in faster productivity improvements. Forgetting indicate that a disruption has affected the learning curve, whereby the current productivity is associated with a lower volume on the present learning curve.

Knowledge depreciation refers to the ongoing erosion in the knowledge stock, and indicates that the experience-based knowledge stock is not accurately represented by the total accumulated production (see Boone et al. 2008 for review of empirical studies in depreciation). Instead, the learning stock, or knowledge, depreciates even while it is being used and is better represented by some fraction of the total accumulated production. The net result of both forgetting and deprecation is that in order to predict productivity at a given point in the production history, the basic learning curve must be modified to show that not all production experience is reflected in productivity.

Higher levels of labor intensity may make professional services more vulnerable to organizational forgetting or knowledge depreciation as a result of several factors. In professional services especially, much of the process knowledge resides in the workers. When a worker leaves, the unique knowledge that he or she possesses leaves too, unless it has been transferred back into the organization. The result is depreciation in the organizational knowledge stock. This may be reflected in lower rates of learning. For example, Darr et al. (1995) found that the learning rate in labor-intensive environment such as a network of pizza restaurants was lower than the 80% learning rate found in most manufacturing firms.

The higher variability of professional services may exacerbate forgetting unless: (a) past projects are sufficiently codified so that the relevant knowledge is captured; and (b) subsequent customers request similar service products, as some time passes before similar service products are reproduced.

KNOWLEDGE DIFFUSION

Knowledge diffusion (or transfer) refers to the productivity improvements that are due to others' production experience. Researchers have examined knowledge transfer among firms within the same industry, across shifts within a single firm, and among stores in a retail chain (Argote et al. 1996; Darr et al. 1995; Epple et al. 1996). In general, the work has focused on transfers between units performing the same tasks. What happens, however, when the tasks are not the same? Or, as in this case, when only some of the tasks are similar and yet the units still work closely together, how readily does knowledge diffuse?

Physically closer networks will experience more a frequent and rapid knowledge transfer (Rothwell 1994; van Dierdonck et al. 1991). Organizations or units that are more similar will also experience greater knowledge transfer. Lane and Lubatkin (1998) found that knowledge transfers more frequently and easily among firms that share similar knowledge bases, organizational structures, and policies. Within a single firm, employees with similar professional experiences and training will communicate more easily than employees with different experiences and backgrounds (Goldhar and Jelinek 1985; Leonard-Barton 1995). Members of a common profession tend to share the same skills, jargon, and knowledge base, all of which make communication easier (March et al. 1991). The result is faster, easier, and richer knowledge transfers among employees with similar professions, backgrounds, and training.

However, knowledge can also be lost in the transfer process. Knowledge transfers incompletely across shifts within a single firm or among stores in a retail chain (Argote et al. 1996; Darr et al. 1995; Epple et al. 1996). Intra-firm knowledge diffusion depends on the ability and willingness of both the recipient and the originator to transfer knowledge; both parties must want to exchange knowledge. The originator must be willing to share knowledge and the recipient must have the capability to adopt the new knowledge (Szulanski 1996).

Knowledge transfer is made more difficult by the formal and informal boundaries between units. Boundaries, whether created formally by organizational structures or informally by alliances and communities, serve to channel knowledge flows. Knowledge flows more easily on one side of an organizational boundary but is more difficult to transfer across boundaries (Brown and Paul 1996). Consequently, specific efforts must be made in order to ensure that important knowledge crosses boundaries.

The characteristics of knowledge will also influence its transferability. Knowledge that can be documented, whether through manuals, blueprints, or memos, is more easily and rapidly transferred (Nelson and Winter 1982; Nonaka 1994; Polanyi 1967). In addition, articulable knowledge is more easily organized and communicated via a variety of means, which may serve to reinforce one another.

Most research has assumed a symmetrical knowledge transfer or spillover throughout industries or within firms. When more than two units are involved, symmetrical models of knowledge transfer do not capture bidirectional knowledge flows among all units. Jarmin (1994) indicated that knowledge is unlikely to diffuse equally to all firms in an industry or to all subunits of a firm. Epple et al. (1996) found that the knowledge transfer between shifts was not symmetric.

Asymmetric models also capture information that indicates a unit's ability to absorb information. While knowledge diffusion is determined by the capabilities of the originating and adopting units, the absorptive capabilities of the adopting unit appear to be more important (Cohen and Levinthal 1990).

MODELS AND RESULTS

FIRM-LEVEL LEARNING CURVE

The traditional form of learning curve is often written as (see Epple et al. 1991):

$$l_t / q_t = A Q_{t-1}^{-\gamma}, \tag{19.1}$$

where l_t is the labor hours worked by the engineers in month t, q_t is the number of drawings produced in month t, and Q_{t-1} is the cumulative number of drawings (knowledge stock) up to month t, or $Q_{t-1} = \sum_{i=1}^{t-1} q_i$, $q_0 = 0$. A and γ are positive constants. The rate of learning can be expressed by the progress ratio $p = 2^{-\gamma}$, which is the percentage decrease in labor hours to create a drawing for every doubling of drawings produced. For estimation, (1) can be rewritten in the commonly used form (from Eppl et al. 1991):

$$\ln(q_t) = \alpha + \beta \ln l_t + \gamma \ln Q_{t-1} + \varepsilon_t, \tag{19.2}$$

where $\alpha = 1/\ln(A)$, β, and γ are all coefficients that need to be estimated, and t is the error term. Table 19.1, column 2, shows the estimated parameters using ordinary least squares for firm-level models of learning and depreciation (see Boone et al. 2008 for details).

Firm-Level Depreciation of Knowledge

The model in Equation 19.3 allows for a depreciation in knowledge stock. If K_t is the stock of knowledge at time t, then $K_t = \lambda K_{t-1} + q_t$, $0 \le \lambda \le 1$, and $K_0 = 0$. The parameter λ represents the proportion of knowledge from previous months that is available during future months. Equation 19.2 can be generalized to:

$$\ln(q_t) = \alpha + \beta \ln l_t + \gamma \ln K_{t-1} + \varepsilon_t, \tag{19.3}$$

where $K_{t-1} = \sum_{i=1}^{t-1} \lambda^{t-i-1} q_i$. Equation 19.2 is a special case of 19.3 when $\lambda = 1$. If $0 < \lambda < 1$, then some of the knowledge gained from accumulated production is lost—that is, it is not all available for use in the current month. Column 3 in Table 19.2 shows the parameter estimates for model (3). The value of λ is 0.9934, with a standard error of 0.077, indicating that while some depreciation of knowledge does occur, it is not significant to reject the hypothesis that $\lambda = 1$.

TABLE 19.1

Estimates of Model Coefficients

Parameter	Model (2)	Model (3)
Constant	−4.009[a]	−4.0354[a]
(α)	(0.509)	(1.740)
Man Hours	0.8831[a]	0.8805[a]
(β)	(0.051)	(0.052)
Experience	0.1811[a]	0.1847[a]
(γ)	(0.0429)	(0.058)
Depreciation	1	0.9934[a]
(λ)		(0.077)
R^2	0.826	0.825
N	85	85

Standard Errors in Parentheses; a = $p < 0.05$; b = $0.05 \le p < 0.10$.

TABLE 19.2

Estimates of Model Coefficients for Symmetric Knowledge Transfer

Parameter	Estimate	Standard Error	t statistic
Man Hours (β)	0.6502	0.0199	32.6213
Focal A	0.3974	0.0473	8.3934
Focal B	0.1383	0.0515	2.684
Focal C	0.2548	0.0318	8.0201
Focal D	0.0895	0.0441	2.0314
Non-Focal A	−0.0036	0.0052	−0.701
Non-Focal B	0.0096	0.0056	1.7059
Non-Focal C	0.0078	0.0053	1.4773
Non-Focal D	0.0309	0.0073	4.2277
Constant (α_A)	−5.6273	0.4723	−11.9142
Constant (α_B)	−3.1266	0.4987	−6.2694
Constant (α_C)	−4.0612	0.3275	−12.4003
Constant (α_D)	−2.5126	0.4324	−5.8113
Depreciation (λ)	1		
R^2	0.5551		
N	462		

INTER-DEPARTMENTAL KNOWLEDGE TRANSFER

The second series of models investigate how knowledge flows within the various departments in the organization. When a client hires the firm, the project is allocated to a "focal" department that leads the project. For example, if the project calls for upgrading the HVAC system in a historical building, the lead or focal department will be mechanical. However, the electrical department will be involved in helping with the wiring design, as will the architects in helping to preserve the historical integrity. So, in this example project, the mechanical department is the focal department; and the electrical and architectural departments are collaborators on the project. In contrast to the "Firm-Level Learning Curve" section (which uses monthly input and effort), we drill the data used in the following models down to the project level (total of N projects). This allows us to explicitly model how the experience gained from previous projects—focused in any discipline—carry over to the project at hand. We will designate the departments in this firm as A, B, C, and D in order to maintain confidentiality.

Symmetric Transfer of Knowledge between Departments

$$\ln q_{i,n_i} = \alpha_i + \beta_i \ln l_{i,n_i} + \gamma_i \ln K_{i,n_i-1} + \sum_{j \neq j}^{J} \psi_j \ln K_{j,i_{n_i}-1} \times A_{ij} + \varepsilon_N, \quad (19.4)$$

where $\ln K_{j,n_i-1}$ is the knowledge stock of department j prior to the start of the nth project i of department i. Further, let $A_{ij} = 1$ if unit j is an active collaborator with unit i on a given project, and is zero otherwise. The departmental subscripts i and j take on the values A, B, C, or D, depending on the department. To simplify the analysis (and also as a consequence of Model [3]), we assume that $\lambda = 1$.

This specification allows the total experience level to date for a collaborating department to influence the productivity of the focal department, but also forces the transfer of knowledge (via the coefficient ψ_j to occur in a similar manner, regardless of who a department is collaborating with).

Table 19.2 shows the model results for this model. Here knowledge transfer is restricted to occurring symmetrically across departments so that the knowledge transfer that a collaborating department brings to a project is constant, no matter with whom the collaborator is working. This specification allows for departmental heterogeneity in two ways. First, by estimating the separate intercept for each department, the effect of differing project sizes across departments is dealt with. As for how learning accrues within a department, this specification allows for differing abilities to translate experience into output by estimating a separate experience coefficient for each focal department. The results show that only the experience associated with collaborating department D is capable of being leveraged for increases in output.

Asymmetric Transfer of Knowledge between Departments

This model is similar in most respects to Equation 19.4 except in one important aspect. Rather than restrict the knowledge transfer between the focal and collaborating departments, to be constant, irrespective of the focal partner, we allow for the asymmetric transfer of knowledge by estimating 12 knowledge transfer coefficients (ψ_{ij}) representing how the focal department i can leverage the experience of the collaborating partner j into output gains. We estimate the following model:

$$\ln q_{i,n_i} = \alpha_i + \beta_i \ln l_{i,n_i} + \gamma_i \ln K_{i,n_i-1} + \delta_i \left(\ln K_{i,n_i-1} \right)^2$$

$$+ \sum_{j=1}^{J} \sum_{i \neq j}^{J} \psi_j \ln K_{j,i n_i-1} \times A_{ij} + \varepsilon_N, \tag{19.5}$$

where $A_{ij} = 1$ if unit j is an active collaborator with unit i on a given project, and is zero otherwise.

Table 19.3 shows our preliminary results for this model and we only report the sign of the significant parameters. Diagonal elements are our estimates of γ_i, and all are positive and significant for every department, indicating that departments are able to capitalize on their own experience. The asymmetric patterns seen in the table are consistent with the findings of Table 19.2 but shed further light on exactly which knowledge transfer collaborations provide statistically meaningful output gains. Of particular interest is that almost all departments (except for A) appear to be able to transfer the experience of department D when collaborations occur. In only one case (when focal department A collaborates with department B), does a partner's experience actually seem to decrease the output.

TABLE 19.3

Some Preliminary Results for Asymmetric Knowledge Transfer

Collaborating Department	Focal Department			
	A	B	C	D
A	+	−		
B		+		
C			+	
D		+	+	+

DISCUSSION

FIRM-LEVEL LEARNING AND DEPRECIATION

Table 19.1 confirms that professional services exhibit learning curves. From column 2 of Table 19.1, the rate of learning is 0.1811. So the progress ratio is $2^{-0.1811} = 0.882$, suggesting that for every doubling of output (drawings), the productivity increases approximately by 12%.

A second significant finding is that there is no appreciable depreciation of learning, at least in this firm. In manufacturing contexts, the percent of knowledge stock retained from month to month varies widely. For example, prior results include monthly knowledge retention rates of 67% in automotive assembly (Epple et al. 1996), 75% in shipbuilding (Argote et al. 1990), 81% in a North American truck plant (Argote et al. 1990), and 96% in aircraft production (Benkard 2000). In mass service settings, past research estimated that only 47% of the knowledge stock at the beginning of the month was carried over to the next month (Darr et al.'s 1995 study on pizza franchises).

The results can be explained by two key observations (see also Boone et al. 2008). First, this firm managed all of its work product electronically. Engineers work collaboratively on CAD designs and these are available to all engineers in the firm through an indexed database. Past knowledge is therefore codified, and it is easy for engineers to access past projects for projects at hand. Second, as in any professional service, a portion of the knowledge is tacit (i.e., with the engineer who worked on the project but not in the CAD design itself)—for example, designs that did not work; unique aspects of certain clients or projects; "wisdom" acquired through years of exploring different options for multiple projects, and so on. While such knowledge is not explicit, engineers in this firm seemed to have a general sense of who "knew their stuff." Engineers often approached those who were recognized informally as "experts" in order to seek advice about projects. Often the expert gives them advice that is not documented, thus enhancing the outcome of the project. While such informal interactions are not explicitly captured by the Models (1) to (3), it plays a major role in sustaining the knowledge in the firm. We theorize that larger firms need more formalized procedures to capture the tacit knowledge that is hidden within the professional service worker.

DEPARTMENT-LEVEL DIFFUSION

The results for symmetric knowledge transfer suggest that all departments benefit from their own experience, although at differing levels. This is a function of how well each department is able to codify explicit knowledge and how effective it is in tapping the tacit knowledge residing in its workers; and as the asymmetric transfer results show, how well it taps into the experience of collaborating departments. The results also indicate that experience-based knowledge transfer among departments is limited when one assumes a symmetric knowledge transfer. Department D is able to increase the productivity of others when it is a collaborator. None of other the departments contribute to increases in productivity as collaborators. This suggests that there are barriers to the knowledge transfer from departments A, B, and C to the rest of the organization. The analysis of symmetric knowledge transfer, while important, does not provide information on bilateral knowledge transfer between department pairs.

The results of Model (5) support Jarmin's (1994) assertion that asymmetric knowledge diffusion models provide a more complete and interesting depiction of intra-organizational knowledge flows. Significant and insignificant knowledge transfers are masked by the model that assumes symmetric knowledge transfer.

Two distinct patterns of knowledge transfer are revealed in the data. First, we identify departments that appear to realize more significant outgoing knowledge transfers than incoming. The accumulated experience-based knowledge for department D is significantly associated with the productivity of departments B, C, and, of course, department D itself. In other words, this department appears to contribute significantly (in addition to itself) to the productivity of B and C. At the same time, the productivity of D is not significantly associated with the experience-based knowledge of the other departments.

Departments B and C show the opposite behavior. Their productivity was significantly associated with the experience-based knowledge of D, although no other department appeared to benefit from its production experience. It also appears that some departments (in this case A) had a negative impact on the productivity of others (B).

While our investigation of knowledge diffusion is ongoing, conversations with managers in the firm indicate that the findings may reflect the intensity of the departmental relationships. Commonalities in department cultures or informal network structures are not captured by the data. Knowledge diffusion depends in large part on organizational context; for example, the routines and the values and norms of the "transferring" and "receptor" groups. In addition, diffusion is affected by how similar two groups are, the formal and social relationships linking group members, the distance between sharing groups, and the barriers that exist between the groups. Although all of the departments reside within the same firm and share a larger organizational context, they may share significantly different cultures. More importantly, the relationships between pairs of departments are likely to be different. However, one would expect to see significant bidirectional knowledge transfer where two departments have stronger ties or networks.

The findings also suggest varying levels of absorptive capacity for the different departments. Absorptive capacity could potentially moderate the effect of network ties on knowledge transfers.

CONCLUSION

In this chapter, we set out to investigate two broad areas of inquiry: (1) can we identify learning curves in professional service organizations? Do they sustain the knowledge accumulated? (2) How does knowledge transfer between departments or units in professional service firms?

Results indicate that experience-based learning is significant; and in contrast to most manufacturing firms, professional firms are able to sustain their knowledge with time. Our results also show that knowledge transfer patterns in this organization are asymmetric. Some departments benefit from the production experience of others without exhibiting significant outgoing knowledge transfer. Other departments exhibit significant outgoing knowledge transfer, without showing any productivity benefits from the production experience of others.

Future research can focus on the following key questions:

- Can tour results be duplicated in other professional service industries? Of particular interest are the management consulting firms that, much like the design firm in this chapter, tackle a wide variety of issues in a multitude of industries.
- Can a generalized set of behavioral patterns explain asymmetric knowledge transfer? Why are some departments' transfer knowledge better than others?
- What can professional service firms do to improve the transfer of knowledge between departments so that clients' problems can be solved at less cost and more speed?

REFERENCES

Adler, P.S., 1990. Shared learning. *Management Science* 36(8): 938–957.
Argote, L., 1996. Organizational learning curves: Persistence, transfer and turnover. *International Journal of Technology Management* 11(7–8): 759–769.
Argote, L., Beckman, S., and Epple, D., 1990. The persistence and transfer of learning in industrial settings. *Management Science* 36(2): 140–154.
Argote, L., and Epple, D., 1990. Learning curves in manufacturing. *Science* 247(4945): 920–924.
Bailey, C., 1989. Forgetting and the learning curve: A laboratory study. *Management Science* 35(3): 340–352.
Benkard, C.L., 2000. Learning and forgetting: The dynamics of aircraft production. *The American Economic Review* 90(4): 1034–1054.
Boone, T., Ganeshan, R., and Hicks, R.L., 2008. Learning and depreciation in professional services. *Management Science* 54(7): 1231–1236.
Boone, T., Ganeshan, R., and Hicks, R.L., 2010. *The transfer of knowledge among units in professional services.* College of William and Mary Working Paper.
Brown, J.S., and Paul, D., 1996. Organizational learning and communities-of-practice. In *Organizational learning*, eds. Michael Cohen and Lee S. Sproull, 58–82. Thousand Oaks: Sage Publications.

Cohen, W.M., and Levinthal, D., 1990. Absorptive capacity: A new perspective on learning and innovation. *Administrative Science Quarterly* 35(1): 128–152.

Darr, E., Argote, L., and Epple, D., 1995. The acquisition, transfer and depreciation of knowledge in service organizations: Productivity in franchises. *Management Science* 41(11): 1750–1762.

Dutton, J., and Thomas, A., 1984. Treating progress functions as a managerial opportunity, *Academy of Management Review* 9(2): 235–247.

Epple, D., Argote, L., and Devadas, R., 1991. Organizational learning curves: A method for investigating intra-plant transfer of knowledge acquired through learning by doing. *Organization Science* 2(1): 58–70.

Epple, D., Argote, L., and Murphy, K., 1996. An empirical investigation of the microstructure of knowledge acquisition and transfer through learning by doing. *Operations Research* 44(1): 77–86.

Goldhar, J., and Jelinek, M., 1985. Computer integrated flexible manufacturing: Organizational, economic and strategic implications. *Interfaces* 15(3): 94–105.

Hirschmann, W.B., 1964a. Profit from the learning curve. *Harvard Business Review* 42(1): 125–139.

Hirchmann, W. B., 1964b. Learning curve. *Chemical Engineering* 71(7): 95–100.

Jaber, M.Y., and Bonney, M., 1997. A comparative study of learning curves with forgetting. *Applied Mathematics Modeling* 21(8): 523–531.

Jarmin, R.S., 1994. Learning by doing and competition in the early Rayon Industry. *RAND Journal of Economics* 25(3): 441–454.

Lane, P.J., and Lubatkin, M., 1998. Relative absorptive capacity and inter-organizational learning. *Strategic Management Journal* 19(5): 461–477.

Leonard-Barton, D., 1995. *Wellsprings of knowledge: Building and sustaining the sources of innovation.* Boston: Harvard Business Press.

Lovelock, C., 1992. *Managing services: Marketing, operations and human resources.* Englewood Cliffs: Prentice-Hall.

March, J.G., Sproull, L.S., and Tamuz, M., 1991. Learning from samples of one or fewer. *Organization Science* 2(1): 1–13.

Morris, B., and Johnston, R., 1986. Dealing with inherent variability: The difference between manufacturing and service. *International Journal of Operations and Production Management* 7(4): 13–22.

Nelson, R.R., and Winter, S., 1982. *An evolutionary theory of economic change.* Cambridge: Harvard University Press

Nonaka, I., 1994. A dynamic theory of organizational knowledge creation. *Organization Science* 5(1): 14–37.

Polanyi, M., 1967. *The tacit dimension.* Garden City: Doubleday.

Rothwell, R., 1994. Issues in user-producer relations in the innovation process: The role of government. *International Journal of Technology Management* 9(5–6–7): 629–649.

Schmenner, R.W., 1986. How can service businesses survive and prosper? *Sloan Management Review* 27(3): 21–32.

Szulanski, G., 1996. Exploring internal stickiness: Impediments to the transfer of best practices within the firm. *Strategic Management Journal* 17(SPI 2): 27–43.

van Dierdonck, R., Debackere, K., and Rappa, M. A., 1991. An assessment of science parks: Towards a better understanding of their role in the diffusion of technological knowledge. *RD Management* 21(2): 109–123.

Wright, T.P., 1936. Factors affecting the cost of airplanes. *Journal of the Aeronautical Sciences* 3(2): 122–128.

Yelle, L.E., 1979. The learning curve: Historical review and comprehensive survey. *Decision Sciences* 10(2): 302–328.

Zander, U., and Kogut, B., 1995. Knowledge and the speed of the transfer and imitation of organizational capabilities: An empirical test. *Organization Science* 6(1): 76–92.

20 Learning Curves in Project Management: The Case of a "Troubled" Implementation

Margaret Plaza, Daphne Diem Truong, and Roger Chow

CONTENTS

INTRODUCTION

In order to survive in a rapidly evolving business environment, organizations are forced to organize more work in "projects" (Meredith and Mantel 2003). In contrast to industrial projects of the past, which usually were repeated many times, projects nowadays are complex ventures that can be significantly affected by learning. Learning causes increased productivity during the course of each activity (Ash and Smith-Daniels 1999), influences the effectiveness of teams, and increases efficiency when performing future activities (Cohen and Levinthal 1990; Edmondson 1999). It has a much stronger impact on performance than even team size or diversity (Sarin and McDermott 2003).

Since organizations compete on timing and timelines (Brown and Eisenhardt 1997), it is critical to accurately forecast the project duration and to deliver according to the promised schedule. Learning has a strong impact on activity duration but, unfortunately, a vast majority of currently deployed project management systems do not consider it at all. Even the earned value method (EVM) – one of the most popular multidimensional project management systems, which has been used successfully since 1960 (Jaafari 1996; Kim et al. 2003; Moselhi et al. 2004; Raby 2000) – is based on the assumption that performance is a constant function over time (Anbari 2003; Fleming and Koppleman 2000). The assumption makes EVM ineffective on projects affected by knowledge transfers and learning; for example, in information technology (IT) implementations (Karlsen and Gottschalk 2003).

This section explores the impact of a team's learning on the management of IT projects. The discussion revolves around the real IT project managed through EVM, which was a critical stage of ERP (an enterprise resource planning software) implementation. The project was classified as unsuccessful although all objectives were delivered earlier than planned. The project manager did not consider the impact of a learning curve and calculated resource requirements from the forecasts provided by the EVM control system. As a result, resource requirements turned out to be inflated; the company hired too many consultants and unreasonably extended the project's budget.

The IT project discussed in this chapter serves as a reminder that learning effects must be considered in project management. In the following sections, the sources of inaccuracies resulting from EVM are explored and a decision support system, EVM+ (an extension to EVM), is discussed. This support system addresses the forecasting issues and improves project control. It also proves to be particularly useful during the stages of projects that are strongly impacted by knowledge transfers.

TROUBLED IT PROJECT (PLAZA 2008)

A Canadian company initiated ERP implementation in early 1999, planning to "go-live" during the first weekend of September that same year. It was decided during the package selection phase that standard ERP functionality would have to be customized in order to accommodate the complexity of the ordering and costing processes. For that reason, an IT project was initiated as an integral stage to the ERP project and a development team was added to the core implementation team. The key objectives of a development team were to establish data conversion procedures, prepare data, develop contingency plans and backup procedures, and to develop a set of customizations that were necessary for the transition period. A development team was comprised of 12 members, with the average experience of each team member ranging from one to two years. The IT project was planned to run parallel to the core project and was to be completed two weeks before the final "go-live" date, which could not be extended under any circumstances. The duration of the project was set as $T_0 = 36$ weeks. The project had to be carefully controlled so the project manager decided to use the EVM system and track schedule performance index (SPI)*:

$$SPI = \frac{EV}{PV}. \tag{20.1}$$

---

* SPI is calculated as a ratio between planned and earned values. Planned value (PV) represents the dollar value of work planned to be completed. Earned value (EV) represents the dollar value of work actually completed at any given point in time.

The project manager measured earned value (EV) at time $t_E = 8$ weeks. Since it was only 80% of the planned value (PV) ($SPI = 0.8$), the project manager concluded that the actual performance (p_1) was much lower than assumed during the planning phases (p_0) and forecast the project duration to be $T_1 = 45$ weeks.* Since the completion date could not be extended, the only possible solution was to increase the number of resources. In an attempt to compensate for the lower productivity levels and ensure that the project would be completed within $T_0 = 36$ weeks, the project manager decided to hire three additional resources at higher experience levels ($RI = 1.25$).[†]

Unfortunately, the assessments obtained from the EVM model were inaccurate and the project team continued to complete all tasks ahead of schedule. Since the IT project had to be coordinated with an ERP implementation, the project manager released additional development work to keep the expensive consulting resources fully utilized. As a result, although the project was completed on time, almost 40% more development work was added to the planned amount. Adding more resources increased the initial budget of 1.3 million Canadian dollars by an additional 30%. It was clear that the EVM model failed to accurately forecast the duration on this implementation.

WHY DID THE EVM MODEL FAIL TO GENERATE ACCURATE ASSESSMENTS?

The EVM is a popular project control system which had previously been very useful on several industrial projects. The major difference between a typical industrial project and IT implementation is the significant learning and integration issues that dominate technology projects (Plaza 2008; Plaza et al. 2010; Plaza and Rohlf 2008). The earned value method assumes a constant performance over time (Raby 2000), which was not a valid assumption on the IT project due to the fact that when the project was launched, the average experience of the project team was at least two to three times below the level of an experienced developer. Although the project team started the project with some level of experience, it was still on the learning curve during the implementation.

If SPI is measured during the early stages of the project and $SPI < 1$, as was the case during the IT project, then the performance is indeed lower than originally assumed. However, due to the learning effects, the performance will continue to increase and might, in fact, significantly exceed the assumed ceiling. If "non-linear" factors are excluded from the analysis and the duration is forecast from the linear factors only, the project manager would incorrectly presume that the performance would stay at a low level until the end of the project and would add more resources in order to compensate for this (Plaza and Turetken 2009).

Both the non-linear performance and the "initial level of experience" must be factored in by a project management method, especially if significant learning effects are present. For example, the project control models based on the logistic equation

---

* According to EVM: actual performance $p_1 = SPI * p_0$ and forecast duration $T_1 = T_0/SPI$.
† Resource index (RI) represents the level of change in the resource base. It is calculated as a ratio between the changed and previous numbers of resources.

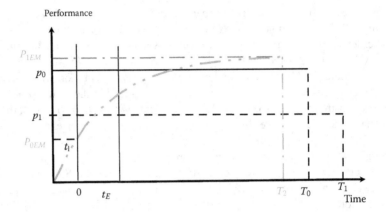

FIGURE 20.1   Learning curve parameters.

(L-curve) and on the assumption that a team begins the project with a very low level of experience, provides a much more accurate prediction of the project completion time than the EVM model (Plaza 2008; Plaza and Turetken 2009). In those models, non-linear factors, which cause performance escalation during project execution, are segregated from the linear factors, which change the performance uniformly during the entire project. The enhanced EVM model (EVM+) is also based on an L-curve and includes the impact of the initial performance level (Plaza 2009). The learning curve parameters utilized by the EVM+ system are depicted in Figure 20.1.

p_{1EM} is the performance ceiling and an asymptote to the learning curve; t_1 represents the duration of initial training before the commencement of a project; and k is a learning curve coefficient, which depicts the slope of a learning curve. The initial performance p_{0EM} can be calculated from an L-curve equation. Note that in Figure 20.1 $p_{1EM} \neq p_0$, which means that the performance ceiling determined during the project execution might be different from the performance assumed during planning. p_{1EM} depends on the combination of *SPI* and "non-linear" factors (learning effects), and can be calculated from Equation (20.2), where the corrective coefficient, *PCI* depicts the reduction/increase of work caused by learning effects plaza (2009).*

$$\frac{p_{1EM}}{p_0} = \frac{SPI}{PCI}.$$ (20.2)

$$PCI = 1 + e^{-kt_l}\frac{e^{-kt_E}}{kt_E} - e^{-kt_l}\frac{1}{kt_E}.$$ (20.3)

Based on the *EV* measured at time, t_E, the project manager can calculate a projected earned value for the remainder of the project, *PEV(t)*, from Equation (20.4), where *C* represents the total weekly cost of resources.

---

* If learning effects can be neglected, "non-linear" factors would be excluded (K → ∞) and *PCI* = 1.

$$PEV(t) = C \frac{p_{1EM}}{p_0}\left[t + e^{-kt_l} \frac{1}{k} e^{-kt} - e^{-kt_l} \frac{1}{k} \right]. \tag{20.4}$$

The actual duration of a project, T_2, can then be forecast from Equation (20.5), which cannot be solved explicitly, so one option would be to use the "goal seek" function from Microsoft Excel (Plaza 2009).

$$PEV(T_2) = C \frac{p_{1EM}}{p_0}\left[T_2 + e^{-kt_l} \frac{1}{k} e^{-kT_2} - e^{-kt_l} \frac{1}{k} \right] = PV. \tag{20.5}$$

In the next section, the impact of learning effects is investigated further. The investigation is structured around three questions, which would be critical for the IT project:

1. How should the team performance be projected?
2. How should SPI be used to forecast the duration of the project?
3. Can the earned value be projected and used as a baseline for the project?

How Should the Team Performance be Projected?

On the IT project C was averaged as \$36,000 per week (\$3,000 per week per resource), and k was assessed as 0.6 (Plaza 2008). Since the team was given one week of training before the commencement of a project, then $t_1 = 1$ week and $p_{0EM}/p_0 = 1 - e^{-kt1} = 0.56$. The planned duration of the IT project was estimated as $T_0 = 36$ weeks. If, after $t_E = 8$ weeks, the project manager measured the earned value and calculated SPI as 0.8 (or 80%), then he or she could have used Equation (20.2) from the EVM+ model to forecast the team performance for the rest of the project. An actual performance forecast from the EVM+ model for a "behind schedule" IT project ($SPI = 80\%$) is depicted in Figure 20.2.

In Figure 20.2, the team performance is also forecast for a project which has the same characteristics as an IT project but is "on schedule" ($SPI = 100\%$). Since the results are sensitive to a variation in the learning curve coefficient (Plaza et al. in press), both forecasts are plotted within ±0.1 tolerance limits for k. The corresponding performances projected from the EVM system (learning effects are not included) are depicted as horizontal lines: $p_1/p_0 = 0.8$ if $SPI = 80\%$, and $p_1/p_0 = 1$ if $SPI = 100\%$.

The forecasts of performance from the EVM+ are much higher than the horizontal lines calculated from the EVM. Although for $SPI = 80\%$ the performance forecast from the EVM+ is below p_0, it would continue to increase, thus remaining above EVM projections. This is why the EVM failed to accurately forecast the duration of an IT project. The important lesson, which can be gleaned from Figure 20.2, is that standard SPI does not accurately forecast the performance on projects impacted by learning and therefore cannot be used to forecast the duration of a project.

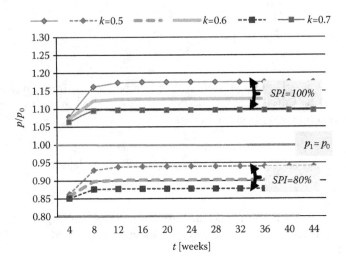

FIGURE 20.2 Performance derived from Equation (20.2) and scaled by the assumed performance p_0 for $SPI = 80\%$ (IT project is late) and $SPI = 100\%$ (IT project is "on schedule").

How Should *SPI* be Used to Forecast the Duration of a Project?

It was established in the previous section that if learning effects are present, then the team outperforms the results offered by the EVM and therefore would complete the project in a shorter period of time than was projected.* In order to improve the accuracy of the duration forecasts, "non-linear" changes in performance caused by learning must be included. The enhanced earned value method model incorporates both "linear" and "non-linear" factors (Plaza 2009), so a project manager can improve the forecasting of the actual project duration T_2 by using Equation (20.5).

The forecasts from the EVM+ model are plotted as a function of *SPI* for the IT project in Figure 20.3. In order to assess the impact of variation in a team's learning abilities, as represented by k, the plots are constructed for the learning curve coefficient $k = 0.6$ and its lower/upper limits.† The duration calculated from the standard EVM is also plotted for a reference.

The results obtained from the EVM+ model are below the durations forecast from the standard EVM for the entire range of *SPI*. It is clear that if learning effects are present, the standard EVM would always forecast the project duration to be too long, which would force the manager to take drastic actions, such as adding resources. As was the case on the IT project, that action proved to be unnecessary and caused significant budgetary problems.

How Should the Earned Value be Projected?

As depicted in Figure 20.2, if performance does not remain constant, planned value does not represent a baseline for the project. Since the EVM+ model allows the

---

* If $SPI = 80\%$ the EVM model would anticipate the duration of the project to be 45 weeks (that is, late by 9 weeks). If $SPI = 100\%$ the EVM model would assess the project to be "on schedule."
† The lower and upper limits for a learning curve can also depict a learning curve assessment error, $kErr$.

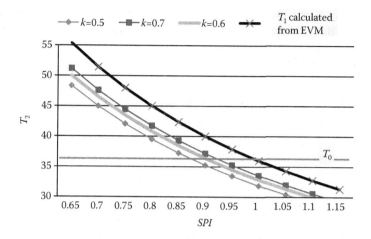

FIGURE 20.3 Project duration plotted as a function of SPI for $k = [0.5, 0.6, 0.7]$.

performance to vary, Equation 20.4 can be used to calculate a projected earned value (*PEV*) in reference to the planned value (*PV*), which becomes a dimensionless project baseline (Plaza 2010). In Figure 20.4 those projections are calculated for $k = 0.6$ and its lower/upper limits.

Since *PEV* includes the learning impact on the performance of a project team, it becomes a much more realistic predictor of a project schedule. For example, according to the EVM+ model, if, after 8 weeks $SPI = 80\%$, then *PV* would be delivered in

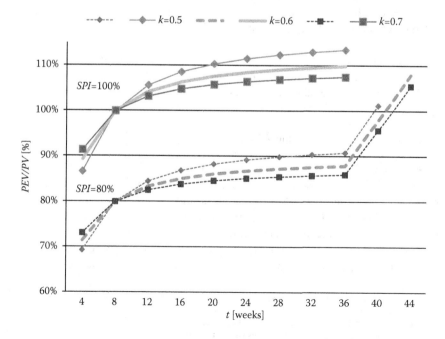

FIGURE 20.4 Earned value projected from EVM+ scaled by planned value.

approximately 40 weeks.* If $SPI = 100\%$, then the project will deliver approximately 110% of PV after 36 weeks.[†] If more resources are added, the actual value delivery during those 36 weeks will be even larger. All those forecasts turned out to be very close to the results observed on the IT project discussed here.

CAN THE EVM+ SYSTEM SOLVE THE FORECASTING ISSUES?

The enhanced earned value method model requires that in order to improve project control, a project manager must assess the team's learning ability before the commencement of the project, establish a learning curve coefficient, and also a performance ceiling. He or she will then be able to forecast the project parameters and improve the accuracy of project planning. The assessments must be verified early on during project execution.

The issues experienced on an IT project could have been prevented if the learning curve coefficient was assessed during the planning stages and the project duration was calculated from Equation (20.5).[‡] If a planned duration turns out to be too long, a manager might try to improve the situation by extending the duration of training (t_1) before the commencement of a project, which would improve the initial performance level and reduce the impact of learning effects. Other options include increasing the number of resources and using more experienced resources (Plaza 2010).

During planning, a manager could evaluate those options in simulations. For example, assume that EI represents the changes in performance levels and RI depicts the changes in the resource base. The combination of both coefficients can be substituted for $SPI/PCI = EI * RI$ in Equations 20.2 to 20.4 (Plaza 2010). A manager can use Equation (20.5) to calculate the project duration (T_2) and the projected earned value in reference to the planned value (PEV/PV) for the following scenarios:

1. extended training/integration period before the commencement of the project,
2. lower than expected performance, and
3. higher than planned number of resources, which are at a higher experience level.

In Table 20.1, the coefficients EI and RI are retrofitted to the situation observed on the IT project and the EVM+ model is utilized as a planning tool.

Base scenario represents the parameters actually assumed during the planning of the IT project and the three scenarios are added to test the impact of other options. The results obtained from the EVM+ model are summarized in Table 20.1 and in Figure 20.5. According to the EVM+ model, the base scenario requires 37 weeks to deliver the planned value. If all other parameters remained the same and only training was extended to 3 weeks (Scenario 1) the project would be completed within 36 weeks. If training remained as planned and the performance turns out to be 10% less than expected

---

* According to the standard EVM, if, after 8 weeks, $SPI = 80\%$, the project will deliver 100% of PV in 45 weeks.
† Depending on the actual value of k the results can be anywhere from 105% up to 115%.
‡ Note that during planning, a manager makes an assumption that p_0 is an asymptote to the learning curve, so $p_{1EM}/p_0 = 1$, and Equation (20.5) is reduced to $PEV(T_2) = C*[T_2 + e^{-kt_1} - (1/k)e^{-kt_1} + (e^{-kt_1})1/k] = PV.$

TABLE 20.1

Planning Scenarios for the Project

	RI	EI	t_1 [weeks]	t_2 [weeks]	PEV/PV [%] After 8 weeks	PEV/PV [%] After 36 weeks
Base Scenario (0)	1	1	1	37	89	97
Scenario 1	1	1	3	36	97	99
Extending training and assuming no change in resource numbers or performance levels.						
Scenario 2	1	0.9	1	41	80	88
No changes in number of resources but performance lower than expected.						
Scenario 3	1.25	1.1	1	27	122	134
Adding more resources at higher experience levels.						

(Scenario 2), the projected $SPI = PEV/PV$ would be only 80% after 8 weeks and the project would require 41 weeks to deliver the planned value. If more resources at higher experience levels were added (Scenario 3), the project would deliver well over 130% of the planned value after 36 weeks, which means that significantly more work would have been completed at the end of the project.

The three scenarios explain the results observed on the IT project and offer more realistic projections than a standard EVM. They also provide some evidence

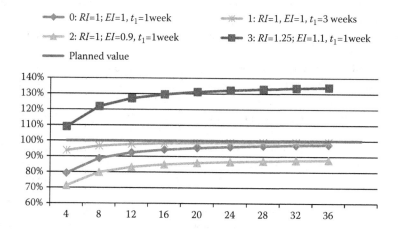

FIGURE 20.5 Earmed value projected under the condition of learning effects as percentage of planned value (assuming $k = 0.6$) for four diferent scenarios: (0) base, (1) longer training, (2) performance lower than expected, (3) more resources at higher experience levels.

that the EVM+ model can minimize the forecasting issues and improve project management.

The EVM+ model requires a significant amount of calculations, which would be extremely cumbersome and considered impractical in a typical project situation. The enhanced earned value method-based system, which is discussed in the next section, streamlines the calculations and facilitates both project planning and execution.

PROJECT MANAGEMENT WITH THE EVM+ SYSTEM

The objective of the EVM+ system* is to assist project managers in developing resource strategies and in tracking the progress of a project. Tooltips are placed throughout the system to describe the key input and output parameters for users with limited knowledge of an EVM+ model. The system is composed of the following components: project details, project resources, project planning, project tracking, and project report.

The *project details* component is the first screen that a project manager views upon initiating or loading a project. This component allows users to define and describe the project by providing the project name, description, start date, budget, and the number of modules that need to be implemented.

The *project resources* component (Figure 20.6) allows the user to define the resource characteristics. The characteristics include performance limit, p_0, learning curve coefficient, k, and learning curve assessment error, $kErr$. The error is used in the project planning component for the sensitivity analysis. If k is not known, the project manager must conduct a test where the team performance is measured as the time required to complete similar tasks at different times during the progression of the project. The test results are entered into the learning curve calculator to determine k (Plaza and Rohlf 2008).

The *project planning* component (Figure 20.7) is used to define resource strategies. Users can also compare multiple strategies using this component. A strategy is defined by setting the number of internal consulting resources, external consulting resources, external training resources, and consulting project managers. Based on the number of resources combined with the resource characteristics set in the project resources component, the system will calculate the key results used to assist users in determining which resource strategy should be used for the project. The key results include the actual project duration, T_2, and project baseline, as represented by *PEV/PV*.

A summary of the scenarios tested during planning are tallied on the left sidebar. Additionally, three types of analysis curves are generated: the performance (learning) curve, the duration versus K, and the cost versus k. The learning curve is a plot

---

* The EVM+ system is a Microsoft Windows Presentation Foundation (WPF) application developed using C# and the .Net Framework 3.5 SP1. Open-source controls and toolkits are utilized to accelerate the development of the tool. These include: WPF Toolkit (Windows Presentation Foundation [WPF] - Release: WPF Toolkit) – used to generate the date picking calendar, Visifire chart controls (Silverlight and WPF chart controls) – used for data visualization, and FluidKit – used to visually enhance the reports.

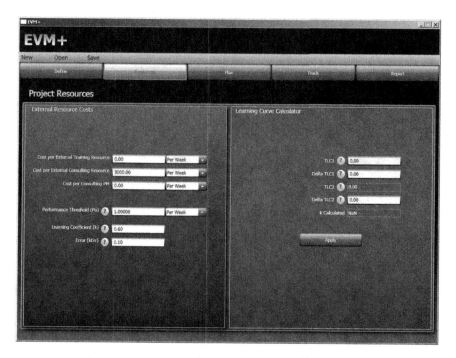

FIGURE 20.6 Project resources.

of performance over time, scaled by the performance ceiling p_0. This helps a manager to forecast the variation of performance in reference to performance assumed during planning. The "duration verus k" and "cost versus k" illustrate the sensitivity of the results to the learning curve coefficient k.

The *project tracking* component (Figure 20.8) allows users to track the progress of the project relative to the active (base) project strategy. Tracking is accomplished by adding statuses. A status contains inputs for the status date (the number of weeks since the project's start date [T_E]), learning curve coefficient and errors ($kErr$), schedule performance index (SPI), and the accumulated actual cost up to the status date.

For each status defined, the system calculates the following EVM+ results:

- Performance correction index (PCI)
- Expected project duration calculated from EVM (T_1)
- Expected project duration calculated from EVM+ (T_2)
- Planned value (PV)
- Earned value (EV)

Similar to the project planning component, a learning curve is also accessible, which shows the team's performance based on the learning coefficients entered. The results are also available for users interested in seeing how the PV, EV and EV/PV change over time. Analysis plots for "T_2 versus k" and "EV versus k" have been included to allow users to view the sensitivity against k.

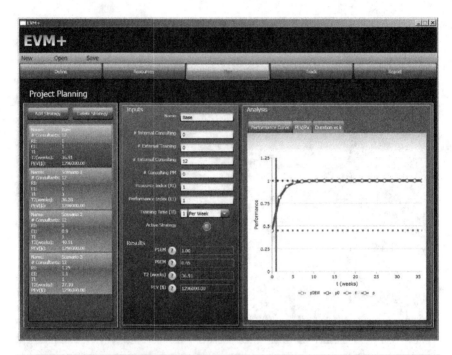

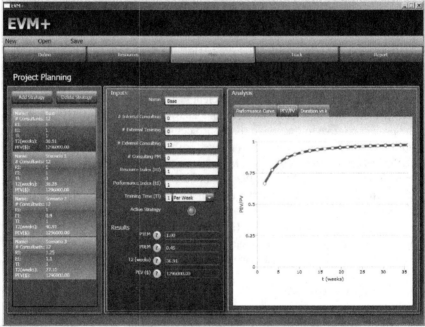

FIGURE 20.7 Project planning.

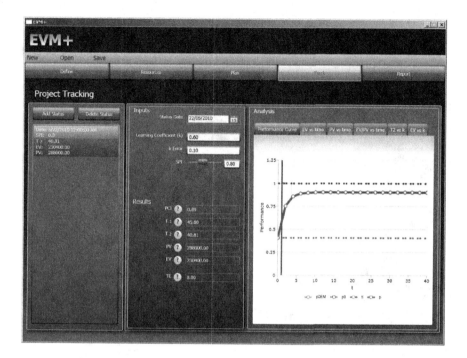

FIGURE 20.8   Project tracking.

The *project reports* component (Figure 20.9) displays a summary of the key results at each status date. For example, the projected earned value to the planned value could be used as a reference from which the progress of the project can be tracked.

CONCLUSIONS

The prototype decision system (EVM+) discussed in this chapter incorporates both "linear" and "non-linear" factors. The system combines a learning curve with the schedule performance index, which improves the accuracy of forecasting and offers a more accurate baseline from which the project progress can be tracked. The next step for our research would be to integrate the EVM+ system with the standard EVM system and incorporate the reports generated by the EVM+ model into a project management methodology.

Although learning curves play an ever-increasing role in project management, the standard project management methods and tools do not consider the impact of learning. Yet, as demonstrated here, if learning effects are not included, the planning becomes unreliable. The models that are currently used do not properly inform the project management, which causes budgetary and scheduling issues.

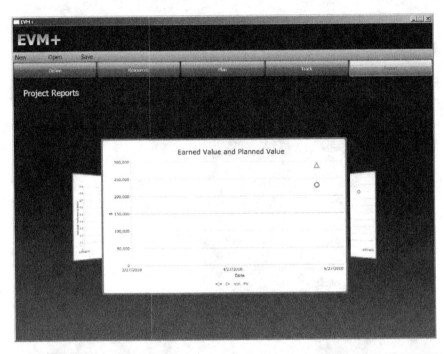

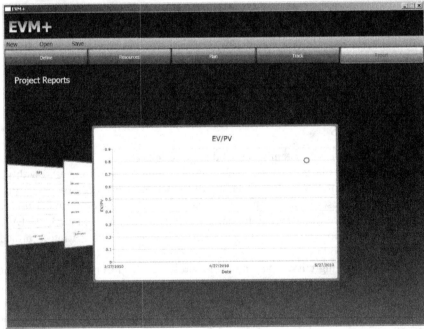

FIGURE 20.9 Project reports.

REFERENCES

Anbari, F.T., 2003. Earned value project management method and extension. *Project Management Journal* 34(4): 12–23.

Ash, R., and Smith-Daniels, D.E., 1999. The effect of learning, forgetting, and relearning on decision rule performance in multiproject scheduling. *Decision Sciences* 30(1): 47–82.

Brown, S., and Eisenhardt, K., 1997. The art of continuous change: Linking complexity theory and time-paced evolution in relentlessly shifting organizations. *Administrative Science Quarterly* 42(1): 1–34.

Cohen, W.M., and Levinthal, D.A., 1990. Absorptive capacity: A new perspective on learning and innovation. *Administrative Science Quarterly* 35(1): 128–152.

Edmondson, A.C., 1999. Psychological safety and learning behavior in work teams. *Administrative Science Quarterly* 44(3): 350–383.

Fleming, Q.W., and Koppleman, J.M., 2000. *Earned value project management* (3rd edition). Pennsylvania: Newton Square, Project Management Institute.

Jaafari, A., 1996. Time and priority allocation scheduling technique for projects. *International Journal for Project Management* 14(5): 289–299.

Karlsen, J.T., and Gottschalk, P., 2003. An empirical evaluation of knowledge transfer mechanisms for IT projects. *The Journal of Computer Information Systems* 44(1): 112–119.

Kim, E., Wells, W.G.J., and Duffey, M., 2003. A model for effective implementation of earned value management methodology. *International Journal of Project Management* 21(5): 375–382.

Meredith, J.R., and Mantel, S., 2003. *Project management – A managerial approach*. New York: Wiley.

Moselhi, O., Li, J., and Alkass, S., 2004. Web-based integrated project control system. *Construction Management and Economics* 22(1): 35–46.

Plaza, M., 2008. Team performance and IS implementations: Application of progress curve to the earned value method during IS project. *Information Systems Frontiers* 10(3):347–359.

Plaza, M., 2009. Integrating learning effects in management of technology projects. *European Journal of Management* 9(4): 21–38.

Plaza, M., 2010. *Decision support for the planning of ERP projects: A methodology for analyzing the impact of training*. Working paper 2007-02-OL-170, Ted Rogers School of Management, Ryerson University, Toronto, ON, Canada.

Plaza, M., Ngwenyama, O.K., and Rohlf, K., 2010. A comparative analysis of learning curves: Implications for new technology implementation management. *European Journal of Operational Research,* 200(2): 518–528.

Plaza, M., and Rohlf, K., 2008. Learning and performance in ERP implementation projects: A learning curve model for analyzing and managing consulting costs. *International Journal of production Economics* 115(1): 72–85.

Plaza, M., and Turetken, O., 2009. A model-based DSS for integrating the impact of learning in project control. *Decision Support Systems* 47(4): 488–499.

Raby, M., 2000. Project management via earned value. *Work Study* 49(1): 6-10.

Sarin, S., and McDermott, C., 2003. The effect of team leader characteristics on learning, knowledge application, and performance of cross-functional new product development teams. *Decision Sciences* 34(4): 707–739.

21 Timing Software Upgrades to Maximize Productivity: A Decision Analysis Model Based on the Learning Curve

Aziz Guergachi and Ojelanki Ngwenyama

CONTENTS

INTRODUCTION

Every year firms spend billions of dollars on upgrading and implementing new information technology (IT) with the goal of improving organizational productivity and end up being unhappy with the outcomes. In 2002, Morgan Stanley reported that U.S. companies wasted $130 billion in the previous two years on IT projects (Ward 2002). Although the evidence from empirical research on IT and organizational productivity improvements is contradictory, managers continue to make large investments in IT. A 1994 Standish Group study reported that 73% of these IT projects did not meet top management's expectations (The Standish Group 1994). However, it has been empirically established that IT can increase organizational productivity if effectively managed (Basu et al. 2001; Bresnahan and Trajtenberg 1995; Brynjolfsson and Hitt 1996; Dewan and Min 1997; Ko and Osei-Bryson 2004a, 2004b; Lee and Menon 2000;

Loveman 1994; Parente 1994). Still under investigation are the conditions under which an organization can maximize its performance improvements (Ko and Osei-Bryson 2004a, 2004b). For example, Loveman (1994) used the Cobb-Douglas production function to study the relationship but found no evidence of a productivity increase from IT investment. On the other hand, Weill (1992) found that transactional IT investments had a positive impact on firm performance, whereas strategic IT or informational IT did not. Prasad and Harker (1997) found that investments in IT labor produced substantially high returns in productivity, whereas investments in IT capital did not. Also, Lee and Menon (2000) found that investments in IT capital were associated with increased productivity in the healthcare industry, whereas investments in IT labor were not. More recently, Ko and Osei-Bryson (2004a, 2004b) found that there is a trade-off point at which investment in IT labor will yield no more improvements in productivity, while conversely, investment in IT capital would. Noticeably missing from all these discussions is the "learning curve (experience curve) effect" and the timing of IT investment and organizational productivity.

It has been observed from empirical studies that when new technologies are implemented in organizations the productivity level drops for a time then rises again and eventually surpasses prior levels (Argote et al. 1990; Argote and Epple 1990; Chambers 2004; Cooley et al. 1997; Dorroh et al. 1986; Jovanovic and Nyarko 1995). One explanation for the initial loss of productivity is employee resistance to the new technology. For example, Tyre and Orkilowski (1994) state that during the initial phase of implementation, organizational members see the new technology as a distinct artifact, completely separate from their daily activities. They have trouble understanding the usefulness of the technology, and often resist its implementation. Argyris (1990) argues that initially "organizational defense routines" develop, which slow down the adoption of new technologies. Another explanation for this initial loss of productivity is the amount of learning required to understand and adapt the technology to work routines (Ghemawat and Spence 1985; Hornstein and Krusell 1996; Li and Rajagopalan 1997). There is a growing literature of empirical studies that have documented the learning curve effect on organizational productivity in different industries (Argote et al. 1990; Brynjolfsson and Hitt 1996; Hornstein 1999; Jovanovic and Nyarko 1995). However, there is little research on applying the learning curve as a basis of analysis for IT implementation decisions. In this paper we address the latter gap in the research. We present a framework and approach to analyzing the timing of IT upgrading and implementation in order to maximize organizational productivity. The rest of the paper is organized as follows: in "The Software Upgrade Problem" section we outline some basic issues of the software upgrading and implementation problem. In "Framework for Analysis" section we present the framework and decision for analyzing the upgrade decision problem. Finally, in the "Concluding Comments" section we conclude the paper with suggestions for future research.

SOFTWARE UPGRADE PROBLEM

Deciding when to upgrade or implement a new IT application is a common problem that all IT managers face. Many organizations use a range of information technologies that have different product life cycles, all of which require upgrading at some point in time. The timing of software upgrade releases is driven by the software vendor's product life cycle and profit maximization goals, rather than the implementing firm's learning and performance curve (Ahtiala 2006). Increasingly, IT managers are pressured into upgrading technologies before their firms have reaped the benefits of their existing technological applications. One strategy that software vendors use to pressure firms into "timely" upgrading is to discontinue technical support for the older version shortly after the new version has been announced. While firms may have no organizational need to implement the newer version of the software, they might feel that by not upgrading at the same time as their competitors, they may risk a loss of competitive advantage. Indeed, software vendors often claim that there will be productivity improvements as a result of upgrading.

The scale of the upgrade problem can also be quite daunting. While the single largest investment in IT that many managers must make involves the initial implementation and upgrading of enterprise resource planning (ERP) systems, other software upgrades such as operating systems can have as significant an impact on the organization's productivity. Take for example, a medium-size organization that has some six thousand computer workstations running various end-user software applications and tens of servers running various ERP-type software applications. Further, assume for simplicity that all the computers are running identical operating systems. A single decision to upgrade the operating systems of all the computers would have productivity implications for the entire organization. Further, upgrading the operating system might also force the upgrading of other software applications, as the newer version of the operating system is often incompatible with the existing application's software. This could lead to a significant cost to the organization. Besides the cost of lost productivity, a variety of other costs must also be considered. First, there is the cost of purchasing the software; then there is the cost of implementation and training; and finally, there is the cost of organizational learning. Numerous vintage technology adoption models claim that due to a variety of sunk costs, it may not always be optimal for firms to immediately invest in the new technology (Lieberman 1987). The timing of the implementation should also be considered. Klenow (1998) argues that a firm should update its technologies when they are operating at peak efficiency and are becoming less efficient than newer technologies that are operated with no experience. But just how would the IT manager determine the appropriate time to upgrade?

FRAMEWORK FOR ANALYSIS

It has long been established that workers in organizations learn and subsequently improve their performance. Empirical studies of this learning-performance relationship have been theorized as "learning curves", (also called "experience curves" or "progress functions") (Adler and Clark 1991; Argote and Epple 1990; Baloff 1971; Chambers 2004; Dorroh et al. 1986; Ebert 1976; Ghemawat and Spence 1985; Hall and Howell 1985; Wright 1936). The learning curve illustrates the commonality of improvement rates in learning among tasks (Dorroh et al. 1986; Hall and Howell 1985; Jovanovic and Nyarko 1995; Jovanovic 1997). Initially, there is a rapid increase in learning potential, followed by lesser rates of improvement. These types of learning patterns have been documented in a variety of activities, including pressing buttons, rolling cigars, performing arithmetic, and writing novels (Ritter et al. 2002). As early as 1936 the relationship between organizational learning and productivity was established. T.P. Wright was the first to publish the organizational learning curve in his study, which focused on the production of airframes (Wright 1936). Wright noted that the total number of labor hours consumed in the production of an airframe was a decreasing function of the total number of airframes already produced. This relationship, defined in the following equation, has been generally used to produce the learning curve in production environments:

$$y = ax^{-b},$$

where y is the units of labor needed to produce the xth unit of output, a is the units of labor needed to produce the first unit of output, x is the aggregate number of units of output, and b is rate of reduction in labor as aggregate output rises.

There has been significant interest among cognitive psychologists, economists, and industrial engineers regarding the process of learning (Adler and Clark 1991; Argote et al. 1990; Argote and Epple 1990; Chambers 2004; Dorroh et al. 1986; Lieberman 1987; Pananiswami and Bishop 1991; Venezia 1985). Bryan and Harter began the formal study of the learning curve in 1899 when they examined the learning patterns of adults who sent and received Morse code. They concluded that after 10 months of experience, the participants were four times more productive in interpreting and forming messages (Arrow 1962). Since then, their conclusions have been adopted for the study of organizational productivity.

The learning curve illustrates that practice will almost always improve performance. It implies that the most dramatic improvements will take place at the beginning of the learning process and then reach a type of plateau. The learning curve also implies that with sufficient practice, people can reach similar rates of performance. As Arrow (1962)—one of the first learning curve researchers—has stated, knowledge increases with time. The process of learning is a product of experience, which allows an individual to obtain the required knowledge. An attempt at solving a specific problem initiates the process of learning, thus increasing the knowledge of the learner (Adler and Clark 1991). More recently, Jovanovic and Nyarko (1995) formulated a Bayesian learning model to illustrate the technology learning effect on productivity. We believe that the learning curve can assist managers in making decisions about the implementation and upgrading of information technologies in organizations. Understanding the organizational learning curve for new software applications and upgrades could yield better productivity outcomes for the organization.

ORGANIZATIONAL CONTEXT

Let us consider the case of Ontario Life, a medium-size insurance company that is heavily dependent on a specific software technology for the execution of its value chain processes. It had been only three years since Ontario Life first purchased this software when the vendor announced an upgraded version. The IT manager, Samuel Walsh, wanted to know if he should go ahead and upgrade the technology right away, or wait until his company had gotten the maximum possible benefits from the version they currently employed. Further, Sam wanted to know if he should postpone the upgrade, and if so, for how long? What are the major parameters that affect the making of such a decision? These are some of the fundamental issues we would like to address in this paper. Learning how to use a technology in an organization is never instantaneous. It will always take some time before the organization reaches the stage where it makes the best use of the technology. To model the technology learning process, we will introduce the parameter r, referred to as the "cost reduction" (dollars/time unit) due to the use of technology.

Basic Concepts

Let T be the task that needs to be carried out in the organization. The costs (units of labor/dollars/time) of executing T, with and without technology, are x_1 and x_2, respectively. The cost reduction r due to the use of technology is by definition equal to $x_2 - x_1$. When the software is initially introduced into the organization, the value of r will be low, due to the fact that the users are at an early stage of learning how to use it. However, as time passes, the value of r will increase until it reaches a plateau, which we define as maximum efficiency (productivity) of use of the software. A typical graph of r as a function of time, t, that describes the organizational learning with the software would have an S-shape curve, such as in Figure 21.1.

Let us now assume that while the organization is in the process of learning how to use the software, the vendor announces an upgrade version of the software at a certain time, t_0. In this paper, we intend to develop a mathematical framework depicting the relationships among the various parameters that affect the making of the decision of whether the organization should upgrade its technology right at time t_0, or wait. To start the development of this framework, we will first look at the ideal case where the organization's members are "instantaneous" learners; that is, where no learning is required when a new technology is introduced into the organization. The analysis of this case is beneficial for two reasons: (1) because of its simplicity, it will be pedagogically useful in pointing out the main parameters involved in the analysis and deriving the basic relationships among these parameters. These relationships should have an intuitive interpretation. (2) This ideal case will serve as a good approximation of a case where the organization is an instantaneous learner—that is, where maximum efficiency in using the software $\{r(t)\}$ is rapidly achieved, as illustrated in Figure 21.2.

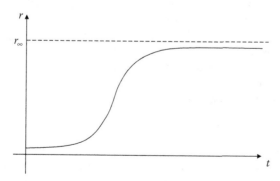

FIGURE 21.1   A normal organizational learning curve.

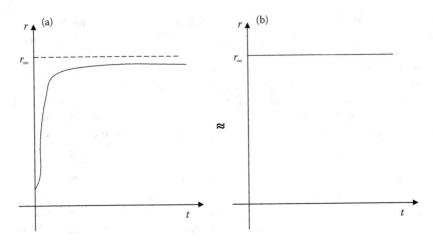

FIGURE 21.2   Case of rapid achievement of maximum efficiency in using the software.

Case of the Instantaneous Learner

In the case where the organization is an instantaneous learner, the curve for $r(t)$ as a function of the time t will not be S-shaped. Since no learning is involved, it will be a straight horizontal line as described by Figure 21.2(b). The organization starts making the best use of the technology immediately after it has acquired it. Therefore, the cost savings that will be generated at a certain time t, as a result of introducing the technology into the organization, can be calculated by the simple equation: $r_{1_\infty} t$. Let us now denote the cost of ownership of technology as C_1. Due to the above savings, the organization will have recovered the cost C_1 at the time instant t_1, defined by the equation: $r_{1_\infty} t_1 = C_1$, which implies that $t_1 = C_1/r_{1_\infty}$.

Assume now that the vendor announces the availability of a software upgrade that is worth acquiring at the time instant t_0. There are two situations that we must consider:

Situation 1: $t_0 \geq t_1$

The vendor announces the availability of the software upgrade *after* the organization has recovered all the costs of ownership of the original version of the software, and has attained maximum productivity from using the software. Consequently, the decision to upgrade is a standard investment decision that is totally independent of the fact that the organization has already owned a previous version of the software. To make such decision, the organization need only look at other factors, such as the availability of cash, the significance of the task T to be executed using the technology, the return on investment, and so on.

Situation 2: $t_0 < t_1$

The vendor announces the availability of the software upgrade while the organization is still in the process of recovering the cost it has paid for the original version of the software and has not yet attained maximum productivity from using it. In this situation, several parameters related to the fact that the organization has already owned

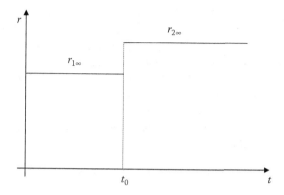

FIGURE 21.3 Technology upgrade takes place at t_0 leading to an increase in the productivity.

a previous version of the technology will affect the decision to upgrade. Examples of such factors are: the proportion of C_1 that has been recovered at the time of the announcement of the availability of the upgrade, the cost of the upgrade, the amount of savings that will be generated as a result of adopting the upgrade, and so forth. In the next paragraphs, we will look at how all these parameters are interrelated.

Let us denote the cost of ownership of the technology upgrade as C_2 (note that this cost could be equal to 0 if the vendor is giving out the upgrade for free), and the new reduction in cost that will result from adopting the technology upgrade as r_2. We implicitly assume that $r_{2_\infty} > r_{1_\infty}$, by definition of upgrade.

Let us assume that the organization does upgrade its technology right at the time, t_0, when the upgrade was announced. Then we will end up with a graph for $r(t)$ as described in Figure 21.3.

The total cost that the organization has paid now is $C_1 + C_2$, and the time instant t_2 at which this organization will recover this cost is determined from the following equation:

$$r_{1_\infty} t_1 + r_{2_\infty} (t_2 - t_0) = C_1 + C_2, \text{which gives } t_2 = \frac{C_1 + C_2 - r_{1_\infty} t_1}{r_{2_\infty}} + t_0.$$

For the upgrade to be worthwhile at the time t_0, we have to make sure that the time t_2 (required to recover the cost $C_1 + C_2$) is less than or equal to the time t_1 (required to recover C_1); that is $t_2 = ((C_1 + C_2 - r_{1_\infty} t_1)/r_{2_\infty}) + t_0 \leq t_1 = C_1/r_{1_\infty}$, which implies that $t_0 \leq C_1/r_{1_\infty} - C_2/r_{2_\infty} - r_{1_\infty}$. Define $\Delta C = C_2 = (C_1 + C_2) - C_1$ as the increase in cost that is required for the upgrade, and $\Delta r = (r_{2_\infty} - r_{1_\infty})$ is the improvement in savings that results from the technology upgrade. Thus, the previous equation becomes: $t_0 \leq C_1/r_{1_\infty} - \Delta C/\Delta r$.

Therefore, we see that, for the upgrade to be worthwhile to the organization, there is a condition on the time t_0 (at which the upgrade takes place) that needs to be satisfied: t_0 should be less than or equal to the time t_{cond} defined by: $t_{cond} = C_1/r_{1_\infty} - \Delta C/\Delta r$.

In other words, the upgrade is beneficial and should be carried out by the organization only at the times t_0 belonging to the interval $[0, t_{cond}]$. Now, for this interval to exist, we need to make sure that t_{cond} is greater than 0: $t_{cond} > 0$; that is, $C_1/r_{1_\infty} > \Delta C/\Delta r$.

This is a basic condition that needs to be satisfied for the organization to entertain the idea of upgrading the software. In business terms, this condition specifies that the relative increase in the cost of ownership should be less than the relative increase in the rate of savings. Intuitively, this condition makes good sense: for the upgrade to be worthwhile, its extra cost should be relatively low, and its contributions to the organization's productivity should be relatively high. Having examined the instantaneous learning case, we can specify two basic conditions that must be satisfied for the software upgrade to be worthwhile before the time t_1. These are: $\Delta C / \Delta C_1 < \Delta r / \Delta r_{1_\infty}$ and $t_0 \le C_1 / r_{1_\infty} - \Delta C / \Delta r$.

Everyday Case: The Organization Has a Learning Curve

In general, organizations do have learning curves. Consequently, achieving maximum proficiency in using new software requires time and effort. However, the shape of the learning curve, the graph of the function $r(t)$, is not normally known a priori. Because of this, it will not be possible to derive simple algebraic relationships among the various parameters involved in the decision, as we did for the ideal case. We will, therefore, limit the discussion below to establishing a relationship that makes use of the integral of the functions $r(t)$.

Let t be a time instant occurring after the technology is introduced in the organization (see Figure 21.4). The shaded area in Figure 4 represents the total amount of money that the organization would have saved up until the time t, as a result of implementing the software. In other terms, it is the return on the investment in the software that the organization would have achieved by the time t. Mathematically, the shaded area is the integral of the function $r_1(t)$ between 0 and t; that is, $\int_0^t r_1(t)\,dt$.

As in the previous illustration, the cost of ownership of software is defined as C_1. Then, as a result of the savings in cost, the organization will recover the cost C_1 at the time instant t_1, defined by the equation: $\int_0^{t_1} r_1(t)\,dt = C_1$.

Again we assume that the software vendor announces the availability of the software upgrade at the time instant t_0. Then, as in the ideal case, there are two possible situations:

Situation 1: $t_0 \ge t_1$

The vendor announces the availability of the software upgrade after the organization had recovered the cost of ownership of the original version of the software and had achieved maximum productivity from its use. In this situation, the fact that the

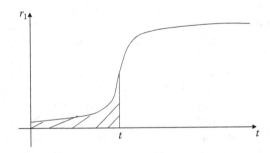

FIGURE 21.4   The learning curve in an organization representing the productivity versus time.

organization has already owned a previous version of the software becomes irrelevant to the decision to upgrade.

Situation 2: $t_0 < t_1$

The vendor announces the availability of the software upgrade while the organization is still in the process of recovering the cost it has paid for the original version of the software and has not yet achieved maximum productivity from its use. In this situation, the fact that the organization has already owned a previous version of the technology has to be taken into account in the decision to upgrade. To analyze the issue, let us assume that the organization goes ahead and carries out the upgrade at time t_0. Such a decision will lead to a discontinuous learning curve for $r(t)$, illustrated in Figure 21.5. Note that the functions r_1 and r_2 in Figure 21.5 represent the reductions in cost corresponding to the original and the upgraded versions of the software, respectively. The values of r_{1_∞} and r_{2_∞} represent the reductions in cost corresponding to both of these software versions, respectively, when the learning process is completely finished. Of course, an assumption that we will have to make is that the value of r_{2_∞} is greater than that of r_{1_∞}, otherwise we would not have a true software upgrade.

The cost C_2 of the software upgrade has two components: (1) the price, P_2, which includes what the organization pays to the vendor (which could be 0, if the vendor gives out the upgrade for free); and the cost of deployment of the new version; and (2) the losses, L_2, that the organization incurs from abandoning the investment in the original version of the software, which is composed of the cost of the original, the cost of deployment, and the cost of learning to t_0, and the new investment in climbing another learning curve for the software upgrade. These losses are represented by the shaded area in Figure 21.6.

The computation of L_2 depends on the shapes of the learning curves of the original, and the upgraded versions of the technology. Mathematically, L_2 is obtained by the following integral:

$$L_2 = \int_{t_0}^{t_e} \left[r_1(t) - r_2(t) \right] dt,$$

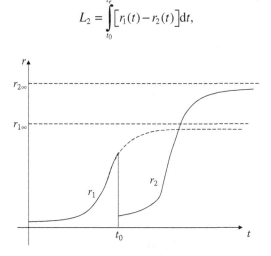

FIGURE 21.5 Discontinuous learning from early upgrading.

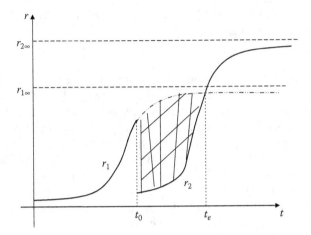

FIGURE 21.6   The region of loss from upgrading too early.

where in this equation t_e is the time point at which r_1 and r_2 intersect. Now in some cases, there may be no losses. This could happen when the organization is able to learn the upgrade version of the software more efficiently than it did with the original. For such an occurrence, the losses would be equal to 0.

POINT OF COST RECOVERY

With an understanding of the above, we can now specify the point of cost recovery. This would be the ideal point beyond which the organization could expect, with reasonable certainty, productivity gains from upgrading the software. This "point of cost recovery" is the time instant t_2 at which the organization will recover the total cost, $C_1 + C_2$, that it has paid for both the original and the upgraded versions of the software. We can determine this point using the following equation:

$$\int_0^{t_0} r_1(t)\,dt + \int_{t_0}^{t_2} r_2(t)\,dt = C_1 + C_2,$$

which can be transformed as follows:

$$\int_0^{t_0} r_1(t)\,dt + \int_{t_0}^{t_2} r_2(t)\,dt - C_1 = C_1 + C_2 - C_1$$

$$\int_0^{t_0} r_1(t)\,dt + \int_{t_0}^{t_2} r_2(t)\,dt - C_1 = \Delta C.$$

The reader will observe that ΔC is the increase in cost required for the software upgrade. By replacing C_1 by its expression, we obtain:

$$\int_0^{t_0} r_1(t)\,dt + \int_{t_0}^{t_2} r_2(t)\,dt - \int_0^{t_1} r_1(t)\,dt = \Delta C$$

$$\int_{t_0}^{t_2} r_2(t)\,dt - \int_{t_0}^{t_1} r_1(t)\,dt = \Delta C$$

$$\int_{t_0}^{t_1} r_2(t)\,dt + \int_{t_1}^{t_2} r_2(t)\,dt - \int_{t_0}^{t_1} r_1(t)\,dt = \Delta C$$

$$\int_{t_0}^{t_1} \left[r_2(t) - r_1(t) \right]\,dt + \int_{t_1}^{t_2} r_2(t)\,dt = \Delta C$$

$$\int_{t_1}^{t_2} r_2(t)\,dt = - \int_{t_0}^{t_1} \left[r_2(t) - r_1(t) \right]\,dt + \Delta C.$$

PRODUCTIVITY IMPROVEMENT CONSTRAINT

While cost recovery is necessary for productivity gains from upgrading the software, it is not a sufficient criterion for ensuring such gains. What is necessary is to set a time constraint for the software upgrade which would ensure productivity gains for the organization. As we stated earlier in our illustration, for the upgrade to be worthwhile at the time t_0, we have to make sure that the time t_2 (required to recover the cost $C_1 + C_2$) is less than or equal to the time t_1 (required to recover C_1); of course, in the case of the instantaneous learner, that time point is t_0, which is unknown and must be derived for the organization that has a learning curve. Therefore, we specify that the condition $t_2 \leq t_1$ must be satisfied. However, since $r_2(t)$ is a positive function, we can state the condition as:

$$\left(t_2 \leq t_1 \right) \leftrightarrow -\int_{t_1}^{t_2} r_2(t)\,dt \geq 0$$

$$\left(t_2 \leq t_1 \right) \leftrightarrow -\int_{t_1}^{t_2} r_2(t)\,dt \geq 0$$

$$\left(t_2 \le t_1\right) \leftrightarrow \int_{t_0}^{t_1}\left[r_2(t) - r_1(t)\right]dt - \Delta C \ge 0$$

$$\left(t_2 \le t_1\right) \leftrightarrow \Delta C \le \int_{t_0}^{t_1} \Delta r(t)\,dt.$$

The reader will observe that $\Delta r(t)$ is the productivity gain, $r_2(t) - r_1(t)$, that results from the software upgrade. Consequently, for the software upgrade to be worthwhile to the organization, the values of ΔC and t_0 must satisfy the condition:

$$\Delta C \le \int_{t_0}^{t_1} \Delta r(t)\,dt.$$

PROOF OF CONSISTENCY

As a proof of consistency of the above results, let us see if we can re-derive the relationship $t_0 \le C_1/r_{1_\infty} - \Delta C/\Delta r$, which we found for the "everyday case", from the above general inequality. Let us assume, then, that the functions $r_1(t)$ and $r_2(t)$ are the constants: $r_1(t) = r_{1_\infty}$ and $r_2(t) = r_{2_\infty}$. The general inequality leads to: $\Delta C \le (r_{2_\infty} - r_{1_\infty})$ $(t_1 - t_2)$. Since $t_1 = C_1/r_1$, we obtain:

$$\Delta C \le \left(r_{2_\infty} - r_{1_\infty}\right)\left(\frac{C_1}{r_{1_\infty}} - t_0\right).$$

By solving for t_0 we get the term that we wanted, which is:

$$t_0 \le \frac{C_1}{r_{1_\infty}} - \frac{\Delta C}{r_{2_\infty} - r_{1_\infty}} = t_0 \le \frac{C_1}{r_{1_\infty}} - \frac{\Delta C}{\Delta r}.$$

CONCLUDING COMMENTS

In the above, we presented and proved a mathematical model for timing software upgrades based on learning theory, specifically the learning curve. Our model is general and applicable to all situations of software—and other information technology—upgrade problems. It is also relatively easy to implement in standard spreadsheet software. As such, it could be used by managers in their decision-making about investments in IT. Further, while our work here is theoretical, it can serve as a basis for empirical studies to determine the general learning curves for different types of software, such as ERP, and for different types of organizations. Such research could assist information technology managers by suggesting estimation curves that could be used to support decision-making on a wide variety of technology-implementation problems.

REFERENCES

Adler, P.S., and Clark, K.B., 1991. Behind the learning curve: A sketch of the learning process. *Management Science* 37(3): 267–281.

Ahtiala, P., 2006. The optimal pricing of computer software and other products with high switching costs. *International Review of Economics and Finance* 15(2): 202–211.

Argote, L., Beckman, S., and Epple, D., 1990. The persistence and transfer of learning in industrial settings. *Management Science* 36(2): 140–154.

Argote, L., and Epple, D., 1990. Learning curves in manufacturing. *Science* 247(4945): 920–924.

Argyris, C., 1990. *Overcoming organizational defenses: Facilitating organizational learning.* Boston: Allyn and Bacon.

Arrow, K.J., 1962. The implications of learning by doing. *The Review of Economic Studies* 29(3): 155–173.

Baloff, N., 1971. Extension of the learning curve – Some empirical results. *Operational Research Quarterly* 22(4): 329–340.

Basu, S., Ferland, J., and Shapiro, M., 2001. Productivity growth in the 1990s: Technology utilization or adjustment? *Carnegie-Rochester Conference Series on Public Policy* 55(1): 117–165.

Bresnahan, T., and Trajtenberg, M., 1995. General purpose technologies: Engine of growth? *Journal of Econometrics* 65(1): 83–108.

Brynjolfsson, E., and Hitt, L., 1996. Productivity, business profitability and consumer surplus: Three different measures of information technology value. *MIS Quarterly* 20(2): 121–142.

Chambers, C., 2004. Technological advancement, learning, and the adoption of new technology. *European Journal of Operational Research* 152(1): 226–247.

Cooley, T., Greenwood, J., and Yorukoglu, M., 1997. The replacement problem. *Journal of Monetary Economics* 40(3): 457–499.

Dewan, S., and Min, C.K., 1997. The substitution of information technology for other factors of production: A firm-level analysis. *Management Science* 43(12): 1660–1675.

Dorroh, J.R., Gulledge, T.R., and Womer, N., 1986. A generalization of the learning curve. *European Journal of Operational Research* 26(2): 205–216.

Ebert, R.J., 1976. Aggregate planning with learning curve productivity. *Management Science* 23(2): 171–182.

Ghemawat, P., and Spence, A., 1985. Learning curve spillovers and market performance. *The Quarterly Journal of Economics* 100(5): 839–52.

Hall, G., and Howell, S., 1985. The experience curve from an economist's perspective. *Strategic Management Journal* 6(3): 197–212.

Hornstein, A., 1999. Growth accounting in technology revolutions. *Federal Reserve Bank of Richmond Economic Quarterly* 85(3): 1–22.

Hornstein, A., and Krusell, P., 1996. Can technology improvements cause productivity slowdowns? *NBER Macroeconomics Annual* 11: 209–259.

Jovanovic, B., 1997. Learning and growth. In *Advances in economics*, eds. D. Kreps and K.F. Wallis. Cambridge University Press.

Jovanovic, B., and Nyarko, Y., 1995. A bayesian learning model fitted to a variety of empirical learning curves. *Brookings Papers on Economic Activity* 1: 247–305.

Klenow, P., 1998. Learning curves and the cyclical behavior of manufacturing industries. *Review of Economic Dynamics* 1(2): 531–550.

Ko, M., and Osei-Bryson, K.M., 2004a. Exploring the relationship between information technology investments and firm performance productivity using regression splines analysis. *Information & Management* 42(1): 1–13.

Ko, M., and Osei-Bryson, K.M., 2004b. Using regression splines to assess the impact of information technology investments on productivity in the healthcare industry. *Information Systems Journal* 14(1): 43–63.

Lee, B., and Menon, N., 2000. Information technology value through different normative lenses. *Journal of Management Information Systems* 16(4): 99–119.

Li, G., and Rajagopalan, S., 1997. The impact of quality on learning. *Journal of Operations Management* 15(3): 181–191.

Lieberman, M., 1984. The learning curve and pricing in the chemical processing industries. *RAND Journal of Economics* 15(2): 213–228.

Loveman, G.W., 1994. An assessment of the productivity impact of information technologies. In *Information technology and the corporation of the 1990s: Research studies*, eds. T.J. Allen and M.S. Scott Morton, 84–110. New York: Oxford University Press.

Pananiswami, S., and Bishop, R., 1991. Behavioral implications of the learning curve for production capacity analysis. *International Journal of Production Economics* 24(1–2): 157–163.

Parente, S., 1994. Technology adoption, learning-by-doing and economic growth. *Journal of Economic Theory* 63(2): 346–369.

Prasad, B., and Harker, P., 1997. Examining the contribution of information technology toward productivity and profitability in U.S. retail banking. Working paper no. 97–09, Financial Institutions Center. The Wharton School.

Ritter, F.E., Shadbolt, N.R., Elliman D., Young, R.M., Gobet F., and Baxter G.D., 2002. *Techniques for modeling human performance in synthetic environments: A supplemental review*. Human Systems Information Analysis Center. Wright-Patterson Air Force Base, OH.

The Standish Group, 1994. *Chaos Report*, http://www.standishgroup.com/sample_research/chaos_1994_2.php

Tyre, M.J., and Orlikowski, W.J., 1994. Windows of opportunity: Temporal patterns of technological adaptation in organizations. *Organization Science* 5(1): 98–118.

Venezia, I., 1985. On the statistical origins of the learning curve. *European Journal of Operational Research* 19(2): 191–200.

Ward, S., 2002. Companies squander billions on tech. *USA Today*, May 20.

Weill, P., 1992. The relationship between investment in information technology and firm performance: A study of the valve manufacturing sector. *Information Systems Research* 3(4): 307–333.

Wright, T.P., 1936. Factors affecting the cost of airplanes. *Journal of the Aeronautical Sciences* 3(2): 122–128.

22 Learning Curves for CAD Competence Building of Novice Trainees

Ramsey F. Hamade

CONTENTS

INTRODUCTION

Despite the hype associated with the advertised "zero learning curve" computer-aided design (CAD) (e.g., All the "zero learning curve" features of AutoCAD®), learning theory predicts that a long time must be spent training and practicing before a CAD novice becomes an expert. Throughout the training, trainees' productivity increases as a result of them being introduced to more complex concepts and capabilities of CAD functions, along with the constant practice of their newly learned skills. This is reflected in higher efficiency and more effective performance while building CAD models by expert CAD operators. This is due to accumulated procedural and honed declarative skills. A distinction can be made between the declarative and procedural components of acquired CAD knowledge. While declarative knowledge relates mainly to learning specific instructions—for example, which icons to press on the graphic user interface (GUI), or which feature attributes to define in order to accomplish a CAD task – the procedural component of CAD knowledge manifests itself through the schema that trainees develop to construct a solid CAD model. This procedure is not system specific, so this train of thought is transferable across different CAD systems. Declarative knowledge, on the other hand, is system specific.

It deals with the specifics of the task at hand, lends itself to description through language and visual aids, and is easy to capture.

Empirical learning curves that describe CAD training do not exist in the open literature. In order to remedy this deficiency, Hamade et al. (2005) set out to apply the principles of the theory of learning to the learning of CAD software. They showed that novices' learning of mechanical CAD software (Pro ENGINEER®, Parametric Technology Corporation, Waltham, Massachusetts) correlates well with established theoretical learning curves that depict learning behavior. For performance assessment purposes, beginner-level test-solid CAD parts were utilized by which the performance time was recorded for each subject. Also recorded was the total number of features-of-size (or, simply, features) needed by each trainee to construct the part. (A feature in Pro ENGINEER® terminology describes a basic solid shape such as a cube, cylinder, hole, and so on). It was found that as procedural and declarative knowledge was accumulated, fewer (but more complex) features were utilized, thus contributing to improved performance. Consequently, a method was developed that enabled the separating of the declarative and procedural components of acquired CAD from the total learning curve without having to record individual menu clicks. This is accomplished through using a procedure based on physically decomposing the total learning curve based on the number of features utilized in the model construction. Formal training of mechanical CAD was found to be highly cognitive, with procedural skills seen to have an even larger cognitive load than declarative knowledge. These findings were further validated by Hamade et al. (2007) by including test parts of a more challenging nature. It was also shown that the method is equally applicable to a newer version of the CAD software. More recently, Hamade et al. (2009) provided further insight into how human performance is affected by cumulative experience. In this chapter, the author presents a summary of the work, outlined in Hamade et al. (2005, 2007, 2009), on learning theory as applied to the mechanical CAD training of novices. In Section 2, the experimental method is explained. The "learning curve" that is applicable to this line of research is described in Section 3; while in Section 4, procedural and declarative knowledge are explained from a CAD perspective. Section 5 presents the method used in separating the performance-time learning curve into its procedural and declarative components. The findings are then summarized in Section 6.

METHOD OF CAD KNOWLEDGE ACQUISITION

TRAINING AND PARTICIPANTS

This study was conducted over a 16-week-long semester in a course for CAD training offered at the American University of Beirut. This elective, senior-level course is taught once a year by the author. Training was carried out using Pro ENGINEER®, a feature-based, associative, and parametric CAD solid modeler. The software's GUI features a combination of icons and pull-down menus. Training was conducted based on two 2-hour hands-on sessions per week, for a total of four hours of lab training. These training sessions included instructions on using the CAD and on dealing with the ever-increasing and challenging commands. Students were also assigned one homework exercise after each lecture, which was consistent with the taught features. Although the supervised practice time did not exceed four hours per week, trainees were allowed to practice out of class as often (and for as long) as they desired. The training was performed in a laboratory that contained 24 powerful PC workstations equipped with fast graphics and memory access capabilities. Each participant had full access to one of these workstations. Once the license was acquired from the server, the users were able to begin using the software without delay or interruption.

The 44 subjects (all male except for one female) that were studied constituted the total enrolled pool of mechanical engineering students over a two-year period. The trainees were found (Hamade and Artail, 2008) to have good technical and science backgrounds—especially in mathematics—and to have already taken introductory courses on mechanical drawing and engineering graphics. They were also found to have already been exposed to 2-D CAD (mostly AutoCAD).

CAD KNOWLEDGE ACQUISITION

The developing skills of the trainees were measured utilizing a simple test part of which four variations—designated Test Parts 1, 2, 3, and 4—are shown in Figure 22.1. These variations were each envisioned to be of the same novice level of difficulty (suitable for beginners), but were different enough so as not to bias the test results by having the trainees build the same model four times. These parts were designed so they could be constructed by an expert using only two or three relatively complex features. A novice, on the other hand, would typically use many more simple features.

The subjects were called to perform Parts 1, 2, 3, and 4 in the laboratory after 2, 4, 7, and 12 weeks from the start of training, respectively. The performance studies were performed in the same laboratory in which the training was conducted. In every exercise, the trainees were asked to build the test part and both the build time and the number of features used by each student were recorded. The object to be modeled was provided as a hard, fully dimensioned drawing that contained several dimensioned views of the test part (including a reference isometric view). Before the trainees were told to start modeling on the computer, they were given one minute offline to understand, visualize, and strategize on how to construct the model online. This short period of time should involve pure procedural cognitive thinking, the goal of which is to comprehend the object and develop a suitable procedure (or a CAD

FIGURE 22.1 Novice-level assessment CAD solid models: Test Parts 1 through 4 (from left to right).

feature-build plan). Although the students were allowed access to the online help menu and instruction manuals, few students, if any, were observed to take advantage of this opportunity over the course of the performance exercise.

LEARNING CURVE

The classical learning curve (Wright's learning lurve [WLC]; Wright 1936) is described by

$$y(n) = y(1)n^{-b}, \tag{22.1}$$

where $y(n)$ is the production time for the nth repetition, $y(1)$ is the time to produce the first repetition, n is the repetition number, and b is the learning slope, with $b = -\log \varphi/\log 2$, and φ ($0 \leq \varphi \leq 1$) being the learning rate. For example, if the time taken to perform the first repetition is 100 minutes, $y(1) = 100$, and the trainee's learning rate is 70%, $\varphi = 0.7$ ($b = 0.514$), then the time to perform the tenth repetition is 30.6 minutes. A trainee with a large learning slope learns faster than one with a smaller slope. Repeating the same example for $\varphi = 0.9$ ($b = 0.152$), the time to perform the tenth repetition becomes 70.5 minutes.

Several studies (see Lieberman 1987) have suggested that learning is a function of time rather than cumulative output, such as:

$$T(t) = T(1)t^{-b_t}, \tag{22.2}$$

where $T(t)$ is the performance time when recall occurs by time t; $T(1)$ is the theoretical production time at first repetition (taken here at week 1); and b_t is the exponent of the learning curve, describing the speed of learning. The time-based model in Equation 22.2 will be followed in this chapter (other multivariate forms were explored in Hamade et al., 2009).

PERFORMANCE-TIME LEARNING CURVE

The first type of learning curve constructed in this study is a plot of performance time (speed in minutes) versus the number of training weeks. This build time variable is a direct measure of the user efficiency in using the CAD tool and, consequently, the resulting learning curve may be thought of as the efficiency learning curve. In building the test models as discussed in Section 2, the trainees' performance times were found to fit the power form in Equation 22.2. Figure 22.2 shows the performance times, with spread bars, for Test Parts 1 through 4 at 2, 4, 7, and 12 weeks of training, respectively. For the entire class, $T(1)$ and b_t were found to be 48.14 and 0.693, respectively, with a learning rate of 61.8%, $\varphi = 0.618$, which is indicative of a highly cognitive learning process. Such a cognitive declarative knowledge is revealed by the trainees' ability to learn and use increasingly more complex commands (that require more menu picks and/or keystrokes and, thus, are more time consuming than their simple counterparts) and execute them at increasing speeds.

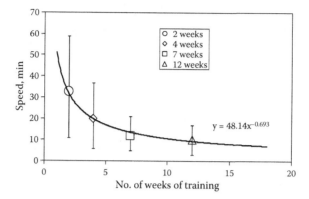

FIGURE 22.2   Performance times, with spread bars, for Test Parts 1 through 4 at 2, 4, 7, and 12 weeks of training, respectively, with power trend-line shown through the empirical data. (From Hamade et al., *Computers & Industrial Engineering*, 56, 1510, 2009.)

MODEL-SOPHISTICATION LEARNING CURVE

Procedural knowledge was acquired as trainees developed more advanced skills in the building of features, allowing them to think up of new strategies involving fewer, more effective features. In building the test models, the trainees' sophistication improved, with model builds approaching the minimum number of features possible. Therefore, the second type of learning curve is a curve where the tracked variable is the number of features (of size) that each trainee used in building the model. This learning curve reflects the sophistication or quality of the constructed CAD models and may be thought of as the effectiveness learning curve. Similar to speed, the number of features (of size) used by each trainee at each assessment interval serves as an indication of the development of procedural knowledge. This measure was also found to fit the power form in Equation 22.2 where $F(t)$ is the number of features used in modeling at time t; $F(1)$ is the theoretical number of features utilized in the model's construction (theoretical modeling sophistication); and b_f, similar to b_t, was derived through monitoring trainees' sophistication progress. Figure 22.3 shows the number of features, with spread bars, for Test Parts 1 through 4 at 2, 4, 7, and 12 weeks of training, respectively. $F(1)$ and b_f for the class of trainees were found to be 12.33 and 0.504, respectively, with a learning rate of 70.5%, $\varphi = 0.705$, which is indicative of a highly cognitive learning process.

PROCEDURAL AND DECLARATIVE KNOWLEDGE: A CAD PERSPECTIVE

Performance and innovation levels improve as knowledge is accumulated over time, with the improvements in performance time and sophistication being mainly due to the accumulation of declarative and procedural skills (Bhavnani et al. 1999; Rodriguez et al. 1998). Specifically, and in the context of mechanical CAD platforms, acquired declarative knowledge may be captured by such knowledge as:

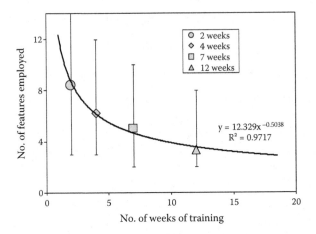

FIGURE 22.3 Features employed, with spread bars, for Test Parts 1 through 4 at 2, 4, 7, and 12 weeks of training, respectively. Power trend-line is shown through the empirical data.

learning how to operate the GUI, learning how to properly assign feature attributes, and learning how to sketch geometry in "sketcher mode." The execution of these commands entails that the operator invokes motor operations (key strokes and menu picks) with an underlying high percentage of additional cognitive skills being used to decipher these commands and their proper execution. Figure 22.4 is an example

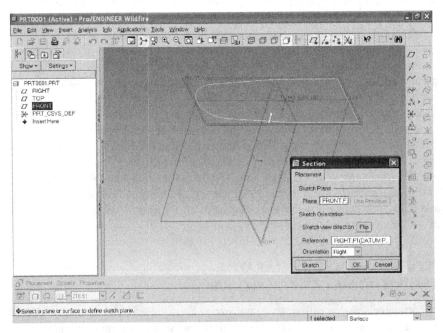

FIGURE 22.4 An illustration of a declarative task in which the sketcher in Pro ENGINEER® is being utilized to sketch geometry for the base feature of Test Part 1. (From Hamade et al., *Computers & Industrial Engineering*, 56, 1510, 2009.)

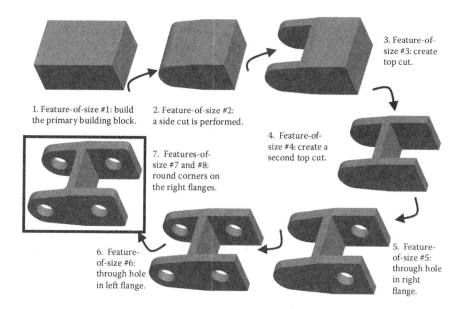

1. Feature-of-size #1: build the primary building block.

2. Feature-of-size #2: a side cut is performed.

3. Feature-of-size #3: create top cut.

4. Feature-of-size #4: create a second top cut.

7. Features-of-size #7 and #8: round corners on the right flanges.

6. Feature-of-size #6: through hole in left flange.

5. Feature-of-size #5: through hole in right flange.

FIGURE 22.5 An illustration of an unsophisticated CAD part build procedure where a large number of relatively simple features (eight in total) is required to construct Test Part 1. (From Hamade et al., *Computers & Industrial Engineering*, 56, 1510, 2009.)

of the latter in which the sketcher in Pro ENGINEER® is being utilized to sketch geometry for the base feature of Test Part 1.

Procedural knowledge operates at a higher level than the details-oriented declarative knowledge. For the creation of a solid model, a build plan is required. Such a plan involves breaking the desired shape into relatively simple shapes. This is an exceptionally cognitive exercise which represents the procedural knowledge component of the total learning curve. Figures 22.5 and 22.6 are illustrations of two such possible build plans (identified as build scenarios 1 and 2). Figure 22.5 outlines the first scenario where there is a large number of relatively simple features (eight in total) required to construct Test Part 1. Such a procedure is likely to be devised by novice CAD users. Figure 22.6 illustrates the second scenario, which requires more sophistication but which results in CAD parts with fewer (though more complex) features.

DISAMBIGUATING PROCEDURAL AND DECLARATIVE PERFORMANCE

The class performance-time learning curve that was constructed above will be decomposed here into its procedural and declarative components. In Hamade et al. (2005), it was shown that the aggregate performance-time learning curve may be written mathematically as the sum of the procedural and declarative curves as follows:

$$T(t) = T(1)t^{-b_t} = \{T_p(1) + T_d(1)\}t^{-b_t} = T_p(1)t^{-b_p} + T_d(1)t^{-b_d}, \qquad (22.3)$$

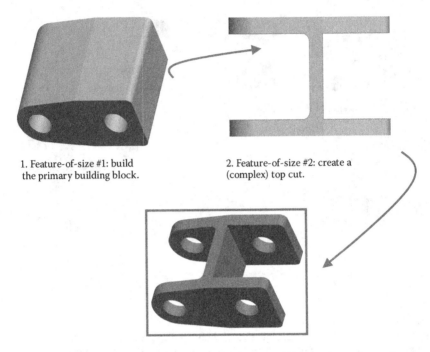

1. Feature-of-size #1: build the primary building block.

2. Feature-of-size #2: create a (complex) top cut.

FIGURE 22.6 Illustration of a sophisticated CAD part build procedure where a few number of relatively complex features (two in total) is required to construct Test Part 1. (From Hamade et al., *Computers & Industrial Engineering*, 56, 1510, 2009.)

where T, t, and b have the same meaning as above, and where the subscripts t, p, and d stand for total, procedural, and declarative, respectively. The total learning rate, b_t, is bound such that $b_d \leq b_t \leq b_p$. The total sum of the procedural and declarative components should add up to the total performance time measured in the experiments. Once the declarative curve has been determined, figuring out the procedural knowledge curve then becomes a simple matter of subtracting the declarative curve from the total curve.

Determining the declarative component of the curve is based on a relation linking the number of features used to the performance time, and is plotted in Figure 22.7. In the figure, a significant downshift is observed in building the model of Part 2 as compared with Part 1. The line that represents Part 3 has a smaller slope than those for Parts 1 and 2. These observations imply that it took increasingly less time to complete the features used to construct the test parts as the number of features used decreased. However, as the features became complex and their number approached that of the absolute minimum possible, the trend is reversed. This can be seen in the slightly larger slope for Part 4, which was constructed as the training drew to a close.

In Figure 22.8, iso-procedural curves are overlaid on the aggregate learning curve. Iso-procedural curves are so called because each of these curves corresponds to a fixed number of utilized features (2, 4, 6, 8, 10 or 12), resulting in no procedural

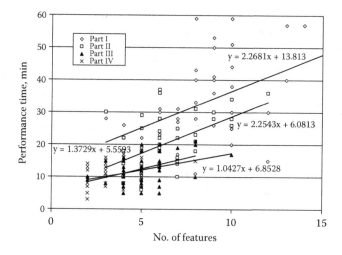

FIGURE 22.7 Performance time versus number of features employed. (From Hamade et al., *Computers & Industrial Engineering*, 56, 1510, 2009.)

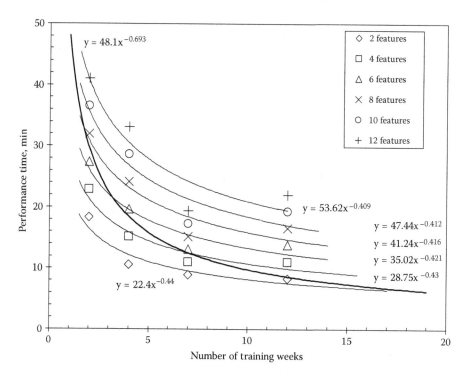

FIGURE 22.8 The total learning (performance) curve superimposed on iso-procedural curves corresponding to building test parts using 2, 4, 6, 8, 10, and 12 features. (From Hamade et al., *Computers & Industrial Engineering*, 56, 1510, 2009.)

content. The data needed to plot these iso-procedural curves is determined from Figure 22.7 by drawing vertical lines at 2, 4, 6, 8, 10, and 12 features, respectively. The intersection of each of these lines with the four straight lines in the figure yields four data points, which are used to construct the iso-procedural curves. Power fits for these curves are shown in the figure and are contrasted with the aggregate learning curve (heavy curve). In Figure 22.8, the bottom curve, with $T(1)$ of 22.4 and a slope of 0.44 (learning rate = 73.7%), is assumed in this study to be the optimum learning curve for the class aggregate. If one were to start out the model build at this best possible two-feature scenario, then one would assume that any improvement can be attributed only to the declarative component. Iso-procedural learning curves corresponding to higher feature counts of 4, 6, 8, 10 and 12 features all have progressively smaler slopes of 0.43 (learning rate = 74.2%), 0.421 (74.7%), 0.416 (74.9%), 0.412 (75.1%), and 0.40 (75.8%), respectively, and have larger $T(1)$ values of 28.75, 35.02, 41.24, 47.44, and 53.62 minutes, respectively. Figure 22.8 reveals that the aggregate learning curve is bound by the iso-procedural curves representing high feature counts and low feature counts at the beginning and the end of the training, respectively.

Figure 22.9 depicts incremental CAD knowledge acquisition. The average trainee starts the learning process along a high feature count iso-procedural curve and continues learning by "sliding along" and acquiring declarative knowledge "speed." This sliding action represents the declarative portion of performance improvements. At the point at which the learning curve intersects the next lower, more effective iso-procedural curve, a procedural jump is made vertically to this lower iso-procedural curve. This is a purely procedural increment. This incremental sliding behavior continues until a plateau is reached at which learning occurs at infinitesimally small steps that require excessively long times.

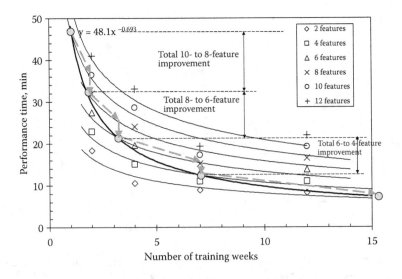

FIGURE 22.9 The incremental depiction of knowledge acquisition.

Quantitatively separating the declarative and procedural components of the total learning curve is illustrated in Figure 22.10. The total knowledge acquisition $T(t)$ may be thought of as the sum of the vertical increments due to declarative, $T_d(t)$, and procedural, $T_p(t)$, knowledge. The summation of the declarative components of the performance time may be mathematically described as:

$$T_d(t) = \ldots + [T(1)t_{10}^{-b_t} - T_{d10}(1)t_8^{-b_{d10}}] + [T(1)t_8^{-b_t} - T_{d8}(1)t_6^{-b_{d8}}]$$
$$+ [T(1)t_6^{-b_t} - T_{d6}(1)t_4^{-b_{d6}}] + [T(1)t_4^{-b_t} - T_{d4}(1)t_2^{-b_{d4}}],$$

(22.4)

where t_{10} represents the intersection point of the total curve and that of the 10-features iso-procedural line, and $T_{d10}(1)$ and b_{d10} are the power law parameters of the 10-features iso-procedural line, respectively. The same carries on for the other numeral subscripts. The first bracket in Equation 22.4 represents the declarative increment due to the step down from the 10-feature line to the 8-feature line, or $\Delta T_{d(10-8)}$. This is the vertical performance difference between the total learning curve evaluated at t_{10} and that of the 10-feature iso-procedural line evaluated at t_8 with a value of $48.1(1.05)^{-0.693} - 47.44(1.8)^{-0.4117} = 9.3$ minutes.

The performance-time increments due to the procedural component may be described by:

$$T_p(t) = \ldots + [T_{d10}(1)t_8^{-b_{d10}} - T(1)t_8^{-b_t}] + [T_{d8}(1)t_6^{-b_{d8}} - T(1)t_6^{-b_t}]$$
$$+ [T_{d6}(1)t_4^{-b_{d4}} - T(1)t_4^{-b_t}] + [T_{d4}(1)t_2^{-b_{d2}} - T(1)t_2^{-b_t}],$$

(22.5)

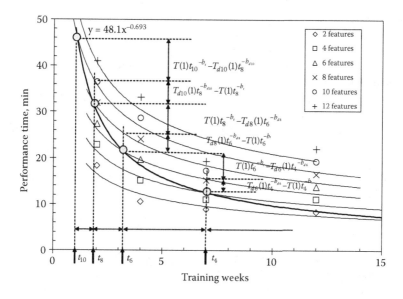

FIGURE 22.10 Quantitative description of incremental knowledge acquisition for both declarative and procedural skills.

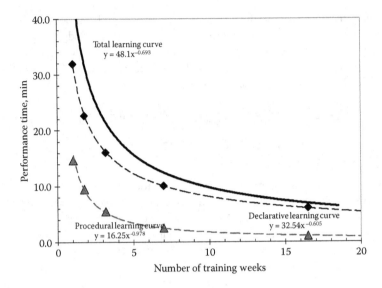

FIGURE 22.11   The total learning curve co-plotted with the declarative and procedural learning components as calculated from Equations 22.4 and 22.5, respectively.

where the first bracket reflects the procedural increment due to the step from the 10-feature line down to the 8-feature line, or $\Delta T_{p(10-8)}$. This increment has a value calculated as $47.44(1.8)^{-0.4117} - 48.1(1.8)^{-0.693} = 5.2$ minutes. Therefore, the total incremental time due to the step down from 10-features to 8-features is the sum of its declarative and procedural components $\Delta T_{(10-8)}$, or 14.5 minutes, as is shown in Figure 22.10. The quantitative values of the subsequent steps can be similarly computed.

The declarative and procedural increments are added up according to Equations 22.4 and 22.5, respectively, resulting in the declarative ($32.5t^{-0.605}$) and procedural ($16.25t^{-0.978}$) learning curves, which are shown plotted along with the total learning curve in Figure 22.11. As training progressed, it is natural to expect that individual trainees would develop both types of knowledge (declarative and procedural), albeit at different rates. Figure 22.11 reveals a much larger slope for procedural (0.978; learning rate = 50.7%) than declarative (0.605; learning rate = 65.7%) knowledge acquisition. Both of these learning rates are extremely fast, indicating that CAD knowledge acquisition is highly cognitive. Only in a few other studies have such fast rates been reported (Dutton and Thomas 1984; Camm 1985; Blancett 2002).

SUMMARY

In this study, a method was laid out for constructing the learning curves associated with CAD knowledge acquisition. Two types of learning curves were devised: a performance-time learning curve and a model-sophistication learning curve. Both of these learning curves were found to follow a power law in a fashion consistent with that of the time-based model in Equation 22.2. The learning rates of both of these curves suggest that both types of knowledge acquisition are highly cognitive. The findings suggest that to attain maximum productivity levels while CAD training, participants have to become efficient by performing a task both quickly and effectively, using a smaller number of features.

Furthermore, a technique was developed to decompose the learning curve into two components: procedural and declarative. Figure 22.11 reveals that the performance-time learning curve is dominated by declarative knowledge. This is especially true at the conclusion of the training. Procedural knowledge plays an important rule only at the start of training, where such top-level cognitive thinking will pay big dividends for those who can see how to construct effective model build plans by utilizing a small number of features. Also, the improvement at the start of training is much larger than progress toward the conclusion of training, at which point the performance-time learning curve predicts a "bleeding edge" where additional training returns very low values of return on investment (ROI).

ACKNOWLEDGMENTS

The financial support provided by the University Research Board of the American University of Beirut is gratefully acknowledged.

REFERENCES

All the "zero learning curve" features of AutoCAD®. http://au.autodesk.com/?nd=class&session_id=2919 (accessed March 2, 2010).

Bhavnani, S.K., John, B.E., and Flemming, U., 1999. Strategic use of CAD: An empirically inspired, theory-based course. *Proceedings of the 1999 Conference on Human Factors in Computing Systems*, May 15–20, Pittsburgh, PA, 183–190.

Blancett, R.S., 2002. Learning from productivity learning curves. *Research Technology Management* 45(3): 54–58.

Camm, J.D., 1985. A note on learning curve parameters. *Decision Sciences* 16(3): 325–327.

Dutton, J.M., and Thomas, A., 1984. Treating progress functions as a managerial opportunity. *The Academy of Management Review* 9(2): 235–247.

Hamade, R.F., Artail, H.A., and Jaber, M.Y., 2005. Learning theory as applied to the mechanical CAD training of novices. *International Journal of Human-Computer Interaction* 19(3): 305–322.

Hamade, R.F., Artail, H.A., and Jaber, M.Y., 2007. Evaluating the learning process of mechanical CAD students. *Computers and Education* 49(3): 640–661.

Hamade R.F., and Artail, H.A., 2008. A study of the influence of technical attributes of beginner CAD users on their performance. *Computer-Aided Design* 40(2): 262–272.

Hamade, R.F., Jaber, M.Y., and Sikström, S., 2009. Analyzing CAD competence with univariate and multivariate learning curve models. *Computers and Industrial Engineering* 56(4): 1510–1518

Lieberman, M.B., 1987. The learning curve, diffusion, and competitive strategy. *Strategic Management Journal* 8(5): 441–452.

Rodriguez, J., Ridge, J., Dickinson, A., and Whitman, R., 1998. *CAD training using interactive computer sessions*. Proceedings of the 1998 Annual ASEE Conference, Seattle, WA.

Wright, T., 1936. Factors affecting the cost of airplanes. *Journal of Aeronautical Science* 3(2): 122–128.

23 Developments in Interpreting Learning Curves and Applications to Energy Technology Policy

Bob van der Zwaan and Clas-Otto Wene
In Memoriam: Leo Schrattenholzer

CONTENTS

INTRODUCTION

In 1999, a group of scientists and analysts from academia, industry, and government agencies who were participating in a workshop arranged by the International Energy Agency (IEA) observed that experience and learning curves "are underexploited for public policy analysis" (IEA/OECD 2000, Appendix B, 112). They recommended that experience and learning curves "are used to analyze the cost and benefits of programs to promote environment friendly technologies" and "are explicitly considered in exploring scenarios to reduce CO_2 emissions and calculating the cost of reaching emissions targets" (IEA/OECD, 2000, Appendix B, 114). The IEA Committee on Energy Research and Technology (CERT) supported the findings of the workshop and initiated an international collaboration.

McDonald and Schrattenholzer (2001) provided the first overview of experience curves for energy technology. Particular attention in their seminal publication was given to the distribution of learning rates for 48 cases of different energy technologies. While the statistical significance of a number of the included learning rates was not particularly high, several robust overall conclusions could be made on the basis of their distribution. One of the main findings was that their distribution had

425

both commonalities and differences with respect to those published by Dutton and Thomas (1984) for (not exclusively energy-based) technologies manufactured by individual companies. Both of these studies found a median learning rate of close to 20%, but McDonald and Schrattenholzer (2001) found a higher frequency of smaller learning rates than Dutton and Thomas (1984) did.

The last 10 years have seen a steadily increasing amount of studies being carried out on technology learning as measured by experience curves, initially on renewable technologies, but now also including fossil, nuclear, hydrogen and end-use technologies (Junginger 2010; Schoots et al. 2008; 2010). Recent high-level policy documents embrace the insights from experience and learning curves into the crucial role of market deployment in order to make low-carbon energy technologies more cost-efficient (IEA/OECD 2006, 2008, 2010; Stern 2006; EESC 2009).

The work in the late 1990s on experience and learning curves at the IEA Secretariat established their political implications and global reach (IEA/OECD, 2000), but this work could rely on earlier pioneering efforts in order to apply to learning curves a political message in the energy field. Maycock and Wakefield (1975) and Williams and Terzian (1993) analyzed the reduction in cost of photovoltaic (PV) cells, and Neij (1999) did the same for wind power. Tsuchiya (1989) showed how niche markets can act as stepping stones for the riding down of the experience curve and inspired the launching of Japan's Residential PV System Dissemination and PV-Roof Program (IEA/OECD 2000, 64–74). Early efforts to introduce learning curves into energy-economic models failed because of the strong non-linearities of the curves and the increasing returns to scale that the introduction of these curves generated in these models (Manne 1994). Technology learning was then introduced in the technology modules of the U.S. National Energy Modeling System, but it proved the model could only be solved iteratively (Kydes 1999). Messner (1997), and Mattsson and Wene (1997) first managed to introduce learning curves in global bottom-up energy system models and solve their models in single-solution runs, while van der Zwaan et al. (2002) were the first to achieve this in a top-down version of energy-economic-environment (EEE) models. Following these efforts, technology learning is now a standard feature in most bottom-up energy systems and top-down EEE-integrated assessment models (Kahouli-Brahmi 2008).

The reason for the political relevance of experience and learning curves is that technology learning is a key factor behind technological change (Sagar and van der Zwaan 2006). The curves are therefore important tools in answering the question of how to retain economic growth while using technology to transform the global energy system so that it can sustain a low-carbon economy. The insights from experience and learning curves suggest that such a transformation can be made, but initial investments in learning will be in the order of hundreds of billions of U.S. dollars (IEA/OECD 2000, 2008; van der Zwaan and Rabl 2004). This raises the questions of how well we understand the technology learning process, and how well we can trust forecasts based on this understanding.

Although high-level policy reports acknowledge the existence of learning effects, they also point to the large uncertainties in the estimates of future learning phenomena. Because of the long time spans involved, these uncertainties translate into large uncertainties about the resources or learning investments that are needed to bring the new technologies to breakeven with incumbent high-carbon technologies. The reports stop short of recommending experience and learning curves as operational policy tools. In their report to the 2006 G8 meeting of heads of state, the IEA finds:

> "Technology learning is the key phenomenon that will determine the future cost of renewable power generation technologies. Unfortunately, the present state of the art does not allow reliable extrapolations." (IEA/OECD 2006, 231)

The Stern review observes that:

> "There is a question of causation since cost reductions may lead to greater deployment; so attempts to force the reverse may lead to disappointing learning rates. The data shows technologies starting from different points and achieving very different learning rates." (Stern 2006, 362, 411–412)

A key criticism is that the curves appear to express purely empirical relations between cost, price or technical performance, and cumulative production or use. Theoretical grounding is needed to explain observed learning rates, limit uncertainties in extrapolations, and legitimize government deployment programs. Another important point of criticism is that the set of technologies with associated learning curves is biased: many technologies exist for which no learning has been observed, for which learning has inadvertently stopped, or for which, in fact, diffusion itself has halted (or has even returned) as a result of unforeseen (and unforeseeable) events (Sagar and van der Zwaan 2006). Furthermore, it is argued that reported learning rates often confuse at least two (and sometimes more) economic phenomena that are fundamentally different in nature. Learning-by-doing versus economies-of-scale is a proper point in case, since the former is realized by cumulative capacity or activity, while the latter is dominated by the size of the plant or sector under consideration at a given moment in time (for a recent example, related to hydrogen technology, in which these phenomena are disentangled, see Schoots et al. 2008, 2010). Several mechanisms have been proposed to explain technology learning and the observed relationships (Abell and Hammond 1979; Arthur 1988; Argote and Epple 1990; Adler and Clark 1991; Nemet 2006). Unfortunately, they generally fail to reconstruct the shape of the learning curves or explain the observed learning rates.

In this chapter we provide some interpretations of experience and learning curves starting from three different theoretical platforms. These interpretations are aimed at explaining learning rates for different energy technologies. The ultimate purpose is to find the role that experience and learning curves can legitimately play in designing efficient government deployment programs and in analyzing the implications of different energy scenarios. The "Component Learning" section summarizes recent work by the authors that focuses on the disaggregation of technologies in their respective components and argues that traditional learning for overall technology should perhaps be replaced by a phenomenology that recognizes learning for individual components. The "Learning and Time" section presents an approach that departs more strongly from the conventional learning curve methodology, by suggesting that exponential growth and progress may be the deeper underlying processes behind observed learning-by-doing. Contrary to this view, the cybernetic approach presented in the "Cybernetic Approach" section sees learning curves as expressing a

fundamental property of organizations in competitive markets and applies the findings from second order cybernetics to calculate the learning rates for operationally
closed systems. All three interpretations find empirical support. The "Conclusions"
section summarizes the pros and cons of the three approaches from an energy technology policy perspective and provides some concluding remarks.

COMPONENT LEARNING

Ferioli et al. (2009) investigate the use of learning curves for the description of
observed cost reductions for a variety of energy technologies, using as a starting point
the representation of energy processes and technologies as the sum of different components. While it is widely recognized that in many cases learning-by-doing may
improve the overall costs or efficiency of a technology, they argue that, so far, insufficient attention has been devoted to studying the effects of single-component improvements, which, taken together, may explain an aggregated form of learning. Indeed,
for an entire technology, the phenomenon of learning-by-doing may well result from
the learning of one or a few individual components only. They analyze under what
conditions it is possible to combine learning curves for single components to derive
one comprehensive learning curve for the total product. The possibility that, for certain technologies, some components (e.g., the primary natural resources that serve as
essential input for their fabrication or use) do not exhibit cost improvements might
account for the observed time-dependence of learning rates as reported in several
studies. The learning rate might also change considerably depending on the (sub)set
of data considered, a crucial issue to be aware of when one uses the learning curve
methodology. These observations may have important repercussions for the extent to
which learning curves can be extrapolated in the future.

Learning-by-doing has often been shown to slow down in the long term. A way
to describe such slowing down is to consider a product, process, or technology as
an aggregate of several components. The cost C of every industrial product can be
expressed as the sum of the costs of its components. If one assumes that the cost of
each component decreases over time according to a power law relation as a result of
learning, it is possible to write the overall cost relation of a generic product as:

$$C(x_t) = \sum_{i=1}^{n} C_{0i} \left(\frac{x_{ti}}{x_{0i}} \right)^{-b_i} = C_{01} \left(\frac{x_{t1}}{x_{01}} \right)^{-b_1} + C_{02} \left(\frac{x_{t2}}{x_{02}} \right)^{-b_2} + \ldots + C_{0n} \left(\frac{x_{tn}}{x_{0n}} \right)^{-b_n}, \quad (23.1)$$

in which the index i represents a given cost component. Each component is, in
principle, characterized by a different learning parameter b_i, and a different initial
cumulative production x_{0i}. Aggregate learning may or may not be broken down into
component learning according to Equation 23.1. At any rate, the value of cumulative
production for each component is at least as important as the individual learning
parameter. The reason for this is that, between components, x_{0i} may have widely
diverging values, and along with b_i, x_{0i} determines how much scope exists for future
learning. For example, the production of wind turbines has a negligible effect on the
historic cumulative production of steel or aluminum, so that not much production

cost reduction for these construction materials (needed for components like the support mast and turbine housing) can be expected by the deployment of wind turbines. On the other hand, continued cost improvements can be expected for the fabrication of (lightweight) rotor blades that so far have reached a much more limited cumulative production. It is therefore necessary to discuss, both in general and for each technology independently, under what conditions the conventional learning curve equation can be broken down into the component learning expression of Equation 23.1. Conversely, one may question when the validity of an equation of the form of Equation 23.1, can be demonstrated, and can it be approximated by the conventional expression for learning curves; that is, with only one term.

For ease of exposition, Ferioli et al. (2009) analyze the properties of a simplified model in which the cost for a product or technology is determined by only two components – one characterized by learning, and one for which the cost is constant in time (i.e., no cost reduction can be observed). If α is the share of the total cost that initially can be attributed to the learning component, then $1 - \alpha$ is the beginning cost share of the non-learning component. The overall cost as a function of the cumulative production of the learning component can, in this simplified case, be expressed as:

$$C(x_t) = \alpha C_0 \left(\frac{x_t}{x_0} \right)^{-b} + (1 - \alpha) C_0, \tag{23.2}$$

where C_0 is the total cost at production level x_0. Equation 23.2 can be considered a special case of the more elaborate model presented by Equation 23.1 and can be useful to highlight some properties of the latter. A theoretical justification for this model is the observation (see, e.g., Schoots et al. 2008; Schoots et al. 2010; van der Zwaan and Rabl 2004) that some parts of a technology, such as the raw materials and labor, may not experience cost reductions or may become more expensive over time. The precise value of α can, in principle, be determined for each technology, which we find may be a valuable avenue for further work. Equation 23.2 assumes that the learning component is the innovative part of the new total product, so that the cumulated production of the learning component and the overall technology are the same. It is also supposed that the component does not improve from simultaneously being part of another technology, so that the capacity of the composite and its learning component evolve synchronously.

One can test whether data that is usually fitted with a traditional learning curve which does not distinguish between different components can be fitted on the basis of Equation 23.2. Figure 23.1 shows that data for the price of gas turbines from MacGregor et al. (1991) can be fitted in two different ways. The data points can be fitted with a learning curve over a range spanning three orders of magnitude of cumulative deployment. The learning rate (LR) is found to be 13% with $R^2 = 0.95$ (Figure 23.1, left plot). It can be observed, however, that the data present an inflection point. In the literature (see notably Seebregts et al. 1999), it has been proposed to fit such data with a piecewise-linear learning curve. Figure 23.1 (right plot) shows a two-piece learning curve applied to the present case: one obtains LR = 19% and LR = 10%, and $R^2 = 0.97$ and $R_2 = 0.94$, for the two learning-curve pieces,

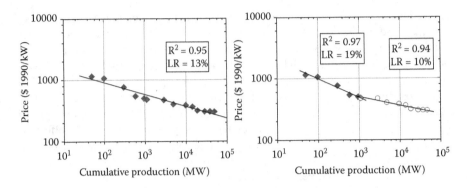

FIGURE 23.1 A set of gas turbine prices fitted in two different ways: linearly and piecewise linearly. (Data from MacGregor et al., *The market outlook for integrated gasification combined cycle technology.* General Electric Company, 1991.)

respectively. Given the empirical nature of learning curves, the fact that the learning rate changes over time leads to methodological issues: a constant learning rate is one of the fundamental assumptions of the learning curve methodology.

The same set of data can also be fitted with an expression of the form of Equation 23.2, as shown in Figure 23.2. For example, the system can be described as composed of a learning component (with LR = 24%) that makes up 80% of the total cost, and a non-learning part (hence with constant cost) that accounts for the remaining 20%. One could think of the costs associated with the steel as being necessary to fabricate the turbine. This fit is clearly better than both of those shown in Figure 23.1, since it represents overall an $R^2 = 0.97$, while applying to the same data set. In any case, if one finds learning curves acceptable when they possess an accuracy of $R^2 > 0.90$, one cannot discard any of these regressions—certainly not the two-component one.

While one may find all three fits to the gas turbine price data of Figures 23.1 and 23.2 acceptable, clear differences occur when one uses these learning curves for price

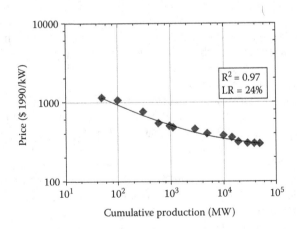

FIGURE 23.2 The fit proposed for the gas turbine price data from Figure 23.1, based on Equation 23.2 with LR = 24% (for the learning component) and $\alpha = 0.8$.

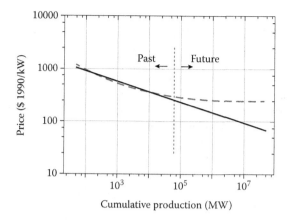

FIGURE 23.3 Two fits for gas turbine prices extrapolated over three orders of magnitude into the future: A linear regression based on the traditional learning curve (——) and one based on Equation 23.2 (– – –).

or cost estimates in the future. Extrapolating cost data over several orders of magnitude of cumulative production can lead to significant errors in the estimates of both the breakeven capacity and learning investment (needed to reach competitivity with the incumbent technology) when one uses the wrong learning model. We point this out in Figure 23.3, in which both the fit of the left plot of Figure 23.1 and that of Figure 23.2 are depicted and extrapolated over three orders of magnitude in the future. Indeed, we see that the two lines diverge rapidly for higher values of cumulative production, with obvious consequences in terms of, for example, the total learning investment needed to reach a given cost level in the future. As the cost of the innovative component is reduced, the non-learning component gains in relative weight in its contribution to the overall costs. Hence the composite learning process is slowed down.

The above argument suggests that technology cost reductions may not continue indefinitely and that well-behaved learning curves do not necessarily exist for every product. In addition, even for diffusing and maturing technologies that display clear learning effects, market and resource constraints can eventually reduce the scope for further improvements in their fabrication or use. It appears likely that some technologies, such as wind turbines and photovoltaic cells, are significantly more amenable than others to industry-wide learning. For such energy technologies, Ferioli et al. (2009) assess the reliability of using learning curves to forecast cost reductions. They argue that due attention must be paid to cost components that do not decrease over time, or may even increase.

LEARNING AND TIME

In another paper, Ferioli and van der Zwaan (2009) analyze the dynamics for the growth and cost reduction of innovative products in the energy sector. They provide a series of examples showing that simple exponential relations can be used to describe growth and cost reduction as functions of time for many types of technologies. These two simple models, for technological growth and progress respectively, when taken together, are shown to be a de facto equivalent to the well-known learning curve. They propose a stylistic computational component-based model that accounts for both these exponential relationships. The main novelty of this model is that it introduces time in the learning curve methodology. While there may be additional explanatory variables, they argue that accounting for time improves the understanding and use of learning curves.

In their article, they first point out that much attention has been paid over the course of the past two decades to how technology development feedback and inducement may be included in theories of economic growth (for an overview see Aghion and Howitt 1997). This "endogenous growth theory" may benefit greatly from the abundance of detailed technology assessment studies in the innovation literature (e.g., Grübler et al. 1999). As these technology assessments confirm, arguably the simplest model to mathematically represent (the early phase of) technological diffusion assumes growth with a fixed percentage, β, each year. Hence, the total output increases by a factor $\alpha = 1 + \beta$ each year, and the output in year $n + 1$, $y_n + 1$, is simply $y_n + 1 = \alpha y_n$ (with y_n being the output in year n). Time can be uniformly normalized by dividing by the unit of one year in order to ensure the dimensional consistency of the equations. Without a loss of generality, an output can be assumed of one in year 1. The annual output as a function of time can thus be written as:

$$y(t) = \alpha^t. \tag{23.3}$$

The cumulative production, $x(t)$, is found by integration of Equation 23.3, which (unsurprisingly) is again an exponential in t (with an extra multiplicative factor and a constant introduced by integration). Therefore the growth model for cumulative production can be approximated by a simplified exponential relation of the form:

$$x(t) = ab^t, \tag{23.4}$$

in which a and b are fitting parameters. The parameter b of Equation 23.4 is thus not exactly identical to the annual growth parameter α in Equation 23.3. For many products, companies and technologies, growth in cumulated production can be approximated by this reduced exponential model. Ferioli and van der Zwaan (2009) give several examples for non-energy technologies, based on data from Goddard (1981). Likewise, they show examples for energy technologies, which include PV modules, wind turbines, sugarcane-based ethanol production in Brazil, and large stationary fuel cells. The historic cumulative production for these technologies and products is fitted with Equation 23.4 based on data from Harmon (2000), Neij (2004), Goldemberg (1996), and Adamson (2006), respectively. The fits prove good to excellent.

Having developed a simple model for technological expansion, they then focus on the dynamics of progress. They assume that cost reduction occurs through increases in productivity, such that the unit cost is reduced by a fixed percentage every year. Hence, the cost in year $n + 1$ equals $C_{n+1} = \gamma C_n$, with $\gamma = 1 - \delta$ and δ the productivity gain. Cost as a function of time can thus be expressed by an exponentially decaying relation of the form:

$$C(t) = df^t, \tag{23.5}$$

in which d and f are fitting parameters ($f < 1$). For several technologies, it is demonstrated that costs as a function of time can be approximated by Equation 23.5, based on data from Goddard (1981), Harmon (2000) and Goldemberg (1996), respectively.

Ferioli and van der Zwaan (2009) show that the elimination of time from Equations 23.4 and 23.5 leads to a power law relation between the cost and cumulative output:

$$C(x) = d\left(\frac{x}{a}\right)^{\frac{\ln f}{\ln b}}. \tag{23.6}$$

Equation 23.6 is a de facto equivalent to the learning curve. Parameters a and d are equivalent to the usual normalization point $(x_0, C(x_0))$, and since $f < 1$ (that is, costs are decreasing over time) and $b > 1$ (output is increasing over time), the exponent in Equation 23.6 is negative, so that Equation 23.6 represents a learning curve (with $LR = 1 - 2^{\ln f/\ln b}$).

They next construct a concise model based on the exponential relations for growth and technical progress (cost reduction) as expressed by Equations 23.4 and 23.5. For the cost part of their numerical model, they consider the overall cost of a product as the sum of the costs of its components (as in Ferioli et al. 2009). For example, the cost of individual parts plus the cost of assembling and marketing may make up the total cost of a given consumer product. As with output growth, they suppose that cost reduction takes place through a stochastic process. Cost reduction occurs in a number of finite steps that reduce the cost of each component by a fraction. The magnitude and number of the cost reduction steps are not fixed, but vary randomly for each step respectively. The purchase of a new automated machine, for example, may in one step reduce the cost of the manufacturing part of a final product by $x\%$. The cost reduction for each component at each step materializes randomly in time, that is, not as a function of the cumulative production. The cost part of the model is graphically depicted in Figure 23.4.

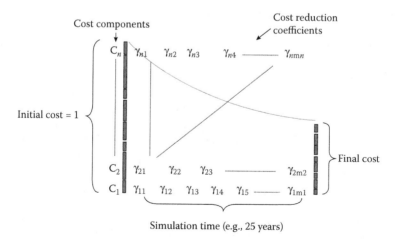

FIGURE 23.4 Illustration of the cost reduction model in Ferioli and van der Zwaan (2009). The overall technology consists of n components, the cost of each decreasing step-wise over time on the basis of a multiplication of positive coefficients $\gamma_{ij} < 1$.

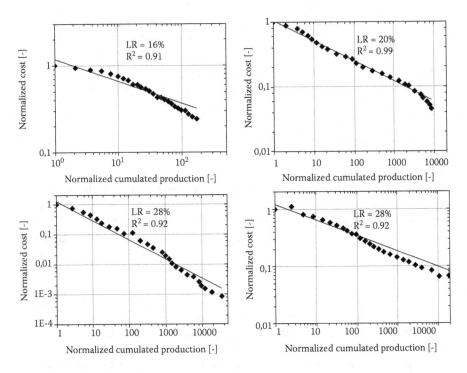

FIGURE 23.5 Cost reduction as a function of cumulative production, as calculated with the simple stochastic model in Ferioli and van der Zwaan (2009) (♦) and best fits with the power law relation of the traditional learning curve (—).

With the two generators of data for technological growth and progress in this simple numerical model, and once the few parameters involved have been appropriately chosen, the calculated trends for the output growth and the cost reduction prove to simulate observed evolutions of these variables for many products and technologies. It is demonstrated that, taken together, these two sub-models are equivalent to the learning curves as reported in the literature, which is confirmed by the results reported in Figure 23.5.

As an additional veracity check to validate their model, Ferioli and van der Zwaan (2009) compare their learning curve simulations directly to cost-versus-output data as available in the literature. Figure 23.6 shows two arbitrary examples of reported learning curve data, for the cost and cumulative output of PV modules (Harmon, 2000) and Brazilian ethanol (Goldemberg 1996) respectively. The numbers for PV, collected over more than two decades, represent an average annual growth rate during this period of as high as approximately 40% per year (Harmon 2000). During this interval the cost of PV modules decreased by more than an order of magnitude. For Brazilian ethanol, the average growth rate was significantly lower, at approximately 10% per year, during a little less than two decades. The ethanol production cost reduced by about a factor of three during these years (Goldemberg 1996). Ferioli and van der Zwaan (2009) contrast these historic data with results from their simulations. The main assumptions in their numerical model concern the average annual

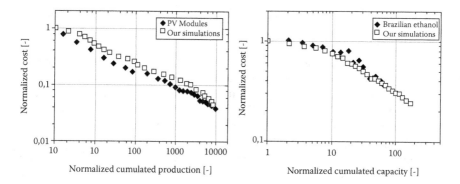

FIGURE 23.6 Comparison of learning curve data for PV (Harmon 2000) and Brazilian ethanol (Goldemberg 1996) (♦) with data generated by the model simulations by Ferioli and van der Zwaan (2009) (□).

growth rate over the whole period under consideration, the standard deviation of this yearly growth, and the final total cost reduction, which are all taken so as to stay close to the real-life values of these quantities. It proves that their model can properly reproduce the learning curves published for these two cases. It can be shown that other learning curves can be simulated in a similar way through using this model.

CYBERNETIC APPROACH

The learning curves for a technology show the performance of a human interactive system producing and/or operating the technology. Is this learning system best understood as an open system reacting to demands and opportunities in the environment, or as a closed system acting autonomously based on its internal structure? Most explanations of learning curves and technology learning focus on the role of environmental interactions in explaining the phenomenon. The cybernetic approach (Wene 2007, 2008a, 2008b) takes the opposite view by assuming that all performance improvements are the result of closed chains of internal operations. Features of the environment, and events and processes in the environment, appear as perturbations, and the system may decide to modify its operations, but these modifications are always conditioned by the internal structure of the system. In this approach, learning rates are the result of the eigenbehavior (Varela 1979; von Förster 2003) of the system. The underlying theory provides a spectrum of *eigenbehavior* values for the unperturbed system in a perfect competitive environment, while the basic eigenbehavior (or eigenvalue) corresponds to a learning rate of 20%. Perturbations from the environment result in the observed broad distribution of learning rate values (see, e.g., McDonald and Schrattenholzer 2001) around the unperturbed eigenbehavior value(s) (Wene 2010).

The cybernetic approach applies theoretical results for biological and social systems (von Förster 1980, 2003; Varela 1979, 1984; Luhmann 2002) to technology learning systems, among others in the energy field. This section exposes the basic features of this approach and provides the results for learning rate distributions. The autonomy of the system is expressed as *operational closure*, meaning that the system forms and controls all its operations. While the system is open to information, and in this case to material and energy flows, the network of internal operations closes on itself. The closure theorem of cybernetics states that in every operationally closed system, eigenbehaviors arise. The task of the analyst is to find the operational loops that represent learning and define the operators whose fixed points provide the values for the eigenbehavior. Guides for this task are the form of the learning curve and the well-known OADI-SMM (observe, assess, design and implement—shared mental model) model for organizational learning (Kim 1993; Espejo et al. 1996). Occam's razor, which is generally a good principle in theory design since it urges for a minimum of elements and assumptions, possesses particular relevance when specifying the loops and operators involved in OADI-SMM.

The OADI mnemonic emphasizes learning as a result of self-reflection; that is, making new designs after assessing the observations of one's own action, which in this case is the implementation of previous design efforts. This is consistent with the assumption of operational closure and is captured in the three feedback loops in Figure 23.7. The hypothesis is that these loops, taken together, provide the elements needed to explain technology learning as expressed by learning curves. The internal and external feedback loops express implementation and observation. The external loop closes over the market and the internal loop over producing. Together, with the self-reflecting, third loop (SRL), they provide the double closure proposed by von Förster (2003) as the process required for an organism or an organization to modify its behavior in order to manage environmental perturbations without losing operational closure. The internal and external loops reflect the double closure over production and sales as analyzed by Baecker (1996). The self-reflection loop represents assessment and design and operates on the internal state, Z. This internal state sets the transfer function of producing. In other words, through its actions on Z, the SRL determines the performance (i.e., the relation between the input and output) of the learning system. The computing block integrates the three loops. The SRL thus drives the learning of the system, but is for its operations totally dependent on the status of the two other loops.

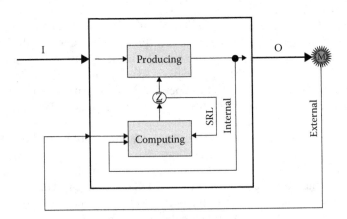

FIGURE 23.7   The elements of the technology learning system.

In order to find the mathematical expression for the operators describing the action of the three loops on the state function, we observe that the learning curve can be written as:

$$P(t) = \text{output/input} = C_0 \, X(t)^E, \qquad (23.7)$$

in which $P(t)$ is the performance of the system at time t. If all inputs are measured in monetary units, then the performance is the output generated from one monetary unit of input (which is equivalent to the inverse of the cost per unit output). In this equation, C_0 is a constant and $X(t)$ is the sum of all output until time t, while E is referred to as the "experience parameter." The progress ratio (PR) and learning rate (LR) are, as usual, given by:

$$PR = 1 - LR = 2^{-E}. \qquad (23.8)$$

Taking the logarithm of Equation 23.7, and differentiating plus rearranging provides:

$$X/dX \times dP/P \times 1/E = 1 \qquad (23.9)$$

Let us consider the unperturbed case. The external loop carries the urge from the perfect competitive market to improve performance by as much as possible. Without perturbations, this loop can be seen as a constant factor. The internal loop tracks output from production, which is represented by the right-hand-side of Equation 23.7. The SRL sets performance via Z, which is represented by the left-hand-side of Equation 23.7. Occam's razor suggests using two operators that work on two independent parts of the state function, which we will call C^+ and C_{SRL}.

One more aspect of the learning curve helps to constrain the choice of operators and initial state function. The learning curve requires that the identity of the output is retained, which can be interpreted as implying that the technical properties of the output pertaining to its original use remain the same. The importance of self-reflection and identity suggests searching for solutions based on complex algebra. For the unperturbed case, the identity requirement can be interpreted as constraining all solutions to the unit circle in the complex Argand plane. C^+ becomes a stereographic operator projecting $X(t)$ onto the unit circle, so that $X(t) \to \infty$ is represented by $-i$ in the complex plane. In the aggregated universe of the learning curve, the C^+ operator corresponds to corporate memory; that is, the SMM in the OADI-SMM.

Equation 23.9 suggests that the C_{SRL} operator is a function of the cumulative output (X) and additional output dX. Wene (2007, 2010) presents arguments for the mathematical form of this operator consistent with Equation 23.9 and with research on individual and organisational learning. Ultimately, however, as in all theoretical research, the unique form of C_{SRL} must be justified by its ability to explain and forecast empirical results.

This theory now yields values for the experience parameter, which correspond to the eigenbehaviors of the technology learning system in the unperturbed case:

$$E(n) = 1/\big[(2n+1)\cdot \pi\big] \qquad \text{with } n = 0,1,2,3,\ldots, \qquad (23.10)$$

while the corresponding learning rates are:

$$LR(n) = 1 - 2^{-E(n)} \qquad \text{with } n = 0, 1, 2, 3, ..., \tag{23.11}$$

and the first four modes of learning are:

$$LR(0, 1, 2, 3) = 20\%, 7\%, 4\% \text{ and } 3\%. \tag{23.12}$$

Changes in eigenbehavior due to perturbations in the external loop can be analyzed without introducing new operators. The double closure between internal and external loops, which modifies the operations in the self-reflecting loop, recursively acts on the system state with an operator matrix with the two basic operators as elements. The result is a distribution of learning rates around the eigenvalues for the unperturbed case. The form and width of this distribution depend on the nature of the perturbations. The latter may appear as negative in the form of environmental restrictions, or as positive in the form of learning spill-overs from other enterprises or industries.

Figure 23.8 from Wene (2010) compares theoretical and empirical distributions of experience parameters. The empirical distributions are from the compilations by, respectively, Dutton and Thomas (1984), based on cost measurements in individual enterprises (Figure 23.8a), and Weiss et al. (2010), for energy supply technologies based on market prices (Figure 23.8b). Negative and positive perturbations are assumed to be additive and Poisson distributed. The assumptions on perturbations are simple but provide good fits and some initial observations.

The Dutton and Thomas (1984) distribution of individual enterprise technologies shows no significant influence from higher order learning modes. The distribution of learning rates for energy supply technologies (Weiss et al. 2010), on the other hand, cannot be reproduced without higher learning modes. The theoretical curve depicted in Figure 23.8b only includes the distribution around the first higher learning rate mode, LR(1), but the fit could be improved by also including higher modes. One can speculate about the cause for the difference between these two distributions. One reason may be that the learning systems for some energy technologies are not properly closed. The network of operations may have side-loops outside of the system; that is, the chosen system boundaries are too narrow to provide full operational closure. A case in point may be wind turbines, for which measurements indicate learning rates of 4–8% that may be representative of higher learning modes (Neij 1999; Durstewitz and Hoppe-Kilpper 1999). Indeed, it proves that increasing the system boundary to encompass the complete wind power plant (Neij et al. 2004) or including the production of electricity from wind power (IEA/OECD 2000; Neij et al. 2004) provides learning rates representative of the zero learning mode. The results of Junginger et al. (2005) for wind farms point to the need to widen the boundaries from national to global learning systems. An analysis of the organisation of the entire wind industry, rather than of just the technology itself, may thus be required to understand the low learning rates observed for wind turbines. This example at least points towards the need for the systematic organisational studies of entire industrial sectors in order to understand learning systems that have enough autonomy to provide operational closure.

Cost analysis requires the assessment of the network of operations within a firm. This aids the investigator in avoiding boundaries which do not provide operational

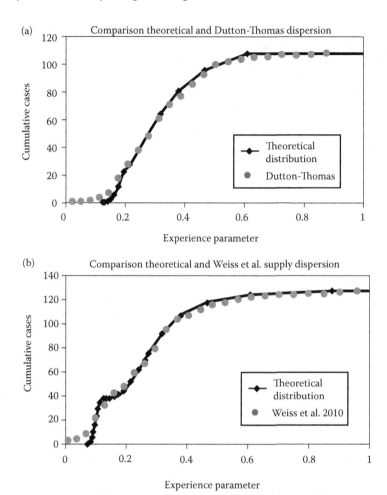

FIGURE 23.8 Comparison of theoretical and empirical distributions. (From Wene, C.O. 2010. Adaptation in technology learning systems. *The proceedings of the 11th IEAA European conference*. 25–28 August, 2010, Vilnius, Lithuania.)

closure. This may be one explanation for the absence of higher learning modes in the Dutton and Thomas (1984) distribution. Operations can, of course, be disturbed by external features, events, and processes. Why then should such disturbances influence the industry but not the firm? This suggests another explanation for the lack of higher learning modes in Figure 23.8a—namely the choice of firms and technologies. Many of the energy technologies in the distribution by Weiss et al. (2010) are regulated— especially fossil and nuclear technologies, which are exposed to insistent but changing external perturbations, related, for instance, to emissions and safety standards. Such perturbations may be too strong to manage through double closure, thus forcing the system into a higher learning mode. Wene (2008b) discusses government spending on public research and development (R&D) as another potentially disturbing factor in the system's environment for producers and users of energy technologies. The purpose of the spending is typically to produce knowledge that increases learning, but systemically this external production of knowledge represents a perturbation that disturbs the internal network of operations. Prospects for radical innovation through public R&D may yield important rethinking or second-loop learning, which may lead to swift cost reductions and improvements in efficiency. External R&D that represents a continuous series of minor improvements, however, may disrupt the system's own learning operations and leave it in a higher learning mode with a lower learning rate.

CONCLUSIONS

This chapter has presented three approaches, which at increasing theoretical depth search to explain the learning curve phenomenon. It is not possible at this point to strike a total balance between the approaches, but following the declared purpose of the chapter we here summarize some of their merits, and present potential deficiencies from an energy technology policy point of view.

Component learning provides a quality-controlled methodology to construct a learning curve for a new technology relying on components for which knowledge exists on learning parameters. A typical example is the fuel cell, but the approach applies in principle to any technology. The usefulness of this method for technology policy is obvious, since it provides refined estimates of learning rates for new technology, and can assist in estimating joint learning for two technologies with similar components. While it provides limited insight into the learning process itself, the component learning approach may increase the reliability of forecasts, particularly in the energy field. In its present form, however, it ignores the possibility of interference between components, which may lead to lower or higher rates of learning.

Factoring the learning curve into market expansion and cost dynamics provides new ways to analyze and extract meaning from market and cost time series. Learning curves for different technologies can be simulated through the use of a small set of fitted parameters. This method widens the theoretical base for learning curves to economic growth and innovation theories, but a further exploration and backing of this base is both desirable and necessary. This may further increase the reliability of technology forecasts based on this approach. In order to provide proper forecasts, however, cost parameters would need to be explained in terms of the measurable properties of new technology, or of the learning system behind new technology.

The cybernetic approach finds learning rates around 20% to be a natural consequence, or an eigenvalue, of the operations of a learning system and independent of any parameter value. This method argues for continuous and reliable learning over the lifetime of a technology. Measured dispersions around the eigenvalues can be reproduced with a few fitted parameters. Like the "learning and time" method, this approach widens the theoretical base for learning curves—in this case to second-order cybernetics and to models for organizational learning—but also here the theoretical foundation should be further investigated. Also, the approach has yet to explain how parameters fitted to the measured distributions of learning curves relate to actual perturbations to the learning system, and how learning parameters for a specific technology can be calculated from the properties of the learning system and its environment.

From the viewpoint of energy technology policy, the three approaches appear to be complementary rather than competing. They strengthen learning curves as policy tools but require further development and grounding in empirical results. Over the last decade, measuring, understanding, and applying learning curves for energy technologies has developed into a large field. Our ambition with this chapter has not been to provide a complete picture of the theoretical developments in this domain, nor have we been in a position to consistently list all the energy technologies for which learning curves have been determined. At the time, McDonald and Schrattenholzer (2001) provided such a listing, although since their seminal publication this methodology, as applied to the energy sector, has progressed substantially. However, we hope that this chapter provides the reader with an insight into at least some of the recent developments, and notably those with which the authors are most familiar. We have undoubtedly left out subjects that were close to the heart of one of our great inspirers in this field, but we are confident—having learned from him before his unexpected early departure—that he would have enjoyed the exploration of these alternative approaches.

ACKNOWLEDGMENT

This chapter was written in memory of Leo Schrattenholzer, to whom both authors are indebted. The authors acknowledge the support of research grants from the Netherlands Organization for Scientific Research (NWO, under the ACTS Sustainable Hydrogen Program, projects 053.61.304, 053.61.305 and 053.61.024) and from the Swedish Energy Agency (project 32179-1).

REFERENCES

Abell, D.F., and Hammond, J.S., 1979. Cost dynamics: Scale and experience effects. In: *Strategic planning: Problems and analytical approaches.* Prentice Hall: Englewood Cliffs.

Adamson, K.A., 2006. *Fuel cell today, market survey: Large stationary applications.* http://www.fuelcelltoday.org/online/survey?survey=2006-11%2F2006-Large-Stationary (accessed May 22, 2008).

Adler, P.S., and Clark, K.B., 1991. Behind the learning curve: A sketch of the learning process. *Management Science* 37(3): 267–281.

Aghion, P., and Howitt, P., 1997. *Endogenous growth theory.* Cambridge: MIT press.

Argote, L., and Epple, D., 1990. Learning curves in manufacturing. *Science* 247(4945): 920–924.

Arthur, B., 1988. Competing technologies: An overview. In *Technical change and economic theory.* eds. G. Dosi, C. Freeman R. Nelson, G. Silverberg and L. Soete. London: Pinter.

Durstewitz, M., and Hoppe-Kilpper, M., 1999. Using information of Germany's '250 MW wind'-programme for the construction of wind power experience curves. In *Proceedings IEA workshop on experience curves for policy making – The case of energy technologies*, eds. C.O. Wene, A., and Voss, T. Fried. 10–11 May 1999, Stuttgart, pp. 129–134, Forschungsbericht 67, Institut für Energiewirtschaft und Rationelle Energieanwendung, Universität Stuttgart.

Dutton, J.M., and Thomas, A., 1984. Treating progress function as a managerial opportunity. *Academy of Management Review* 9(2): 235–247

EESC, 2009. Towards an eco-efficient economy – Transforming the economic crisis into an opportunity to pave the way for a new energy era. Exploratory Opinion of the European Economic and Social Committee, TEN/398-CESE 1700/2009, Brussels, 5 November 2009.

Espejo, R., Schuhmann, W., Schwaninger, M., and Bilello, U., 1996. *Organisational transformation and learning – A cybernetic approach to management.* Chichester: Wiley.

Ferioli, F., Schoots, K., and van der Zwaan, B.C.C., 2009. Use and limitations of learning curves for energy technology policy: A component-learning hypothesis. *Energy Policy* 37(7): 2525–2535.

Ferioli, F., and van der Zwaan, B.C.C., 2009. Learning in times of change: A dynamic explanation for technological progress. *Environmental Science and Technology* 43(11): 4002–4008.

Goldemberg, J., 1996. The evolution of ethanol costs in Brazil. *Energy Policy* 24(12): 1127–1128.

Grübler, A., Nakicenovic, N., and Victor, D., 1999. Dynamics of energy technologies and global change. *Energy Policy* 27(5): 247–280.

IEA/OECD, 2000. *Experience curves for energy technology policy.* International Energy Agency/Organisation for Economic Co-operation and Development, Paris.

IEA/OECD, 2006. *Energy technology perspectives 2006—In support of the G8 plan of action, scenarios and strategies to 2050.* International Energy Agency/Organisation for Economic Co-operation and Development, Paris.

IEA/OECD, 2008. *Energy technology perspectives 2008—In support of the G8 plan of action, scenarios and strategies to 2050.* International Energy Agency/Organisation for Economic Co-operation and Development, Paris.

IEA/OECD, 2010. *Energy technology perspectives 2010—In support of the G8 plan of action, scenarios and strategies to 2050.* International Energy Agency/Organisation for Economic Co-operation and Development, Paris.

Junginger, M. (ed.), 2010. *Technological learning in the energy sector: Lessons for policy, industry and science.* Cheltenham, U.K.: Edward Elgar. (forthcoming)

Junginger, M., Faaij, A., and Turkenburg, W.C., 2005. Global experience curves for wind farms. *Energy Policy* 33(2): 133–150.

Kahouli-Brahmi, S., 2008. Technology learning in energy-environment-economy modeling: A survey. *Energy Policy* 36(1): 138–162.

Kim, D.H., 1993. The link between individual and organizational learning. *Sloan Management Review* 35(1): 37–50.

Kydes, A.S., 1999. Modeling technology learning in the national energy modeling system. In *Proceedings IEA workshop on experience curves for policy making – The case of energy technologies*, eds. C.O. Wene, A. Voss, and T. Fried. pp. 181–202, Forschungsbericht 67, Institut für Energiewirtschaft und Rationelle Energieanwendung, Universität Stuttgart, Stuttgart, Germany.

Luhmann, N., 2002. *Theories of distinction.* Stanford: Stanford University Press.

MacGregor, P.R., Maslack, C.E., and Stoll, H.G., 1991. *The market outlook for integrated gasification combined cycle technology.* General Electric Company.

Manne, A.S., 1994. Private communication.

Mattsson, N., and Wene, C.O., 1997. Assessing new energy technologies using an energy system model with endogenized experience curves. *International Journal of Energy Research* 21(4): 385–393.

Maycock, P.D. and Wakefield, G.F., 1975, *Business analysis of solar photovoltaic energy conversion.* Photovoltaic Specialists Conference, 11th, Scottsdale, Ariz., May 6–8, 1975, Conference Record. (A76-14727 04–44) New York, Institute of Electrical and Electronics Engineers, Inc., 1975, p. 252–255.

McDonald, A., and Schrattenholzer, L., 2001. Learning rates for energy technologies. *Energy Policy* 29(4): 255–261.

Messner, S., 1997. Endogenized technological learning in an energy systems model. *Journal of Evolutionary Economics* 7(3): 291–313.

Neij, L., 1999. Cost dynamics of wind power. *Energy* 24(5): 375–389.

Neij, L., Andersen, P.D., Durstewitz, M., Helby, P., Hoppe-Kilpper, M., and Morthorst, P.E., 2004. *Experience curves: A tool for energy policy assessment.* IMES/EESS Report No. 40, Department of Technology and Society, Lund University, Sweden, see www.iiiee.lu.se

Nemet, G.F., 2006. Beyond the learning curve: Factors influencing cost reductions in photo-voltaics. *Energy Policy* 34(17): 3218–3232.

Sagar, A.D., and van der Zwaan, B.C.C., 2006. Technology innovation in the energy sector: R&D, deployment, and learning-by-doing. *Energy Policy* 34(17): 2601–2608.

Schoots, K., Ferioli, F., Kramer, G.J., and van der Zwaan, B.C.C., 2008. Learning curves for hydrogen production technology: An assessment of observed cost reductions. *International Journal of Hydrogen Energy* 33(11): 2630–2645.

Schoots, K., Kramer, G.J., and van der Zwaan, B.C.C., 2010. Technology learning for fuel cells: An assessment of past and potential cost reductions. *Energy Policy* 38(6): 2887–2897.

Seebregts, A.J., Kram, T., Schaeffer, G.J., Stoffer, A., Kypreos, S., Barreto, L., Messner, S., and Schrattenholzer, L. 1999. *Endogenous technological change in energy system models.* ECN-C--99-025, ECN, Petten, The Netherlands.

Stern, N., 2006. The economics of climate change. *The Stern Review.* Cambridge University Press: Cambridge

Tsuchiya, H.,1989. Photovoltaic cost based on the learning curve. *Proceedings of the international solar energy society clean & safe energy forever symposium*, Kobe City, Sep. 4–8, 1989, p. 402.

van der Zwaan, B.C.C., Gerlagh, R., Klaassen, G., and Schrattenholzer, L., 2002. Endogenous technological change in climate change modeling. *Energy Economics* 24(1): 1–19.

van der Zwaan, B.C.C., and Rabl, A., 2004. The learning potential of photovoltaics: Implications for energy policy. *Energy Policy* 32(13): 1545–1554.

Varela, F., 1979. *Principles of biological autonomy.* Elsevier-North Holland, New York.

Varela, F., 1984. Two principles for self-organization. In *Self-organization and management of social systems.* eds. H. Ulrich and J.B. Probst, p. 25–32. Berlin: Springer.

von Förster, H., 1980. *Observing systems.* The systems inquiry series. Intersystems: Seaside.

von Förster, H., 2003, *Understanding.* Berlin: Springer.

Weiss, M., Junginger, M., Patel, M.K., and Blok, K., 2010. A review of experience curve analyses for energy demand technologies. *Technological forecasting and social change* 77(3): 411–428.

Wene, C.O., 2007. Technology learning system as non-trivial machines. *Kybernetes* 36(3/4): 348–363.

Wene, C.O., 2008a. A cybernetic perspective on technology learning. In *Innovations for a low carbon economy: Economic, institutional and management approaches*, eds. T.J. Foxon, J. Köhler and C. Oughton. Cheltenham: Edward Elgar.

Wene, C.O., 2008b. Energy technology learning through deployment in competitive markets. *The Engineering Economist* 53(4): 340–364.

Wene, C.O., 2010. Adaptation in technology learning systems. *The proceedings of the 11th IEAA European conference.* 25–28 August, 2010, Vilnius, Lithuania. Available at http://www.iaee2010.org/?q=node/84

Williams, R.H., and Terzian, G., 1993. *A benefit/cost analysis of accelerated development of photovoltaic technology.* PU/CEES Report No. 281, Princeton University, N.J.

Index

Q